D1354586

Geological Maps

Geological Maps

An Introduction

JOHN WILEY & SONS

Chichester · New York · Weinheim · Brisbane · Singapore · Toronto

Other Wiley Editorial Offices

John Wiley & Sons, Inc., 605 Third Avenue,
New York, NY 10158-0012, USA

WILEY-VCH Verlag GmbH, Pappelallee 3,
D-69469 Weinheim, Germany

Jacaranda Wiley Ltd, 33 Park Road, Milton,
Queensland 4064, Australia

John Wiley & Sons (Asia) Pte Ltd, 2 Clementi Loop #02-01,
Jin Xing Distripark, Singapore 129809

John Wiley & Sons (Canada) Ltd, 22 Worcester Road,
Rexdale, Ontario M9W 1L1, Canada

A catalogue record for this book is available from the British Library

ISBN 0-471-97696-2

Typeset in 10/12pt Times from the author's disks by Dorwyn Ltd, Rowlands Castle, Hants.
Printed and bound in Great Britain by Redwood Books, Trowbridge, Wiltshire.
This book is printed on acid-free paper responsibly manufactured from sustainable forestry, in which at
least two trees are planted for each one used for paper production.

Contents

Preface to the First Edition

A recent survey of geology students in the UK indicated that although they saw the need for a basic training in mapwork, the three-dimensional aspects it involved formed the single most difficult part of a beginning geology course, and that it was generally taught in a way both abstract and dull. At the same time, there was no book which puzzled students could turn to for explanations; no book which told them more about real geological maps. This book is an attempt to fill that need. It is based on the view that in these days of increasing specialisation the geological map remains the vital coordinating document, and that the proliferation of computer methods of handling three-dimensional data makes a firm understanding and appreciation of mapwork more imperative than ever.

The book is designed for first year undergraduates. An elementary knowledge of rocks and geological processes is assumed, together with a basic understanding of topographic maps. Geological maps, however, are introduced from first principles, so that some of the material may appeal to anyone with an interest in geology; on the other hand, some of the information may be of use to the more advanced student. Those figures that contain formulae, methods, and information for reference have a frame to enable rapid location. The reference list, in addition to citing the sources of the maps included in the book, indicates much further material of relevance to geological mapwork.

In a subject so fundamental and yet so varied, every geologist will have his or her own views on geological maps – the matters needing emphasis, the best methods of interpretation, good examples of maps, and so on. Instructors may therefore urge in their taught courses different priorities from those given here, and, although a wide range of maps and map exercises is included, will prefer to continue to use their own 'pet' examples. But this is meant primarily to be a book for the student – to turn to for clarification, for further information, and simply to learn a little more about geological maps.

I acknowledge the years of undergraduate students at the University College of Wales, Aberystwyth, from whose wishes this book was born, and the individual students who commented on early versions of it. The constructive criticism and advice are also gratefully acknowledged of: Dr. Mark Bentley, of Shell Expro; Peter Hill, of BP; Dr. Dave Wilson of the BGS; colleagues at UCW, Aberystwyth, in particular, Dr. Dennis Bates, Dr. Bill Fitches, Warren Pratt, and Antony Wyatt; and my wife, Jo. For help in the production of the volume I thank Richard Baggaley and Sue Hadden of the Open University Press, and Valerie Grant and Arnold Thawley of the Department of Geology, UCW, Aberystwyth. Finally, I thank my family – Jo, Alastair, and Emily – for putting up with all my nights at the office.

Preface to the Second Edition

Although geology is undergoing great changes, geological maps remain as fundamental as ever. After all, they embody the very matters that set geology apart from other disciplines: dealing with real earth materials in three dimensions and through time. And although maps have always been tremendously effective scientific documents, computer methods are now adding yet further to their power and versatility. So students still need to know about geological maps, but – just as before – some students find the principles difficult and need careful explanations. The first edition of this book was founded on these principles, and they are consolidated here. The advent of ever more powerful computer methods emphasises even more the need for a good understanding of how to deal with the three-dimensional arrangement of earth materials. 'Pity the geologist', said the first edition, 'who does not understand what it is that the computer is doing'. Although the days of using pencils and protractors to work with paper maps may be starting to fade, the principles behind manipulating maps and cross-sections on a computer screen are identical: the need for a clear understanding of the principles is more vital than ever.

Therefore, the early chapters of this book, which introduce and explain the basic geometric principles of mapwork, have been changed little for this new edition. Certain points on which some students appear perennially to get stuck have been expanded slightly and further illustrated. I have tried to improve the effectiveness of some of the examples. A little more emphasis has been given throughout to thickness contours – so-called isopachs – as this device is now being used in an increasingly wide range of applications. But the basic approach is the same: to explain basic map principles using illustrations drawn from real examples.

One significant change from the earlier edition is the provision, in response to many requests, of worked solutions to the exercises. This may help provide further explanation of how geological map methods work, as well as allowing the book to be used more effectively for self-tutoring. The suggested solutions are collected at the end of the book.

Of all the changes in geology since the first edition was prepared, none are more noticeable than those reflecting the burgeoning concern for the natural environment. New uses of geological maps have rapidly grown, to promote better environmental understanding and to foster an integrated approach to the planning of future land use. These were mentioned in the first edition, but they have exploded in range and value. Consequently, I have added a chapter which outlines the use of geological maps in environmental geology. Another recent change in geology concerns its teaching: a much greater number and wider range of students than a decade ago now sample some geology classes. Many of these new students are interested less in a rigorous workout in the technicalities of the subject than a feel for the basic ideas and a vision of the wider applications. Geological maps provide a fine vehicle for explaining how geology underpins so much of the landscape around us – landscape in its broadest sense – and a chapter has been added which introduces this thinking.

In line with the two innovations just mentioned, throughout this new edition the treatment of superficial deposits has been strengthened. Geologists have traditionally tended to confine their enthusiasm to bedrock, even scorning the loose material that so often covers it. Yet these deposits have a tremendous influence on our lifestyles. After all, it is on them, rather than on actual bedrock, that we carry out most of our activities! These covering materials are variously referred to as superficial or Quaternary deposits, or, rather mysteriously, 'drift'. Many of the principles of map interpretation are the same, but I have now tried to emphasise these similarities more, as well as highlighting the differences.

Despite their scientific power, geological maps can be fun. Working with attractively produced maps can be a pleasure. They have aesthetic dimensions, and the attempt in the first edition to glimpse the human stories behind some maps was welcomed by many readers. I have therefore taken the opportunity to expand this theme somewhat. However, several sections on map production and availability have been eliminated, as they are so rapidly being supplanted by computer methods. I have no longer included, for example, such things as addresses from where maps and map catalogues can be obtained, not as a discouragement – quite the opposite – but because this kind of information is now so easily and effectively acquired on the World Wide Web. And this, surely, is where much of the future of geological maps lies. Mapwork of the future may be on a screen rather than on a sheet of paper, but the principles are the same, the documents are still geological maps, and all the signs are that they will remain fundamental to geology.

I have many people and institutions to thank. As in the first edition, this book avoids hypothetical, idealised maps. The examples involve real maps from real places. Consequently I am indebted to all the learned societies, governmental organisations, geological companies, commercial publishers and individuals around the world that have allowed me to draw on their published examples of geological maps. Specific sources are acknowledged in the text. Many of the changes to the book were carried out during my sabbatical leave at the University of Tsukuba, Japan. I thank that university for its generosity, and its members – Professor Yujiro Ogawa in particular – for their hospitality.

Once again I warmly acknowledge the assistance of colleagues at the University of Wales, Aberystwyth. Not least of these are the undergraduate students, who prompted many of the changes herein and then had to act as guinea pigs! Ian Gulley, Mike Jones, and Anthony Smith helped with the drafting. Professor David Gilbertson, Drs. Dick Cave and Antony Wyatt kindly read sections of the book. Al Bolton and Malcolm Peters checked the solutions to the map exercises. My wife, Jo, as usual checked everything. Her unswerving interest, support and understanding remain invaluable. I have failed to excite any interest in geological maps in my two children, Alastair and Emily, but I love them anyway.

List of Exercise Maps

List of Plates

CHAPTER 1

Some Fundamentals of Geological Maps

1.1 Introduction

Geological maps show the distribution at the earth's surface of different kinds of earth materials. To geologists, maps are a fundamental tool. The patterns on the maps show the relationships between the rocks, from which much can be deduced about how they formed, and their three-dimensional arrangement underground. Hence, it becomes possible to predict what occurs some way *beneath* the land surface, and this has tremendous commercial relevance. The three-dimensional methods are used, for example, in working with oil reservoirs, aquifers, ore bodies, land subsidence, and coal seams. Even so, knowledge simply of what is where at the earth's surface also has wide applications. It can explain much about the landscape – in its widest sense – and hence geological maps find applications in subjects such as archaeology and geography. And the blossoming of interest in the natural environment has given a new importance to being able to understand geological maps. This book is aimed, among other things, at helping you develop this understanding.

A geological map may be a geologist's first introduction to an area being visited; it may also represent the culmination of his or her investigations. Maps are commonly used to assemble new information as it is obtained; they are also a highly effective way of communicating the new data to other geologists. A geological map can act as a synthesis of current knowledge on the geology of an area. Indeed, because so many facts and principles are communicated in a single document, Rudwick (1976) called geological maps 'the visual language of geologists'.

Because in nature most geological features are arranged in three dimensions, acquiring a familiarity with how to handle this is an essential part of the training of any geologist. The geological map, despite

its being a flat piece of paper, remains the single most convenient way of representing and working with the spatial arrangement of rocks. It is a central concern of many of the following chapters. At the same time, maps are much used to help reconstruct the geological histories of areas and the geological conditions that existed in the past. This conveying of information in additional dimensions – underground, and back in geological time – sets geological maps apart from other kinds of maps. Another fundamental difference from most other kinds of maps is that geological maps are themselves based on interpretation: much of what is portrayed on the map is not actually visible for the surveyor to see. Consequently, the completed map tends to reflect how well the geology of the area is understood at the time. As North (1928) put it, the map acts as 'an index of the extent and accuracy of geological knowledge at the time of its production, and it is the basis of future research'.

Geology is increasingly becoming a collection of specialised studies, and more and more specialised kinds of maps are evolving. Nevertheless, the conventional geological map continues to provide a common thread. Most specialisations somewhere involve a traditional geological map. However, geological maps themselves are suddenly undergoing very great changes, especially in the way new technologies are being employed in the production of maps and in manipulating map information. This adds tremendous flexibility to the ways we can use maps, but it also makes an understanding of the basic principles behind them more important than ever.

This book is largely concerned with these fundamental principles. However, it also attempts to give glimpses of why many geologists, in addition to understanding the functional significance of geological maps, have a fondness for them, and a respect for the heritage they represent. This first chapter introduces

SCALE:	*1:10 000 000* and smaller	
USE:	Maps of entire continents, oceans, or planets, on single sheets.	

SCALE:	*1:5 000 000* and *1:1 000 000*
USE:	Synoptic views of continents or countries, sometimes on several sheets.

SCALE:	*1:500 000*
USE:	Maps of countries, provinces, states (depending on size); little detail but of use for general planning and overviews.

SCALE:	*1:250 000*
USE:	Regional geology, e.g. the conterminous U S in 472 sheets (2° long. × 1° lat. quadrangles); Australia in 544 sheets; Canada in 918 sheets; U K and adjacent shelf in 106 sheets. Usually have topographic base.

SCALE:	*1:50 000, 1:25 000,* and thereabouts
USE:	The standard scales for reasonably detailed published geological maps of well-investigated countries, e.g.: the previous 'One-Inch' maps of the B G S at 1:63 360; the 'Classical areas' maps of the B G S at 1:25 000; U S G S 15' quadrangles at 1:62 500; U S G S 7½' quadrangles at 1:24 000.

SCALE:	*1:10 000* and larger
USE:	1:10 000 the standard scale for B G S field surveying and detailed investigations. Generally unpublished, apart from coalfields, but copies available to the public. Larger scale maps or plans (true dimensions shown) of sites of scientific or commercial interest: mines, quarries, etc.

Figure 1.1   Some notes on typical scales of geological maps.

the basic features of geological maps, expanding on some of the points mentioned above. We begin with a brief consideration of the topographic base on which geological maps are drawn.

1.2   The Topographic Base Map

Normally the geological information has been added to a topographic base map in order that the geological features can be located. The base map may consist simply of a few recognisable features, such as the shape of a coastline or the position of major towns, or the geology may be superimposed on a complete topographic map. In the USA a single government body is responsible for producing both the topographic and the geological maps of the country. It is called the United States Geological Survey (USGS) and is one of the world's largest geological organisations. UK geological maps are published by the British Geological Survey (BGS). (Such organisations are usually called by their initials. The map on the cover of this book, for example, was produced by the Bureau of Mineral Resources in Australia, widely called the BMR, which has now been reorganised into the Australian Geological Survey Organisation: AGSO.) BGS arranges for its information to be added to the

topographic maps of the Ordnance Survey. Therefore, a first requirement for working with geological maps is a familiarity with the principles of topographic maps, as discussed in standard textbooks on cartography. The most important aspects of topographic maps for geological purposes are summarised in the following sections.

1.2.1   Scale

The **scale** of geological maps is highly variable: from very small-scale maps of entire continents or even planets, to very large-scale maps that show fine details of a particular locality, perhaps one of special scientific or commercial interest. Scale is most usually specified as a ratio, for example 1:100 000, where one unit on the map represents 100 000 of the same units on the ground. Thus one centimetre on a map at this particular scale would be equivalent to 100 000 centimetres, that is, 1000 metres or 1 kilometre. Examples of the kinds of scales typically used for geological maps are given in figure 1.1.

Older, non-metric maps were sometimes referred to by a comparative scale, such as 'one inch equals one mile'. USGS maps are commonly called 'quadrangle maps', as they show a quadrangular area defined by lines of latitude and longitude. The spacing of the lines implies the scale of the map (figure 1.1). Maps

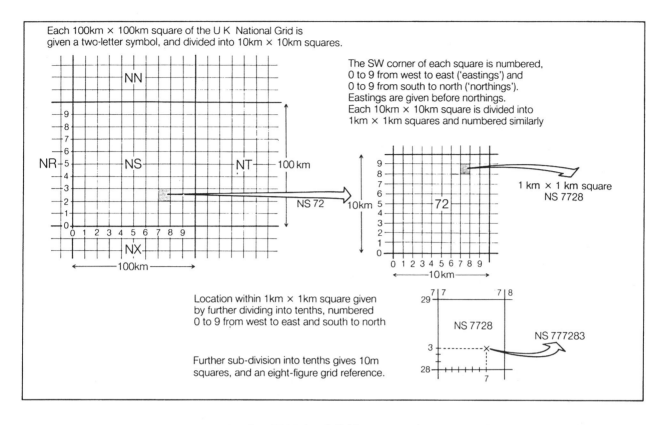

Figure 1.2 Finding locations on maps using the UK National Grid.

may also have a linear or graphic scale, that is, a bar or line divided into segments which correspond to specified distances on the ground. This kind of scale is useful in these days of rapid enlargement and reduction of maps by photocopying machines and computers, because the scale will still be valid at the modified size. Be careful to avoid the common mistake of calling large-scale maps those that cover large areas and that have big numbers in the scale ratios! Large-scale maps show things 'at a large scale', that is, they show things in detail, but the area covered usually is small and the scale ratio likely to be no bigger than 1:50 000 or so. Conversely, small-scale maps show little detail, but cover large areas at a big ratio of scale (1:1 000 000, for example).

1.2.2 Map projection

In small-scale maps, say at 1:500 000 and smaller, the way in which the curved surface of the earth has been projected on to the flat paper is important because of the distortions of angles and areas that can result. You can often find summaries of the different projection systems in the introductory pages of a good atlas. Bugayevskiy and Snyder (1995) provided a comprehensive account. However, the maps normally used for quantitative geological work are at a suffi-

ciently large scale for the effects of projection to be negligible for most purposes.

1.2.3 Grid systems and location

The direction of north is specified on most maps and is normally towards the top of the sheet. Many maps are divided by a **grid system** running north–south and east–west, to aid in locating particular features. Small-scale maps commonly employ latitude and longitude; large-scale maps may involve some arbitrary but standardised system. For example, the UK uses a 'National Grid', summarised in figure 1.2. In the USA the most frequently used method of specifying localities remains the 'township and section' system (figure 1.3). There are, however, increasing attempts to apply the metric Universal Transverse Mercator (UTM) system. This grid already appears on USGS 7½ minute quadrangles and, together with the National Grid, on BGS 1:250 000 sheets.

1.2.4 Relief

Representation of the relief or topography of the land surface is usually omitted from small-scale geological maps. The systems of colour shading commonly employed in small-scale relief maps, for example in many

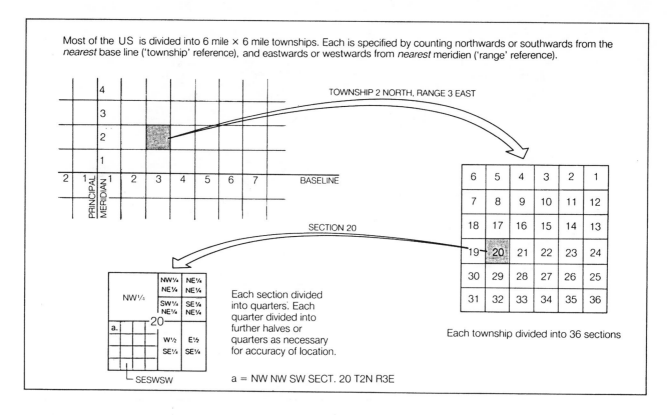

Figure 1.3   Finding locations on maps using the US township and section system.

atlases, would interfere with the colours or ornaments used to depict the geology. However, it may be possible to gain some idea of the relief on such maps from associated features such as river drainage patterns and lakes.

On larger scale maps it is extremely useful to indicate the topography. Older maps employed **hachures**, which can look attractive when executed well and if they do not interfere with the geology. However, they are not quantitative. Hachuring gives a visual impression of relief without specifying the altitude or steepness of slopes. **Spot heights** give very localised information on altitude and are employed on some small-scale maps.

The most successful method of representing relief is by **topographic contour lines** (figure 1.4). These join together points of equal height above some datum, normally sea level. The contour interval is the height difference between adjacent contours. We can estimate the altitude at any point on the map by interpolating between the contour lines. The spacing of the lines indicates the slope of the land. Closely spaced lines reflect steep gradients, curved lines indicate rounded slopes, and so on. Figure 1.5 gives examples. It is vital that you do not memorise a series of 'rules' about the patterns of contour lines, but mentally visualise the relief they are depicting. With a little practice

the ups and downs of the land surface should be apparent in your mind's eye simply by looking at the contour shapes. If more precise information is needed, topographic cross-sections are easy to construct from contour lines. Figure 1.6 shows the method.

It is important to realise that the topographic contour lines are being employed to represent the shape of the land surface, which is three-dimensional, on a flat, two-dimensional piece of paper. In geology, the problem commonly arises of depicting some three-dimensional geological feature on a two-dimensional map, and we use contours for this also. In this book the lines used for the relief of the land surface will always be referred to as **topographic** contours, in order to avoid any confusion with the contour lines to be introduced later for various geological surfaces.

1.2.5   Key, explanation, or legend

There may be further cartographic information provided on the geological map, for example some details of the surveying and production of the topographic part of the map, but it is not normal to provide a full topographic key. The user is assumed to be familiar with the portrayal of roads, political boundaries, rivers and the like. Most of the map key is given over to geological matters.

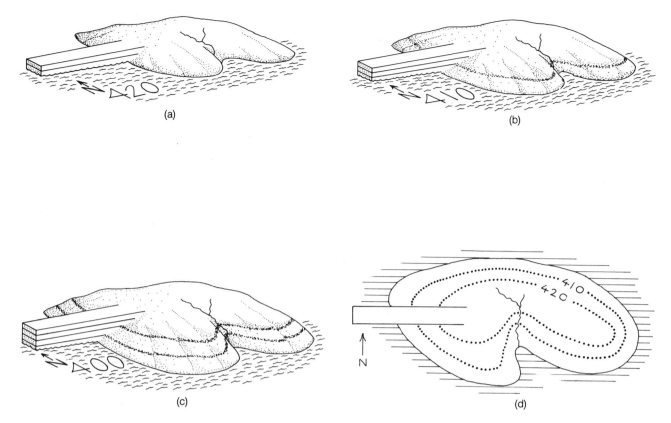

Figure 1.4   The concept of topographic contours, illustrated by a small island in a lake with dropping water level. (a) Lake level at 420 metres altitude. (b) Lake level dropped to 410 m. (c) Lake level dropped to 400 m. Note that the previous lake levels are represented by strand lines on the island. These, like the lake levels, are horizontal, and at 420 and 410 m. (d) Topographic map of the island, lake level at 400 m. Contour interval 10 m. The topographic contour lines, being horizontal, coincide with the strand lines shown in (c).

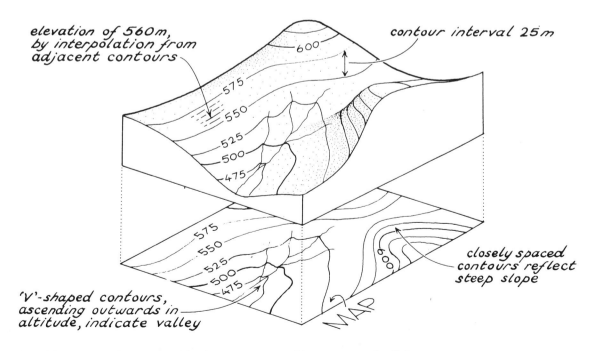

Figure 1.5   Diagram showing relationships between topographic contours and relief.

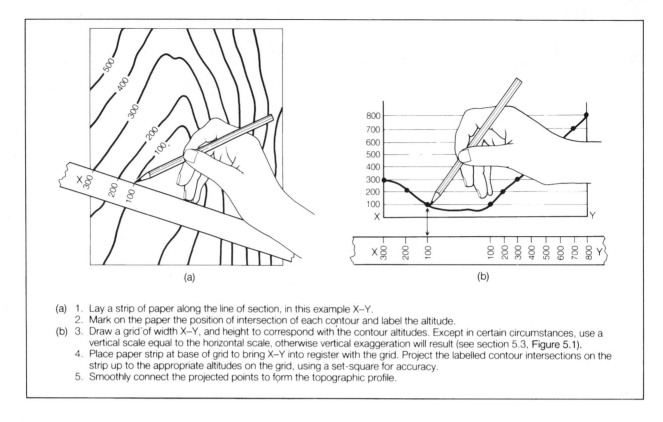

(a) 1. Lay a strip of paper along the line of section, in this example X–Y.
    2. Mark on the paper the position of intersection of each contour and label the altitude.
(b) 3. Draw a grid of width X–Y, and height to correspond with the contour altitudes. Except in certain circumstances, use a
       vertical scale equal to the horizontal scale, otherwise vertical exaggeration will result (see section 5.3, Figure 5.1).
    4. Place paper strip at base of grid to bring X–Y into register with the grid. Project the labelled contour intersections on the
       strip up to the appropriate altitudes on the grid, using a set-square for accuracy.
    5. Smoothly connect the projected points to form the topographic profile.

Figure 1.6   Instructions for drawing a topographic profile from a map.

1.3 Geological Aspects

1.3.1 Key, explanation, or legend

Probably the most striking thing about a typical geological map is its numerous patches of colour. Uncoloured maps have equivalent areas of black and white **ornament**. These colours and ornaments indicate the distribution of the **map units** into which the earth materials have been divided for the purpose of the map. The map **key**, also referred to as a **legend**, **explanation**, or **index**, specifies the geological meaning of the colours and ornaments, together with any symbols used on the map. It can also contain much additional information and should be one of the first things to consult when you examine a map.

On some maps, particularly those at a small scale, the map units represent rocks of different stratigraphic ages, and on some large-scale maps the units are named according to particular fossils that the rocks contain. However, the majority of maps have units divided according to the type of material – the **lithology**. The unit may comprise a single kind of rock or a convenient grouping of different rock types. We commonly refer to a map unit as a **formation** or as 'beds' of rock, and this will be done here, even though

these words may not correspond with their strict stratigraphic definitions. The narrow line that separates two colours or ornaments on a map represents the surface of contact between two adjacent units. The map key may well provide information on the stratigraphic age of the formations, and may make some additional remarks on fossil content, mode of origin, etc. of the beds, but the basic separation of the map units is normally made according to the type of material.

The units may be presented in the key as spaced rectangles, or arranged in a kind of stratigraphic column, sometimes drawn to scale to reflect the thicknesses of the formations and the relationships between them. Whatever the design of the key, it is conventional to show, as far as possible, the oldest units at the bottom and successively younger formations above. The differences between the rock units on the map may be further clarified by adding symbols, usually letters or numbers, to the ornament. Symbols are commonly chosen which convey added information, for instance a letter that acts as an abbreviation for the stratigraphic age of the rock unit. Stratigraphic information is commonly not available for igneous and metamorphic units and so these are listed separately, at the foot of the key.

1.3.2 Superficial and bedrock maps

The loose geological materials, such as sand and peat, that overlie bedrock are known as superficial deposits. These are generally shown on larger-scale geological maps if they are developed substantially, say a metre or more thick, or if they have some particular geological interest. Alluvium, in particular, tends to be shown, in order to emphasise river courses and hence drainage patterns. In general, though, geological maps tend to emphasise the distribution of the different types of bedrock, and certainly any cover of non-geological material such as agricultural soil or concrete is omitted. Maps that exclude superficial deposits altogether are properly called bedrock maps or, by the BGS, solid maps. Conversely, maps that de-emphasise the bedrock are known as superficial deposits, Quaternary, or drift maps (see section 10.5). The BGS has traditionally produced its maps in different versions, variously called 'solid', 'solid and drift', and 'drift' editions according to which aspects are given most emphasis.

The principles of map interpretation that are developed in the following chapters are explained mainly with reference to bedrock formations. They apply, however, to superficial deposits also, with a couple of provisos. These are discussed in section 10.5. Certainly, a familiarity with how to deal with superficial deposits on geological maps is important, for it is, after all, on these materials rather than on bedrock itself that most of our activities are conducted.

Where a map unit of bedrock or superficial deposits reaches the land surface it is said to '**outcrop**', even though it may actually be covered by a thin veneer of material not shown on the geological map, such as soil or a lake. Hence, it is outcrops that are shown on geological maps. Any parts of the outcrop that are not covered at all but are seen as bare material are known as **exposures**. Superficial deposits are exposed in places like river bluffs and sand-pits; exposures of bedrock can be seen in crags and cliffs. Individual exposures are not, however, normally shown on completed geological maps. Note that although some geologists use the terms 'outcrop' and 'exposure' interchangeably, for geological mapwork it is useful to distinguish the two meanings.

1.3.3 The third dimension – geological cross-sections

Geologists often have to deal with the fact that the rocks and structures of the earth are arranged in three dimensions. The distribution of rocks at the earth's surface, which is responsible for the shapes and patterns that are such a striking aspect of geological maps, is simply a function of how this three-dimensional configuration happens to intersect with the present-day

earth's surface, that is, how it outcrops. With practice a geologist can visualise from the outcrop patterns on a map how the rocks are arranged in three dimensions. He or she can picture how the rocks lie below the earth's surface, and how they would once have been above the present land surface, before erosion.

As well as the horizontal map surface, the geology can be depicted in a vertical plane by means of a geological **cross-section**. Most geological maps are accompanied by cross-sections, and the two together are a powerful means of communicating the three-dimensional arrangement of the geology. Maps and sections are two facets of the same thing – the spatial arrangement of the rocks. *Many of the general statements made in this book concerning 'maps' really refer to 'geological maps and cross-sections'.*

Geological maps and sections are so closely interrelated that although cross-sections can be interpreted from maps, in practice the two are usually developed together, and in some cases the geological sections are obtained first and the map is derived from them. An example of this is the exploration for oil below the sea. In this situation there is no accessible bedrock to plot on the map from which we can draw sections! The seismic and drill-core data yield information readily in the vertical plane, enabling a series of geological sections to be built up, and from these sections the geological map is constructed.

1.3.4 The interpretive nature of maps

Although much of this book is concerned with deducing information from completed maps, it is important to understand at the outset that the geological map itself is a highly interpretive document. Numerous interpretive steps are involved in its production. Right from the moment the geological surveyor stands at an exposure of bedrock at the earth's surface, or examines a piece of drill core, he is exercising his own, subjective judgement. The surveyor has continually to decide to which formation, eventually to be a particular colour on the completed map, each rock exposure will be assigned. The map units may be somewhat arbitrary, and because rarely will the bare rock be observable at the land surface, the locations and courses of the geological boundaries between the units will have to be judged. Even the nature of the boundary may be questionable. When you look at most other kinds of maps, for example a road map, you can be sure that something like a road intersection is where the map says it is. Go there and it will be there. Most objects on a geological map, however, represent the surveyor's best estimation of their nature and location. On digging or drilling the map may be proved somewhat wrong.

How the boundaries are depicted on the map has implications not only for the three-dimensional relationships, as mentioned above, but also for portraying what is thought to be the geological evolution of the area. The completed map therefore reflects the surveying team's state of understanding of the map area; even to some extent the state of geological science at the time (Harrison 1963). So, if an area is mapped by two surveyors with contrasting geological ideas, differing maps are likely to result. Oldroyd (1996) reproduced two maps produced at almost exactly the same time by two groups of people with conflicting interpretations of the geology of the area. 'The rocks themselves did not speak', as Oldroyd put it, 'saying how they should be mapped', and consequently the results are quite different. The map units as well as the boundary lines differ. 'The rocks themselves did not change at all', he went on, 'it was their representations that were different'. This is why the geological surveying of a country can never be finalised. Because it is an interpretive document, there can never be an ultimate geological map of an area, as long as geological knowledge continues to improve.

1.3.5   Aesthetics

Geological maps can embody a tremendous amount of observational data, and at the same time have the capacities mentioned above of enabling projection into three dimensions and into past geological times. Nevertheless, despite being such a powerful scientific document, a geological map should be visually pleasing to work with. A good geological map is both scientifically sound and artistically attractive. Indeed, when looking at a map it is often the colours and the design of the map that make the first impact. However, there is no ideal or universally agreed way of presenting information on a geological map. This is one reason why many of the maps reproduced in this book look so different from one another.

Willats (1970) chose to express the aesthetic aspect of maps in a poem, entitled 'Maps and Maidens':

They must be well-proportioned and not too plain;
Colour must be applied carefully and discreetly;
They are more attractive if well dressed but not over dressed;

They are very expensive things to dress up properly;
Even when they look good they can mislead the innocent;
And unless they are very well bred they can be awful liars!

Summary of Chapter

1. Geological maps show the distribution of different earth materials at the earth's surface.
2. Normally the geological data are added to a topographic base map.
3. On a large-scale map it is useful to depict the relief of the land surface, which is best done by topographic contours.
4. The key or legend to the geological map explains the ornament and symbols used to represent the geology, and can contain much information.
5. From the outcrop patterns on the geological map the three-dimensional arrangement of the rocks can be interpreted.
6. Geological cross-sections are complementary to maps in helping portray the three-dimensional arrangement of rocks.
7. The geological map is itself an interpretive document.
8. Although a powerful scientific device, a geological map should be pleasing visually.

Selected Further Reading

Thompson, M.M., 1988. *Maps for America*, 3rd edition. US Geological Survey, 265 p.
  A well-illustrated review of the map products of the USGS, including a short section on geological maps.
Warn, C., 1985. *The Ordnance Survey Map Skills Book*. Arnold-Wheaton and Ordnance Survey, 96 p.
  The book reviews the very basics of working with the topographic products of the UK Ordnance Survey, information now also covered by 'Map skills on CD-ROM', Pebble Shore Information Services, Lewes, Sussex, UK.

CHAPTER 2

The Nature of Geological Maps:
The 'Ten Mile' Map of the UK and the
1:2 500 000 Map of the USA

2.1 Introduction: Cartographic Matters

This chapter will use portions of two real examples of geological maps, one of the UK and one of the USA, to introduce some aspects of map interpretation. It provides a preliminary glimpse of the kinds of interpretations that can be made, before the various concepts are examined more closely in succeeding chapters. We begin by noting some of the cartographic matters; first, the scale. Despite its time-honoured name, the 'Ten Mile' UK map (Plate 1) is actually at a scale slightly larger than 10 miles to the inch, at 1:625 000. There is a north sheet and a south sheet to cover the whole of Great Britain. The USA map (Plate 2) is at 1:2 500 000 and comes in three sheets: the east and west halves of the map and a separate sheet showing the legend.

The UK sheets show degrees and minutes of latitude and longitude at the margins of the map, and also the 10 kilometre squares of the UK National Grid. The USA map shows degrees of latitude and longitude. Note that the maps use an equal area projection, which makes the parallels of latitude more widely spaced than the meridians of longitude. There is no attempt on the relatively small-scale USA map to indicate topography directly; the UK map shows some spot heights, in feet. For example, at [SN7986] the summit of the hill Plynlimon is shown as 2470 feet.

Superficial deposits are included on the USA map, although on the portion reproduced in Plate 2 they are simply termed 'Quaternary'. Plate 1 shows only bedrock, because the UK 'Ten Mile' map is published in two editions, one of which deals specifically with superficial deposits. The 'Quaternary' edition shows such things as boulder clay, glacial gravel, raised beaches, river terrace gravels, lake deposits, and peat.

Sections 10.5 and 11.5 discuss these materials. It is a portion of the solid version that is reproduced here.

The key on the UK map is called an 'Index and Explanation' and is hybrid in nature – it uses different kinds of map units for different kinds of rocks. Intrusive igneous rocks are divided on the basis of lithology – gabbro, granite, diorite, and so on – whereas extrusive igneous rocks are divided by rock type and stratigraphic age. Thus basalt, for instance, appears several times on the key, according to its geological age. The sedimentary rocks are arranged in ascending stratigraphic order, almost wholly using time-stratigraphic names as far as the Devonian period, where there comes in a mixed system involving time, rock type, and the location of the deposits.

The USA map is divided into units on the basis of stratigraphic age. Within some of the stratigraphic intervals the igneous and metamorphic rocks are named according to their lithology. There is, in addition, an attempt to divide the sedimentary rocks of a particular stratigraphic age according to their overall environment of formation, that is, whether they are 'eugeosynclinal' (marine, with volcanic material) or 'continental' deposits. We therefore have to pay close attention to the letter symbols and subscripts of each of the numerous subdivisions. This also explains why, given the size of the country involved, the legend has to occupy its own entire sheet!

2.2 Interpretation of the Maps: Geology and Relief

Let us begin our attempts to see how these geological maps are more than just arrays of fine colours by interpreting some aspects of how geology and topography interact. Starting with the USA map (Plate 2), and the region centred on lat. 43° 0', long. 109° 30', first note the drainage pattern. Many of the river courses are highlighted by the Quaternary material, labelled Q on the map. Presumably this consists here of river alluvium. The areas to the NE are drained by rivers that flow towards the NE, and the SW district is drained by SW-flowing rivers. This drainage divide suggests a central area of upstanding relief, and because many of the rivers originate in small, adjacent but isolated lakes, it is probably an area of irregular topography.

The reason for an area having much higher relief than adjacent ground may well be something to do with contrasting rock types. Here, the apparently higher area is made of granites and gneisses (Wg and Wgn), which are likely to be more durable than the surrounding Tertiary (Tec) materials. However, toughness alone might not account for the relative elevation of such ancient, Precambrian, rocks, which could have been subject to erosion for a very long time. The thick black lines on the map indicate major faults, and it would seem likely that some of these, such as the ones to the SW of the granite–gneiss area, indicate sites of uplift of the masses of resistant rocks. Note that one major fault has been shown as a dotted line where its presence is suspected but covered by a veneer of Tertiary and Quaternary deposits. (In fact, this region consists of the heavily glaciated Wind River Mountains, which are thought to have been up-lifted in Tertiary times along a very major fault.)

In contrast to the last instance, the region shown in pink (Qv) centred on lat. 43° 15', long. 113° 0', shows no drainage at all, except at its margins. Volcanic rocks occupying such a large area are likely to be bedded volcaniclastic rocks or lava flows. Such recent rocks, especially in view of the drainage being off the flanks of the area, may well be lying horizontally. Any rainfall presumably dissipates through pore spaces and underground fissures, which may be common if the rocks are lavas. A picture emerges from the map of a very flat, featureless, volcanic plain, lacking in surface water. (It is the Snake River Plateau. It is no coincidence that this flat empty area of lava flows was used for carrying out the early experiments on nuclear power.)

2.3 Map Patterns and Geological Structure

2.3.1 Dipping formations

On the UK map (Plate 1) divisions 70, 71, 72, 73, and 74 are units of successively younger Ordovician and Silurian ages. They are sedimentary rocks, and so were deposited, back in Lower Palaeozoic times, one on top of the other in a roughly horizontal arrangement (figure 2.1a). During the time since the Silurian, the sediments have been buried, lithified, and eventually brought back to the earth's surface. But if we were to journey from, say, the town of Llandovery [SN7734] to Trecastle [SN8829] we would travel not across the one unit that happens to be at the present erosion level but successively across all five units. The most likely explanation for this is that the rock units have been tilted, as depicted in figure 2.1b. The drawing shows which way the beds of rock must be inclined, for any other direction of tilt would not explain the arrangement seen on the map. It is a general rule that rocks become successively younger in the direction towards which they are inclined. We do not know anything yet, though, about the magnitude of the tilting. The rocks could be gently or steeply inclined, or even rotated to vertical.

On a journey from Abbeycwmhir [SO0571] to Knighton [SO2971] the same five units are also successively crossed, but over a longer distance. It seems that each unit occupies a greater area at the land surface. One immediate explanation for this is simply that the units are thicker here than to the SW – there could have been more of the sediment deposited. The continuing increase in the width of each formation as we look further to the north could suggest a progressive increase in sediment thickness northwards. However, this could be illusory. The units west of Knighton have to be tilted, as explained in the previous paragraph, but the angle might only be small. If the units near Abbeycwmhir are inclined less steeply than those near Llandovery, then, as figures 2.1b and 2.1c show, their outcrop will be broader, even if they have a similar thickness. It may be that both factors are operating; the Abbeycwmhir units could be both thicker and less inclined. It is difficult to gauge on small-scale maps the relative importance of each factor.

If we continued the journey eastwards between Bosbury [SO6943] and Great Malvern [SO7745], the units would be crossed in a very short distance. The reduced width at the surface is so marked that it would seem likely that both the effects mentioned above are operating. The formations are probably relatively thin, and may well be steeply inclined. That the environment of deposition of the sediments was somewhat different here is supported by the development of a limestone unit (coloured turquoise on the map) between divisions 73 and 74. Notice also that the units are here crossed in the reverse sequence from the previous traverses – they become successively younger from east to west. This does not account for

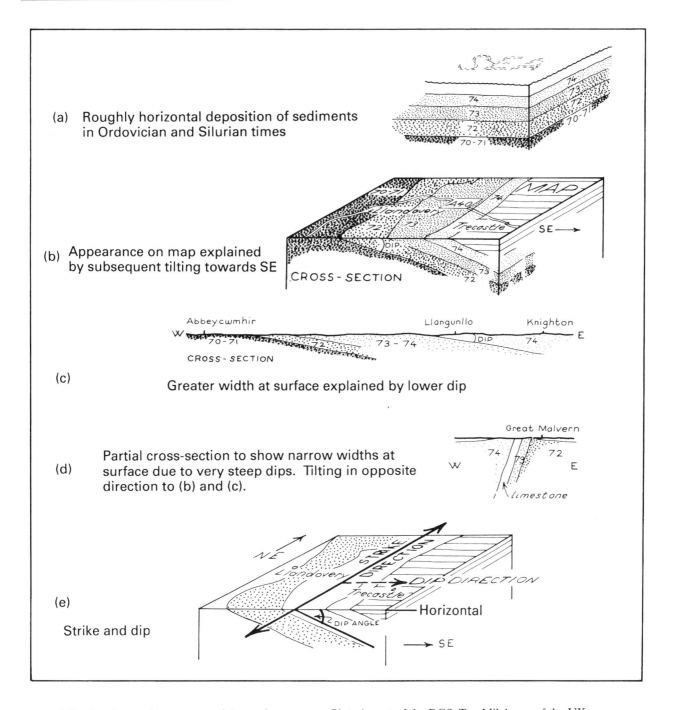

(a) Roughly horizontal deposition of sediments in Ordovician and Silurian times

(b) Appearance on map explained by subsequent tilting towards SE

(c) Greater width at surface explained by lower dip

(d) Partial cross-section to show narrow widths at surface due to very steep dips. Tilting in opposite direction to (b) and (c).

(e) Strike and dip

Figure 2.1 Sketches to show aspects of the geology seen on Plate 1, part of the BGS 'Ten Mile' map of the UK.

their narrowness but it does tell us that the beds here are inclined towards the west (figure 2.1d).

Any tilting from the horizontal which beds of rock show is referred to as the angle of dip, and the direction towards which the beds are inclined is known as the direction of dip. At right angles to the dip direction is the direction of strike, often just called the 'strike' of the beds. These important concepts of strike and dip will be defined and examined more rigorously

in section 4.2. For the moment, we can simply note that the strike direction is reflected by the outcrop patterns, at least on small-scale maps. For example, the rock units just discussed around Great Malvern [SO7745] dip to the NW, and the strike, at right angles to the dip, is N–S, as shown by the N–S arrangement of the outcrops. At Llandovery [SN7734], the outcrops have a NE–SW pattern, indicating a strike in that direction (figure 2.1e). The dip is to the SE.

2.3.2 Unconformities

On Plate 2, along an irregular line running SW from lat. 44° 15', long. 110° 30', NW–SE striking outcrops of rocks varying in age from Precambrian (W) through Upper Palaeozoic to Cretaceous (K), are obliquely cut across by the Quaternary volcanics (Qf) of Yellowstone Park. This appearance – referred to as a discordant relationship – came about in this case because the latter rocks 'are younger, and were laid down, as a series of volcaniclastic rocks and lava flows, on a land surface that already consisted of the NE–SW striking Cretaceous and older rocks. The junction therefore represents a period of geological time (between the Cretaceous and the Quaternary) for which no rocks are present.

A discordant junction that represents a period of non-deposition is known as an unconformity (figure 2.2a). It is commonly recognisable on a map as an irregular line along which younger rocks appear to truncate older rocks. The rocks above an unconformity are markedly younger than those below, and were laid down on a surface developed on the older rocks. In the present example the surface was the landscape of Yellowstone at the time the volcanic material was deposited. Where that preserved surface meets today's landscape, a linear trace is formed, as always with the intersection of two surfaces. Thus on the map we see the trace of unconformity.

A further example of an unconformity occurs on the NE flanks of the Wind River Mountains discussed earlier (section 2.2), around lat. 43° 30', long. 109° 30'. The Tertiary deposits (coloured yellow) appear to be truncating the Cretaceous and older NW–SE-trending units (coloured green, blue, and pink). Just here the Cretaceous and Palaeozoic sequence must be dipping NE, because the units become younger in that direction (see section 2.3.1). The Tertiary material was presumably deposited horizontally on top of the older rocks, after they had been tilted and eroded. The resulting discordant junction therefore represents a span of geological time unrepresented by any rocks (figure 2.2a).

Notice that these lines of unconformity are drawn on the map with the same thickness as ordinary geological boundaries. Apart from the truncating aspect of the above examples, the only difference from normal junctions between sedimentary rock units is the implication of a time gap, a substantial period for which no rocks exist. The reader has to deduce this from the map and its key. We will define and consider unconformities further in chapter 7.

2.3.3 Folded rocks

From Yellowtail Reservoir in the Big Horn Mountains of Wyoming and Montana (Plate 2, lat. 45° 10', long. 108° 10'), a route NE would take us from Upper Palaeozoic rocks across Lower Cretaceous and a succession of Upper Cretaceous units. They are therefore dipping towards the NE. However, SW of Yellowtail Reservoir the units dip in the opposite direction, towards the SW. The simplest explanation of this rather sudden reversal of dip direction is that the units are flexed into an upwarp, the reservoir being situated in the middle (see figure 2.2b). This idea is supported by the fact that to the NW of the reservoir the same units curve round on the map to link up the two oppositely inclined sequences. This curving outcrop pattern, with the oldest rocks in the central part of the arc, is characteristic of beds that are upwarped. Where the curvature of the outcrops is complete, to give a circular aspect to the map pattern, the beds may well be forming a dome.

Continuing SW from Yellowtail Reservoir into the area around the Greybull River, we reach rocks of Tertiary age (Tec). But then the sequence reverses again and further SW units of Palaeocene (Txc) and progressively older Cretaceous age (uK3, uK2) are crossed. These units, too, curve around, SE of Worland, to produce an arcuate pattern, but here with the youngest rocks in the middle. This is the kind of outcrop pattern associated with downwarped units (figure 2.2b). We refer to downwarps on this scale (tens of kilometres or more across) as **basins**, and the upwarps as **arches** or **domes**. Just as with a dome, a completely formed basin will give a roughly circular pattern, concentric around the centre of the structure. A small-scale example occurs at lat. 43° 30', long. 109° 0' and there are several incomplete examples in the vicinity.

This warping of rocks, particularly on a smaller scale, say a few kilometres and less, is known as **folding**. Folds with the oldest rocks in the middle are called **anticlines** and those with the youngest rocks in the middle are called **synclines**. We can detect them on maps by the roughly symmetrical reversal of dip direction (figure 2.2c). If the crest or trough of the fold structure is inclined to the ground surface, it will produce arcuate outcrop patterns like those mentioned above in the Big Horn Mountains.

To give an example from Plate 1, in E Wales around Llanfyllin [SJ0816], the curving outcrop patterns of Silurian rocks (units 70, 71, 72, 73, and 74) are due to a series of anticlines and synclines, the crests and troughs of which are inclined towards the SW. Folded rocks are examined further in chapter 8.

2.3.4 Faulted rocks

Faults are fractures in rocks, along which the rocks have been displaced. Materials of different ages can therefore be brought next to each other. Rather like

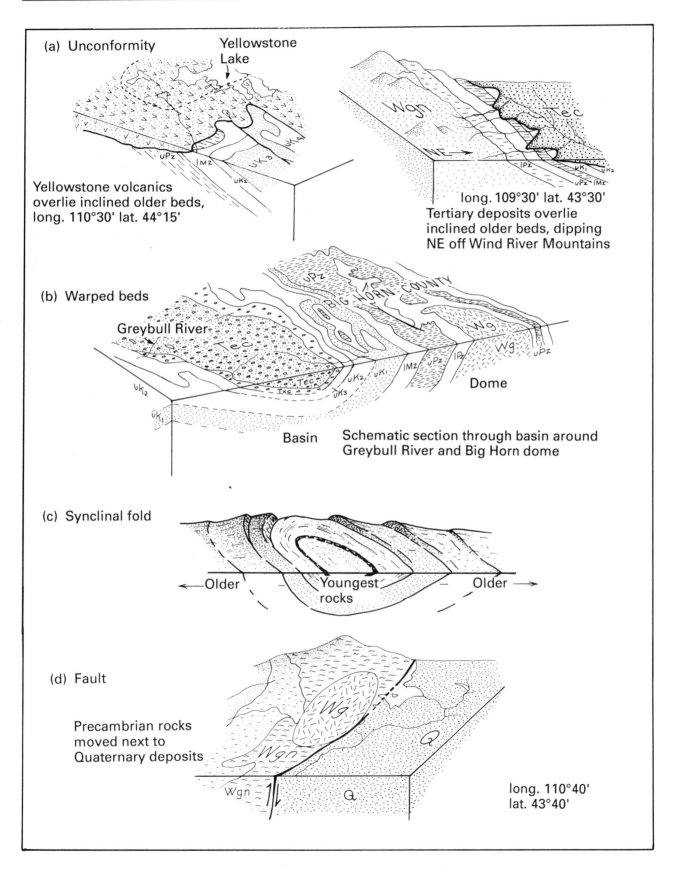

Figure 2.2 Sketches to show aspects of the geology seen on Plate 2, part of the USGS 1:2 500 000 map of the USA.

unconformities, the effect of a time gap is produced, indeed the two kinds of structures can be difficult to tell apart. However, where the geological surveyor has decided that the junction between two different units is a fault, it is usual for this to be indicated on the map by some special symbolism. The UK and USA maps both do this, on the former by a dot–dash line, and on the latter by a heavy line, dotted where concealed by younger deposits.

The fault in Teton County, Wyoming (Plate 2, lat. 43° 40', long. 110° 40') brings Precambrian rocks (Wg) next to Quaternary deposits (Q). In view of the rock displacement required to explain this age contrast, this rather isolated fault must be a major one. More-over, unlike most of the boundaries of the Quaternary outcrops, which are shown to represent normal depositional surfaces, the fact that this boundary is a fault means that the Quaternary deposits themselves have been displaced (figure 2.2d). It is therefore a very young fault and may conceivably be still active. (In fact this fault, the Teton Fault, has displaced the Precambrian rocks by over 8 km, frequently produces tremors today, and has a profound effect on the landscape of this area.)

Eighty miles to the south, around the Idaho–Wyoming state line (Plate 2, lat. 43° 0', long. 110° 45'), a closely spaced system of faults interrupts the upper Palaeozoic (blue, uPz) – lower Cretaceous (green, lK) succession. Most of the normal geological boundaries that are shown indicate an overall dip to the west (going westwards, Upper Palaeozoic units give way to ones of Mesozoic age). But towards the west of the area, instead of the Upper Palaeozoic rocks being at depth and therefore not shown on the map, a number of these faults have displaced the Palaeozoic rocks back to the surface. Such systems of closely spaced faults that can bring older rocks up from depth are typical of what are called thrust belts, which we will look at more closely in section 9.7. Notice two points regarding the age of these faults: (a) in places the faults curve, together with the outcrops of the rocks, suggesting that the faults themselves have been folded; and (b) these faults pass underneath deposits of Tertiary and Quaternary age, indicating that movement along the fractures had ceased before Tertiary times. (These are actually contraction (thrust) faults; part of the important Cretaceous-age Idaho–Wyoming thrust belt.) We will look at faults again in chapter 9.

2.3.5 Igneous rocks, and geological histories

On Plate 1, the long, narrow strips of basalt and dolerite around Dolgellau [SH7317] parallel the nearby sedimentary and volcaniclastic rocks and are therefore likely to be the concordant igneous bodies known as **sills**. In contrast, the igneous rocks on Plate 2 around lat. 46° 10', long. 110° 20', outcrop with no relation to the surrounding rocks. These must be the discordant sheets called dykes. In this example, they are **radial dykes**, arranged like spokes around a hub of igneous material.

It is common with igneous rocks to be able to deduce something from a map about their relative ages. For example, the dykes in the last example cut across Tertiary rocks, and must therefore have been intruded at some more recent geological time. The principle that a geological feature that cuts another must be the younger of the two, often called **cross-cutting relationships**, is a fundamental one in deducing geological histories. It is by no means confined to igneous rocks. For example, the concentric outcrops centred on lat. 42° 10', long. 106° 20' become younger outwards, and are therefore folded in a basin form. However, they are also faulted, the movement of the units being recorded by the offsets of the outcrops along the fault lines. But it is the folded units that are faulted, that is, the folding must have preceded the faulting of the rocks. There was an example at the end of the previous section of how we deduce information on the relative ages of faults. Interpretation of the sequence in which events took place is an important aspect of geological map work.

2.4   Conclusion

Even at this preliminary stage, we have begun to see the wide variety of things that we can interpret from a geological map. From the relationships between rocks and structures of different ages, we can interpret something of the geological history of the area. Some idea of the three-dimensional nature of the rocks can be obtained, especially by noting particular outcrop patterns. All these concepts will be explored more fully in succeeding chapters. We begin by considering in more detail one of the most powerful and useful aspects of large-scale geological maps: the accurate representation of rocks in three dimensions.

Summary of Chapter

1. Non-contoured geological maps may yield information on geology and topography, especially from drainage patterns.
2. Rocks become successively younger in the direction in which they are dipping.
3. Relatively narrow outcrops may reflect steeper dips or thinner units, or some combination of both.

4. Unconformities represent missing stratigraphy, and may appear as discordant junctions between units of different geological ages.

5. Folds produce symmetrically repeated outcrop patterns. The units become younger outwards from the centre of domes and anticlines, and older outwards from the centre of basins and synclines.

6. Faults, fractures along which the rocks have been displaced, are commonly depicted on maps by a particular line symbol.

7. We can deduce the form of igneous bodies, such as dykes and sills, from maps.

8. Aspects of the geological history of an area can be interpreted from maps.

CHAPTER 3

The Three-Dimensional Aspect: Structure Contours

3.1 Introduction

The problem of representing three-dimensional things on a flat piece of paper has exercised minds for many years, nowhere more so than with maps. Many atlases begin by discussing the question of how to represent the spherical earth in a book. A similar problem is the portrayal of the undulations – the relief – of the earth's surface. Early map-makers attempted to depict relief by drawing humpy little hills, often wildly exaggerated in height and steepness. Better pictorial methods gradually evolved, such as shading and hachuring, but in general these are unsuited to geological maps. By far the most successful means yet devised are topographic contour lines. These are now common on larger scale geological maps, say 1:100 000 or larger. Figures 1.4 and 1.5 illustrated the concept of topographic contours, and figure 1.6 showed how to construct topographic profiles from them. It is important that you are completely familiar with topographic contours.

This chapter is concerned with applying all these principles to *underground* surfaces. It begins by emphasising the similarity between topographic contours and those drawn for underground surfaces, called structure contours. The chapter explains how structure contours are derived, and illustrates their use. Although these days the routine construction and manipulation of structure contours are increasingly being carried out by computer methods, the understanding of the three-dimensional principles behind them remains fundamental.

3.2 The Nature of Structure Contours

Contour lines can be used to represent on a piece of paper any three-dimensional surface – it does not have to be the relief of the land. The contours drawn in figure 3.1a could equally represent the shape of the earth's surface or, say, the surface[1] of a rock formation. Although the formation may be underground, it still has an altitude, and the contour lines simply join the points on either its top or bottom surface that have equal height.

It is possible, therefore, to draw on a map contour lines which portray not the land surface but the position and undulations of some underground surface. Such contour lines are called **structure contours**. Without labels, the lines on figure 3.1a could be topographic contours representing a hill, but they could equally well be structure contours depicting a map unit that has been upwarped into a dome. The structure contours sketched in figure 3.1b are of a surface that has the form of a basin, and in figure 3.1c they depict a dome. Note that because the dome in figure 3.1c is deeply buried, the altitudes are negative with respect to sea level.

If the structure contours of a surface are known, a cross-section can readily be constructed, in an exactly analogous way to a topographic profile (figure 1.6). Instead of marking on a strip of paper where the *topographic* contours meet the line of the section, the position and altitude of the *structure* contours are marked, and transferred to the section grid. This has been done to produce the cross-sections shown in

---

[1] Note that the word 'surface' has two slightly different meanings in the map context. In addition to meaning the *land* surface – the outer surface of the earth – the term also applies to the boundary of *any* geological body or curving geological plane, and these can be underground. Thus geologists talk both about 'the beds outcropping *at* the land surface', the ground on which we live, and 'the outcrop *of* a surface', such as a fault or the boundary of a map unit.

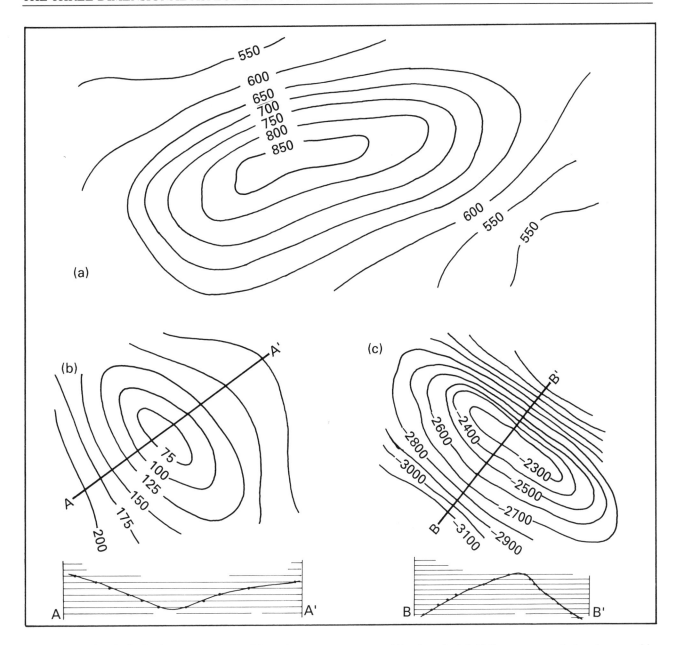

Figure 3.1 The similarity between topographic and structure contours. The lines in (a), if they are regarded as topographic contours, represent a hill. Equally they could be regarded as structure contours representing a dome-shaped geological surface, buried beneath some higher land surface. The lines in (b), without information on whether they are topographic or structure contours, could represent either a depression in the land surface or a buried structure of basinal form. The lines in (c), although similar in form to topographic contours representing a hill, are likely, in view of their negative altitudes, to be structure contours representing a subsurface dome. In practice, different line symbolisms are used on maps to differentiate the two kinds of contours.

figures 3.1b and 3.1c. Some published maps include structure contours, normally of a surface that is considered representative of the structure of the area, but it usually falls to the map reader to construct them. Structure contours can be drawn for any geological surface, for example, an unconformity or the boundaries of an igneous intrusion.

3.3 Examples of Structure Contours on Maps

Figure 3.2a is a structure contour map of the Ekofisk oilfield, the first of the giant oilfields to be discovered in the North Sea. The surface for which the structure contours are drawn is the top of the rock unit which contains most of the oil. The contours show this

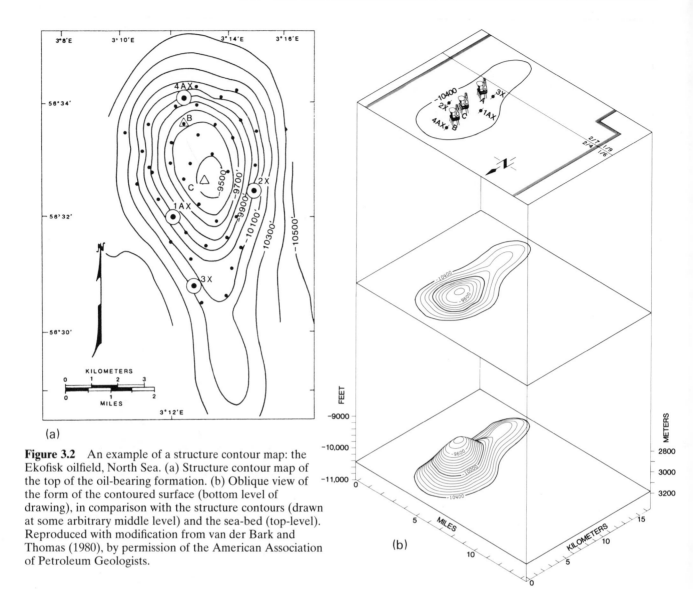

(a)

Figure 3.2 An example of a structure contour map: the Ekofisk oilfield, North Sea. (a) Structure contour map of the top of the oil-bearing formation. (b) Oblique view of the form of the contoured surface (bottom level of drawing), in comparison with the structure contours (drawn at some arbitrary middle level) and the sea-bed (top-level). Reproduced with modification from van der Bark and Thomas (1980), by permission of the American Association of Petroleum Geologists.

formation to be in the form of a deeply buried dome, slightly elongate in a N–S direction. It is in the upper part of this dome that the oil is trapped.

Figure 3.2b is an inclined view of the structure which may help you see how the structure contours are representing the dome. It is vital that you become used to visualising structure contours in three dimensions. Map 1 provides an exercise. Figure 3.2b also shows the location of the main oil wells, positioned to penetrate the oil in the crest of the dome. Perhaps it is becoming apparent to you why structure contours are of such value in applied geology. In fact, it was for practical reasons that the device was originated – in the anthracite fields of Pennsylvania.

Figure 3.3 illustrates an unusual practical use of structure contours, as well as a very irregular contour pattern (see section 4.2 for the reasons for the irregularities). In the area south of Bordeaux, France, the

quality of the grapes depends upon the soil, which in turn depends upon the depth to a particular limestone, known as the Calcaire à Astéries. Better wine is likely to be made where the limestone is buried by no more than a few metres. The structure contours for the top surface of the limestone allow, by comparing with a map showing topographic altitudes, determination of the depth to the limestone at any locality of interest. This provides an initial guide to the likely value of a specific locality for wine-making.

3.4 Structure Contours Derived from Borehole/Well Information

But how can we know where the structure contours run, if the surface they represent is out of sight

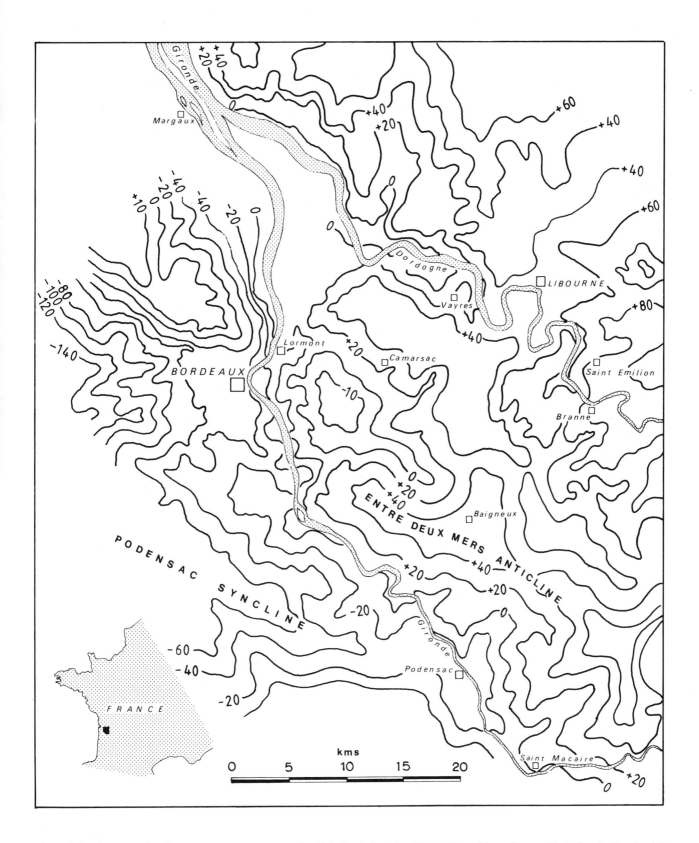

Figure 3.3 An example of a structure contour map: the Calcaire à Astéries, Entre-Deux-Mers, France. Note the shallow burial levels of the contoured surface. Reasons for the irregularity of the contours are mentioned in section 4.2. Based on Vigneaux and Leneuf (1980).

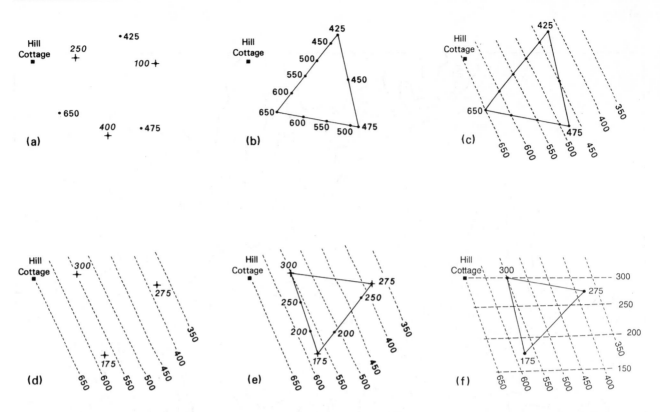

Figure 3.4 Drawing structure contours using interpolation: an illustration involving estimating the depth to an aquifer below a property, given a restricted amount of information. (a) The information provided: a map showing the location of the property, called Hill Cottage, three topographic spot heights, the locations of three boreholes (crosses) and the depths at which the top of the aquifer was encountered in each one. The problem: does the aquifer occur beneath Hill Cottage and, if so, at what depth? (b) Interpolation between the spot heights in order to derive further altitudes, such as those at 600, 550, 500 metres (borehole data omitted here). The distances between the known heights are divided in proportion, to derive additional, round number control points. For example, a point halfway along the right-hand side, halfway between 425 and 475 m, will give the 450 m point; each of seven equal divisions along the base of the triangle represents 25 m (650 – 475 = 175, divided by 7 = 25), therefore the 500 m point is one division along from the 475 m corner, two more divisions gives the 550 m point, etc. (c) Linking the points interpolated in (b) to give topographic contours (borehole data omitted here). Contour interval = 50 m. The land surface appears to be, on the basis of the information available, a smooth slope inclined towards the ENE. (d) Converting the depths to the aquifer, from the borehole data given in (a), to altitudes. For example, drilling at the borehole nearest to Hill Cottage commenced (at the land surface) at an altitude of 550 m and at 250 m depth hit the aquifer, which therefore itself has an altitude of 300 m (550 – 250 = 300); the borehole which encountered the aquifer at 100 m depth commenced at an altitude of 375 m (by interpolation between the two nearest topographic contours), hitting the aquifer at an altitude of 275 m (375 – 100 = 275). (e) Interpolation between the aquifer altitudes in order to derive altitudes at 200 and 250 m. (f) Linking the altitudes interpolated in (e) to give structure contours for the top surface of the aquifer. The surface appears to be gently inclined towards the south. The difference in value between the topographic contour and the structure contour at any specific point gives the depth of the aquifer there. At Hill Cottage the depth is 350 m (650 – 300 = 350). (Although the value could have been obtained by drawing depth contours for the aquifer, the form of the surface would be difficult to visualise from depth values alone, and so it is more normal to contour altitudes.)

underground? The most common method of deriving them, especially in industry, is to use information obtained by drilling boreholes, called wells in the hydrocarbons industry. If the altitude of the land surface where the drilling starts is known, the depth at which the bedding-surface of interest is encountered can be measured, and, by subtraction, its elevation derived (figures 3.4a and d). If at least three elevations are known, it is possible to interpolate values and estimate the structure contour pattern (figures 3.4e and f). In oilfields, there may be a number of wells providing

depth information, and hence greater control on the form of the buried surface. Also, because hydrocarbon work commonly involves surfaces at considerable depths of burial, the altitudes with respect to sea level may well be negative; they are often still quoted in feet. Figure 3.5 is a worked example, and Map 2 provides an exercise.

Any other available information will be added in to help control the accuracy of the structure contours. In oil exploration, especially, the data from seismic sections will be included.

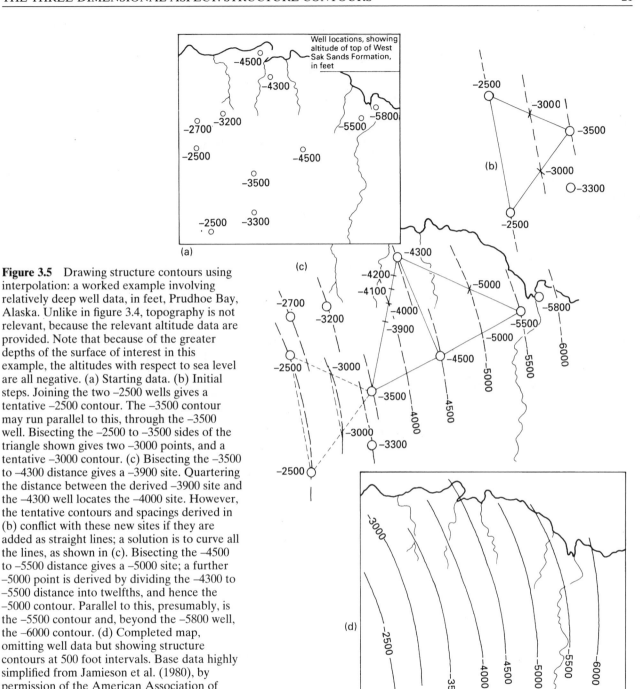

Figure 3.5 Drawing structure contours using interpolation: a worked example involving relatively deep well data, in feet, Prudhoe Bay, Alaska. Unlike in figure 3.4, topography is not relevant, because the relevant altitude data are provided. Note that because of the greater depths of the surface of interest in this example, the altitudes with respect to sea level are all negative. (a) Starting data. (b) Initial steps. Joining the two –2500 wells gives a tentative –2500 contour. The –3500 contour may run parallel to this, through the –3500 well. Bisecting the –2500 to –3500 sides of the triangle shown gives two –3000 points, and a tentative –3000 contour. (c) Bisecting the –3500 to –4300 distance gives a –3900 site. Quartering the distance between the derived –3900 site and the –4300 well locates the –4000 site. However, the tentative contours and spacings derived in (b) conflict with these new sites if they are added as straight lines; a solution is to curve all the lines, as shown in (c). Bisecting the –4500 to –5500 distance gives a –5000 site; a further –5000 point is derived by dividing the –4300 to –5500 distance into twelfths, and hence the –5000 contour. Parallel to this, presumably, is the –5500 contour and, beyond the –5800 well, the –6000 contour. (d) Completed map, omitting well data but showing structure contours at 500 foot intervals. Base data highly simplified from Jamieson et al. (1980), by permission of the American Association of Petroleum Geologists.

3.5 Structure Contours Derived from Topography – The Theory

It is possible to construct structure contours for near-surface rocks without borehole information, if the formation outcrops. For this, it is necessary to know the various topographic altitudes at which the unit reaches the land surface. This, of course, is exactly what is shown on a geological map. Therefore, this technique is much used, both in academic research and commercial work. In fact, the three-dimensional thinking that the method involves is relevant to many kinds of subsurface work. However, beginners can find the concepts difficult. Therefore, the following explanation starts with a broader consideration of the problem and develops the method step by step.

Topographic contours and structure contours both work on the principle of connecting points of equal

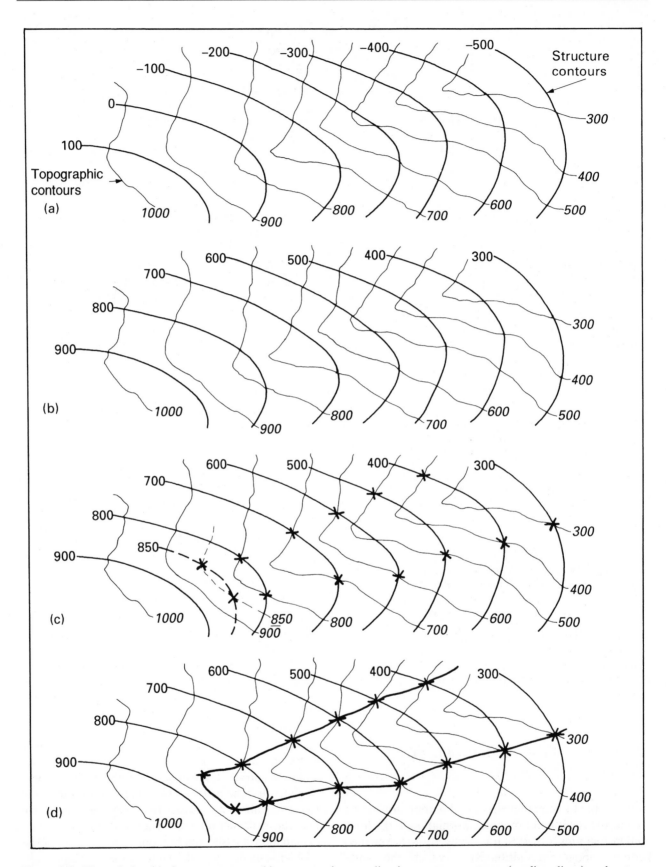

Figure 3.6   The relationship between topographic contours (narrow lines), structure contours (medium lines) and outcrop (broad lines). Section 3.5 explains this figure in detail.

elevation, differing merely in the particular surfaces they represent. They can be shown together on a single map. In figure 3.6a the topographic contours are depicting a land surface ranging between 300 and 1000 metres in elevation, whereas the structure contours are showing a curved formation boundary at an altitude of between –500 and 100 metres. The formation surface is therefore underground. You should look at figure 3.6a and make sure you can see in your mind's eye the two separate surfaces represented: the land surface and the formation at depth.

Figure 3.6b shows a rather similar situation – exactly the same land surface, and structure contours for a curved formation surface. But there is an important difference from figure 3.6a. The particular formation boundary shown, although of the same shape as before, is at a higher altitude and so is less deeply buried. In fact, in some places the topographic contours and the structure contours intersect at precisely the same value of altitude. Where this is the case, the formation surface cannot be buried at all. The formation will actually intersect with the earth's surface: it will outcrop. Figure 3.6c is exactly the same as 3.6b except that it shows the points where the contours are showing that the formation boundary must outcrop.

The values of the contours given on the map depend on the contour interval; there may well be intermediate elevations where the bed also reaches the surface. Figure 3.6c shows in addition the interpolated 850 metre contours, and hence further points where the bed would be expected to outcrop. Figure 3.6d shows the logical extension of this – *all* the points where the bedding surface should reach the land surface. These merge together to form a linear trace. This line is the trace of where the bedding intersects with the land surface: it is the *outcrop* of that bedding surface. Thus, by knowing the structure contours for a geological surface and the topographic contours we have predicted what should be the outcrop of that surface. This is a procedure that finds some use in field surveying (section 15.2), but it is a variation on this approach that allows us to make the more valuable construction of deriving structure contours from topography.

For most purposes it is much more useful to construct not the outcrop from known structure contours and topography, but structure contours from known outcrop and topography, for these last two factors are available at the land surface. It is possible for just the same reasons as developed above. Figure 3.7a shows a small part of the outcrop of a geological surface and it shows topographic contours. Figure 3.7b shows locations through which the 200, 300, and 400 metre structure contours must pass, and figure 3.7e represents the only way in which the structure contours can be satis-factorily drawn in this example. Section 3.6 discusses further why this is the only possible path. (An alternative interpretation involving all the boundaries being wholly vertical is also possible here, but would normally be identifiable if more of the map were seen.)

A large-scale geological map will usually supply the information on topography and outcrop, and so in principle we can construct on it structure contours for surfaces of interest. We are therefore in a position to predict *from a map*, the three-dimensional arrangement of an outcropping formation. Imagine the practical applications of this: a hydrogeologist can derive the location and form of an aquifer. A mining geologist can estimate the length of tunnel or drill-hole necessary to reach the material of value. An engineering geologist can assess the nature of the rocks he is considering excavating for a building foundation. If we derive structure contours for the top and bottom of a formation of commercial value, we will be able to calculate its volume, that is, we can estimate reserves. We have arrived at one of the great practical uses of geological maps.

3.6 Structure Contours Derived from Topography – The Practice

Begin by locating on the map the outcrop of the surface that you are going to contour. If you are interested in the top of a formation, make sure you are dealing with the top surface and not the base! The top will be adjacent to the next youngest formation, and the dip direction of the unit, if it is known, will be towards it (section 2.3.1).

Look for places where the outcrop of the surface crosses topographic contours, and start your construction in an area where there are plenty of intersections. Leave until last those areas where there are few intersections, and therefore least control on the route of the structure contours. Where the outcrop crosses or meets a topographic contour, you know the surface must be at the same altitude as that topographic contour. If you can locate two or more reasonably close intersections with the same altitude, you can tentatively connect them to produce the structure contour for that altitude.

It may seem at first that there are several courses the structure contour could take through the point of known altitude (e.g. see figure 3.7). However, only one of them will correspond with the actual outcrop that is shown on the map (figure 3.7d). For example, if the structure contour (drawn for the *top* of a unit) of a certain altitude crosses an intersection into ground of a lower altitude, then the formation that is being

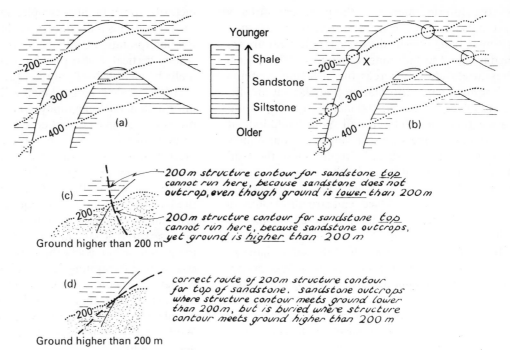

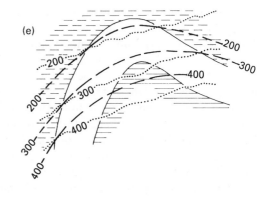

Figure 3.7 Constructing structure contours from topography and outcrop. (a) Portion of a hypothetical geological map, showing topographic contours (dotted) and the outcrop of a sandstone unit. The top of the sandstone is to be contoured. (b) Preliminary steps. Reference to the stratigraphic sequence shown in the key enables the *top* of the sandstone to be located on the map. Circles indicate where the altitude of the top of the unit is known, from intersections with topographic contours: the structure contours will pass through these circles. Consider the circle at X. At first it may seem that there are two possible routes for the 200 metre structure contour to pass through the outcrop/200 m topographic contour intersection, as illustrated in (c) and (d). However, the route shown in (c) is not compatible with the map information, and only the route shown in (d) can be correct. (e) shows the 200, 300 and 400 m structure contours completed from the map information.

contoured will be outcropping there. On the other hand, if the same structure contour enters ground of higher altitude, then the outcrop there will be of material stratigraphically above the contoured formation.

The structure contour of a surface can only cross a topographic contour of the same altitude at a point where that surface outcrops. There is nothing wrong with it crossing topographic contours of higher altitudes, provided the surface at those places is buried. Conversely, it can cross topographic contours of lower altitudes, provided the map shows the surface being contoured to have been eroded away at those points.

Another help in drawing the course of a structure contour is to sketch in lightly some interpolated intermediate altitudes to obtain more control points. In most cases the structure contour will curve smoothly; if it makes violent twists it is likely to be wrong, or there has been faulting of the rocks. Experience

counts a lot in drawing satisfactory structure contours and there is usually a fair amount of trial and error.

When the tentative structure contour seems to be obeying all the topographic and outcrop information in the starting area, it can be firmed up. You can then add in further structure contours in that vicinity, using the same methods. Adjacent structure contours tend to be parallel, so that once one is drawn with confidence, it serves as a guide for the nearby ones. They are likely to be evenly spaced. Structure contours can only touch each other where the surface is vertical. These latter guides take priority over any interpolated points, which, after all, are only hypothetical. Developing several adjacent contours together usually gives better results than completing each line in turn. There is normally little point in adding structure contours of *higher* altitude than the present-day land surface, that is, representing where the contoured surface used to be before erosion. On the other hand, adding *subsurface*

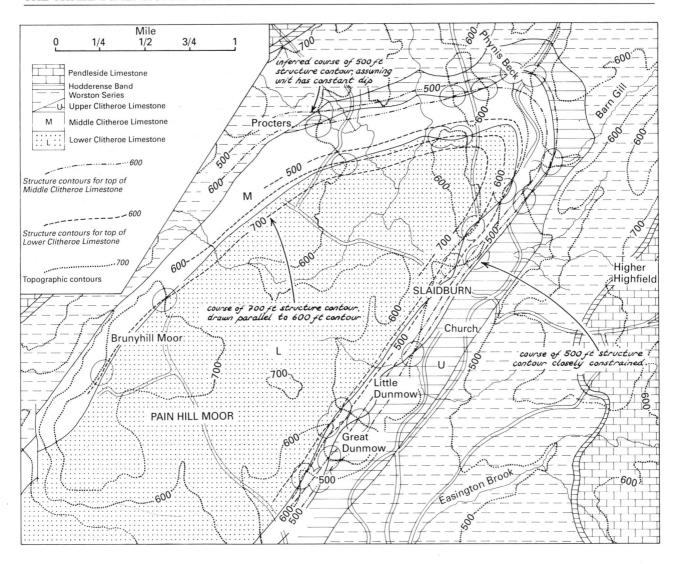

Figure 3.8 Deriving structure contours from outcrop and topographic contours: a worked example, Slaidburn, Yorkshire. The ellipses enclose areas of information particularly useful in drawing the structure contours. With the contour patterns established, additional subsurface contours, say 400, 300 and 200 feet, could be added to allow subsurface predictions. Geological map redrawn from Parkinson (1936), by permission of the Geological Society, London.

contours is of immense practical use, as mentioned earlier, in predicting the underground location of materials.

It is usually easiest to work progressively outwards from the starting area, but with some maps it is necessary to sketch the structure contours for several separated areas where there is good topographic control, and then to extrapolate between them. A look at the outcrop patterns on the map should give you some idea of the form of the rocks (section 2.3) and therefore the kind of overall shape the structure contours are likely to have. With practice you will develop your own preferred way of tackling these constructions.

Figure 3.8 shows a worked example of structure contours derived from the intersection of outcrops and topography on a real geological map. Some explana-

tory comments are added. The important thing when drawing structure contours is not to try to apply a series of memorised rules, but to *understand* the procedure. Always try to visualise in three dimensions what you are doing. Working geologists do not spend vast amounts of time carrying out these constructions, especially in these days of assistance from computers, but an understanding of how the methods work is paramount.

3.7 Structure Contours Derived from Topography and Boreholes

Deriving structure contours from outcrop and topography is useful in near-surface operations, but the

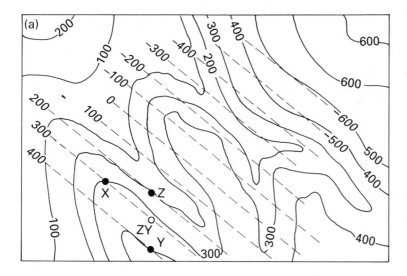

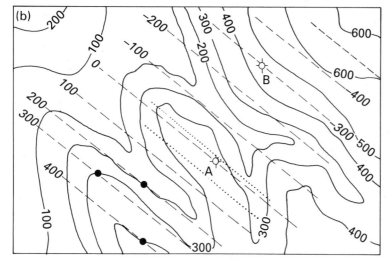

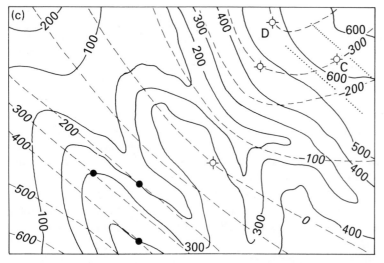

Figure 3.9 An example of the use of borehole/well information to constrain the routes of structure contours drawn from topography. (a) Topographic contours (solid lines) and known outcrops of coal seam at X, Y and Z (solid black circles). At location ZY, halfway between Z (200 metres) and Y (400 m), altitude of seam is presumably 300 m, enabling 300 m structure contour to be drawn through ZY and X. Parallel to this, 400 m structure contour is drawn through Y and 200 m structure contour through Z. Additional structure contours are equidistant, assuming uniform dip of the seam.
(b) Borehole at A encounters coal seam not at –80 m as predicted from (a) (0 and –100 m structure contours of (a) shown as dotted lines), but at 0 m. This suggests dip decreases northeastwards; structure contours in (b), with decreased spacing, reflect this new information. Borehole at B confirms seam at –300 m, as predicted. Structure contours of (b) are best interpretation of information from three outcrops and two boreholes.
(c) Further borehole at C fails to encounter seam at –460 m as predicted from (b). (–400 and –500 m structure contours of (b) shown as dotted lines), but at –300 m. This could indicate a reversal of dip direction in the NE of the area (i.e. seam dips SW), in which case borehole D should encounter seam at about –320 m. However, seam at D is met at –400 m, suggesting that structure contours are not parallel, leading to the refined interpretation shown in (c).

reliability of underground predictions falls off as increasing extrapolation becomes necessary. It may become too approximate for commercial work on more deeply buried rocks. Then it becomes necessary to supplement the map information with some direct underground data. Drilling is expensive, but a carefully sited borehole or two can greatly constrain where the structure contours can be drawn. Figure 3.9 gives an example. Map 5 exercises the use of outcrop and borehole information in conjunction.

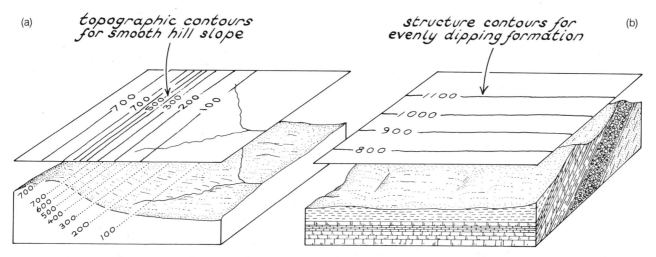

Figure 3.10 The significance of straight contours (strike lines). (a) Straight topographic contours resulting from smooth hillslope with consistent direction. The varying gradient is represented by the spacing of the contours. (b) Straight structure contours representing a smooth geological surface; the even spacing reflects a uniform angle of dip.

3.8 Straight Structure Contours

Structure contours – to repeat – are exactly the same as topographic contours except that they represent some underground surface. They can, however, look a bit different. Figure 3.10a shows the topographic contours for some hypothetical smooth hillslope of fairly even gradient. The topographic contours are straight and evenly spaced, merely becoming closer where the gradient is steeper. Rarely are topographic contours actually like this on maps because natural hillslopes usually have various irregularities due to erosion. Bedding surfaces, however, can have this appearance on large-scale maps, if the inclined plane is smooth and non-undulating (figure 3.10b). Here, the structure contours will appear as straight lines. They are, however, unlikely to be dead straight. You should not construct structure contours with a ruler; natural planes are not that smooth.

Straight structure contours are sometimes referred to as **strike-lines**. This is because structure contours everywhere parallel the strike of the surface they are representing, which is conspicuously constant in direction if the lines are straight. The strike direction is therefore readily visualised and measured from them. Knowledge of the strike direction is essential in assessing the orientation of geological surfaces. The idea of strike and dip was introduced in section 2.3.1 and it is a basic working concept in geology. It enables us to *measure* orientations, and it is to this business of quantifying things from geological maps that we now turn.

Summary of Chapter

1. Structure contours are similar to topographic contours, but instead of representing the land surface they portray some specified underground surface such as the boundary of a map unit.
2. They depict in map view the position and form of the underground surface, and are therefore a highly useful construction.
3. They can be constructed from borehole/well information, by interpolating the elevations of the surface between the holes.
4. Structure contours for outcropping surfaces can be constructed from the topographic elevations at which they outcrop.
5. The drawing of structure contours from outcrop elevations is more closely controlled if there is borehole information in addition.
6. Structure contours for smooth, uniformly inclined surfaces are straight, and are also called strike-lines.

Selected Further Reading

Badgley, P.C., 1959. *Structural Methods for the Exploration Geologist.* Harper & Brothers, New York, 280p.

Chapter 4 of this advanced book is about structure contour maps. It includes a list of properties of structure contours and constructing hints.

Ragan, D.M., 1985. *Structural Geology. An Introduction to Geometrical Techniques*, 3rd edition. John Wiley, New York, 393p.

Chapter 18 is a brief treatment of structure contours.

Tearpock, D.J. and Bischke, R.E., 1991. *Applied Subsurface Geological Mapping.* Prentice-Hall, Englewood Cliffs, New Jersey, 648 p.

Chapter 8 is an advanced treatment of structure contours as used in the oil industry.

Map 1   Raton, New Mexico, USA

Location of
map 1

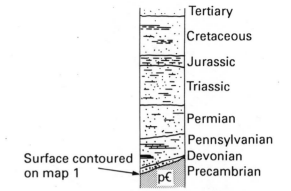

Tertiary
Cretaceous
Jurassic
Triassic
Permian
Pennsylvanian
Devonian
Precambrian

Surface contoured
on map 1

pЄ

In northeastern New Mexico, between the towns of Las Vegas and Raton, is a thick sequence of Palaeozoic and Mesozoic clastic sedimentary rocks. Gas has been extracted from some of the Cretaceous rocks. In fact, there are signs that hydrocarbons have been widely generated in the area, but suitable traps have proved elusive. In the search for oil and gas traps, numerous structure contour maps have been constructed for various stratigraphic horizons.

It turns out that the overall control on the form of the sedimentary basin is the Precambrian basement. On its surface the sedimentary pile accumulated. The form of this surface is depicted in the structure contour map opposite, reproduced with slight modification from Woodward (1984), by permission of the American Association of Petroleum Geologists.

Where in the area is the thickest accumulation of sediments likely to be? If the present-day land surface at that site is 2000 metres above sea level, how thick are the sediments there? Describe in words the form of the Precambrian basement in the vicinity of that site.

Away from the structure just discussed, where is the next thickest sedimentary accumulation likely to be? How does the structure here differ from that described above? What kind of structure separates the two areas? Describe its orientation.

Where in the map area is the highest point on the Precambrian surface? Where does it show the steepest gradient? Where is it least steep?

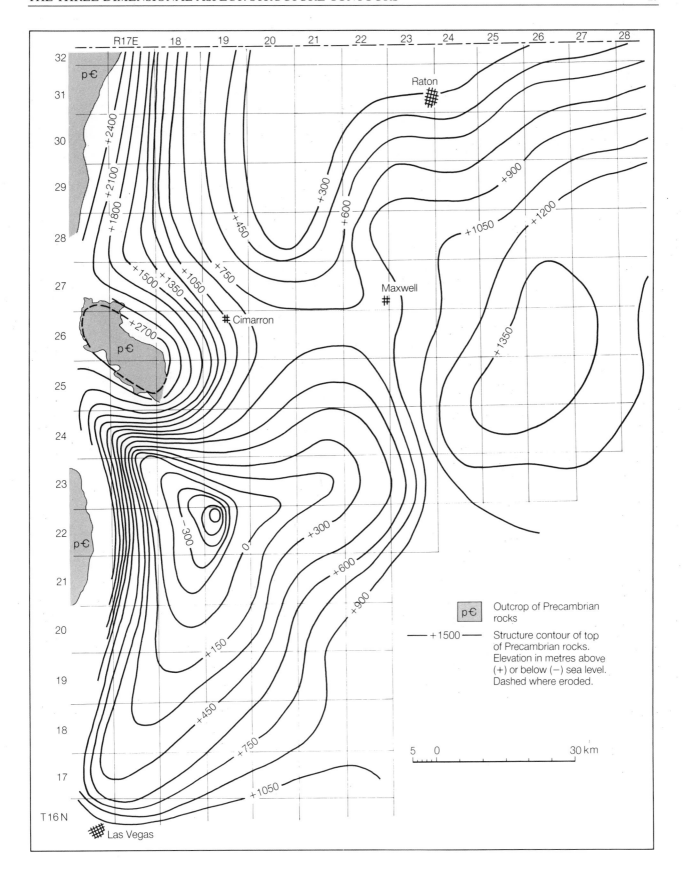

Map 2   Lacq gas field, Aquitaine, France

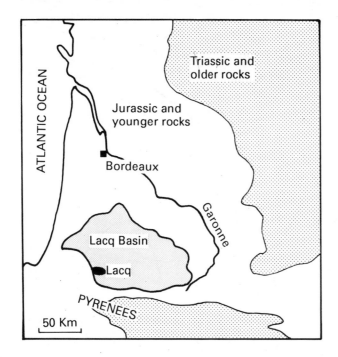

Molasse	Sandy shale	
Tertiary	Shale and Sandstone beds	
Upper Cretaceous	Argillaceous Limestone and Dolomite	
Albian-Aptian		
	Marls, Limestone and Dolomite	
	Limestone	
	Marls (Ste Suzanne)	
Neocomian	Limestone and Anhydritic Shale	
Upper Jurassic		
	Limestone and Dolomitic Beds	
	Ammonite marls	
Dogger	Crystalline Limestone	
Lias	Shale and Limestone	
	Shale	
	Dolomite, Anhydrite and Shale	

During the decade 1950–1960, extensive drilling defined what was to become France's largest gas field. On the map opposite, slightly modified from Winnock and Pontalier (1970), by permission of the American Association of Petroleum Geologists, the locations of some of the wells are shown. The table gives the depth at which each well encountered the top of the Neocomian rocks (a division in the lower Cretaceous), within which gas is trapped. The depths are also plotted, in rounded values, on the map.

From these data draw a structure contour map of the upper surface of the Neocomian. Describe in words the form of the Lacq structure. Sketch a NE–SW cross-section, say through wells 118 and 126, to illustrate the structure.

Lacq gas field, Aquitaine: well depths.

Well number	Depth*	Well number	Depth*
3	11 070	117	14 420
101	11 250	118	11 960
102	12 300	119	11 790
103	11 390	120	11 040
104	10 730	121	13 150
105	10 300	122	11 590
106	12 070	123	12 200
107	13 010	124	12 640
108	12 360	125	12 490
109	14 300	126	14 000
110	13 880	127	12 740
111	12 950	128	11 810
112	11 540	129	11 370
113	10 500	130	11 790
114	12 850	131	11 690
115	12 950	132	12 940
116	13 460	133	10 680

*Depth quoted is to top Neocomian, in feet below sea level.

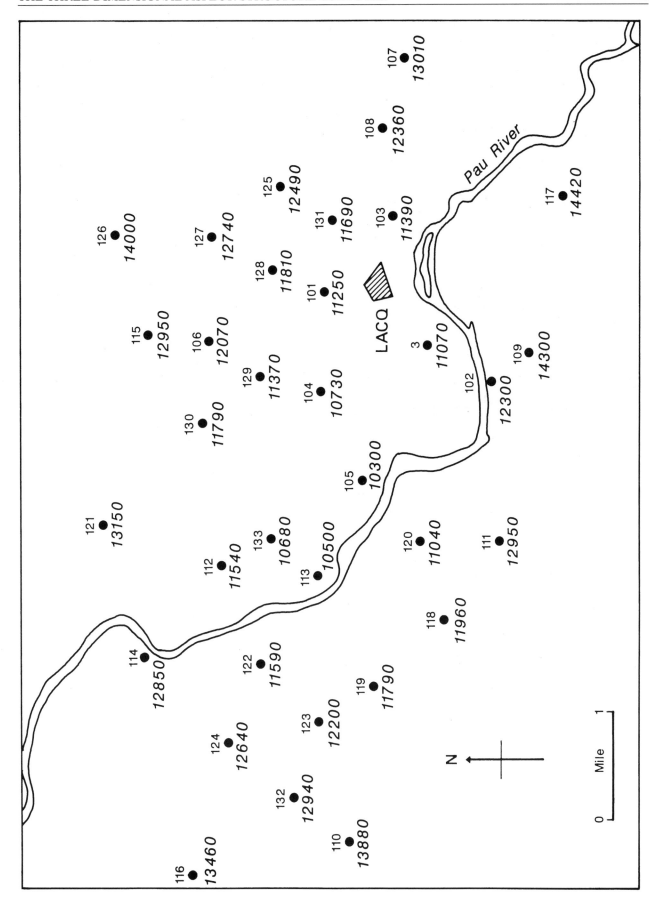

Map 3   Bear Hole, Montana, USA

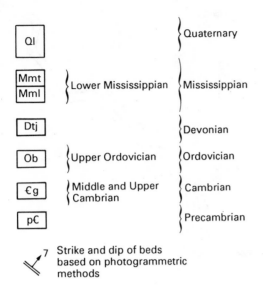

Ql		Quaternary
Mmt / Mml	Lower Mississippian	Mississippian
Dtj		Devonian
Ob	Upper Ordovician	Ordovician
€g	Middle and Upper Cambrian	Cambrian
pC		Precambrian

7 Strike and dip of beds based on photogrammetric methods

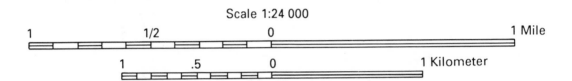

Scale 1:24 000

1   1/2   0   1 Mile

1   .5   0   1 Kilometer

The map opposite is of an area in south-central Montana, in the Big Horn Mountains. The land is part of the Crow Indian Reservation. The geology also appears, in very simplified form, on Plate 2 (note the values of latitude and longitude). The map is redrawn, by permission, from part of the USGS 1:24 000 Preliminary Geological Map of the Bear Hole Quadrangle, Map MF-1885. The original map was produced using a combination of field reconnaissance and air photo interpretation, together with a computer-assisted method of determining the strike and dip of units from the air photos. The technique was feasible because the units are uniformly dipping and of reasonably consistent thickness. The formations in the area range in age from Precambrian to Mississippian (Lower Carboniferous), as indicated on the key.

Identify the Devonian by adding colour to its outcrop. Locate its top and its base. Draw structure contours, say the 8200, 8300, 8400, 8500, and 8600 feet values, for the top and bottom surfaces of the Devonian. Comment on the form and spacing of the structure contours, and hence the form of the Devonian.

The formations appearing in the NE corner of the area are unlabelled. Deduce what they are likely to be.

Map 4   Maccoyella Ridge, Koranga, New Zealand

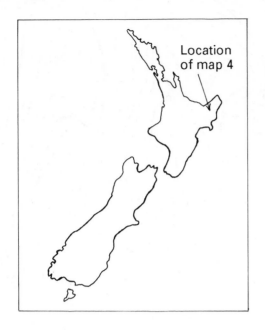

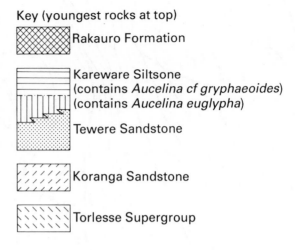

Key (youngest rocks at top)

Rakauro Formation

Kareware Siltsone
(contains *Aucelina cf gryphaeoides*)
(contains *Aucelina euglypha*)

Tewere Sandstone

Koranga Sandstone

Torlesse Supergroup

The facing map is based on part of the New Zealand Geological Survey PTS Sheet N87/9, N88/7: Geology of Koranga, Raukumara Peninsula, by permission of the New Zealand Geological Survey. It is enlarged here from the 1:15 840 (four inches to a mile) of the original to approximately 1:10 000. The grid reference numbers can be used in an analogous way to the UK National Grid.

The map shows a sequence of sedimentary rocks of Jurassic–Cretaceous age. There is some variation in thickness, shown particularly by the Koranga Sandstone, which in some places is absent altogether. Also, the units are slightly folded, so that structure contours will tend to curve, and their spacing may vary, reflecting differing amounts of inclination.

From the age relations given in the key, in what overall direction are the units dipping?

On the map, is the top surface or the base of the Te Were Sandstone further towards the SE? Carefully draw structure contours for the base of the Te Were Sandstone.

Describe in words the form of this surface.

Assuming all the other units dip by the same amount as the Te Were base, sketch a cross-section across the area to show the overall geological structure.

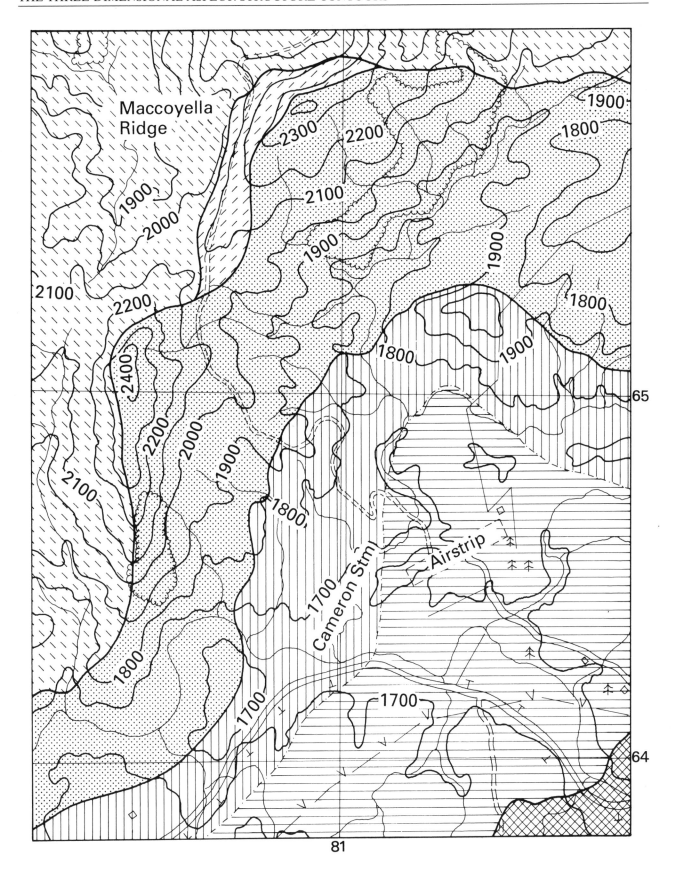

Map 5   Coalbrookdale Coalfield, Shropshire, England

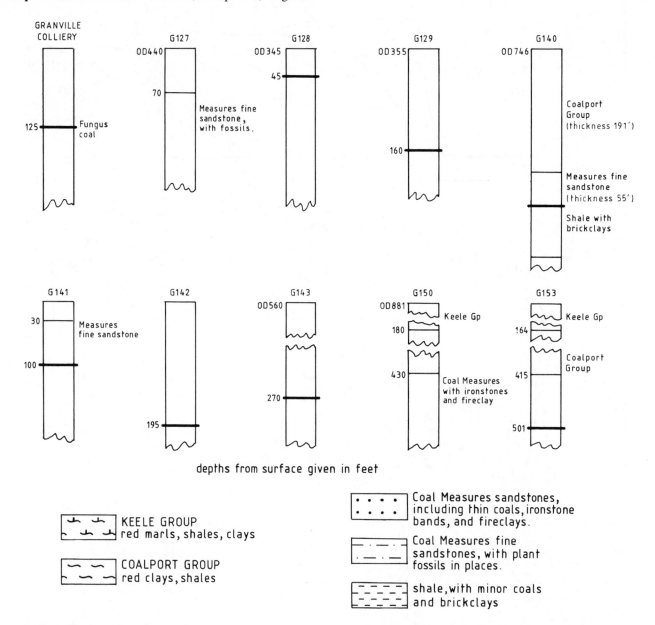

depths from surface given in feet

The Coalbrookdale Coalfield lies between Shrewsbury and Wolverhampton. It is small but because of the nearby ironstone is one of the places where the Industrial Revolution began. Like almost all other UK coalfields, the rocks are of Carboniferous age. Locate it on the part of the 'Ten Mile' map reproduced here as Plate 1. What is the oldest unit that the coal-bearing rocks (Lower Westphalian) are in contact with?

Opposite is a map of the kind of geology found in the Granville Colliery area. What is the overall structure of the Fungus Coal in this area?

A new mine-shaft is being considered in the area where borehole G130 has been sunk. Make a vertical column (like those in the key) of what you predict this borehole should contain. In particular, state the depth at which you predict the Fungus Coal will be reached. (Note that all these problems are best tackled by first drawing structure contours for the Fungus Coal. This is done most accurately by combining the topographic and the borehole information on its elevations. You can ignore the thickness of the seam for this purpose, i.e. the top and bottom surfaces have the same altitude.)

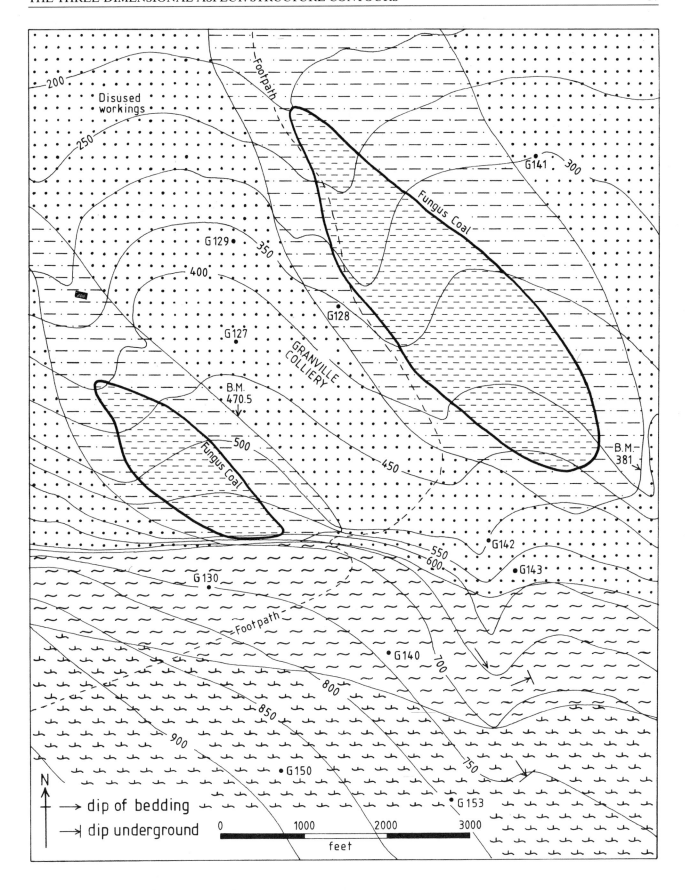

CHAPTER 4

Measurements in Three Dimensions: Strike and Dip, Formation Thickness and Depth

4.1 Introduction

We are beginning to see how it is that geological maps are such a powerful and convenient means of conveying information about the three-dimensional configuration of rocks. Nevertheless, it is often necessary to specify the arrangement in words or numbers. Geologists do this by using the concept of strike and dip. The general idea was introduced in section 2.3.1; the first part of this chapter explains it in detail.

The second part of the chapter expands on methods of subsurface projection and some useful measurements that can be made from maps. These techniques are of use in applied geology, where for many purposes, the work will have to be done as accurately as possible, especially if sums of money are at stake. Some of the corrections that may have to be borne in mind for this kind of mapwork are introduced.

4.2 Strike and Dip

Strike and dip are used to specify the orientation of a geological surface such as the top of a bed of sedimentary rock. The **strike** of bedding is the direction of an imaginary horizontal line running along a planar bed. It has no *position*, just direction. It therefore does not matter *where* on the plane the strike is measured. It is usually given as a compass direction, either loosely in words, say, NE–SW, or in degrees measured clockwise from north and quoted as three figures, say 045. The **angle of dip** is the maximum inclination of the bed, in degrees from the horizontal. To avoid confusion with strike it is quoted as two figures, say 08° or 30°, and is always given after the strike value. In addition to the angle of dip, there is the direction towards which the

surface is inclined, called the **dip direction**. This will always be at right angles to the strike.

Notice that it is not some arbitrary decision by geologists to define the dip direction as perpendicular to strike, or vice versa. It is a property of any tilted plane that the line at right angles to the maximum inclination will be horizontal. Consider the sloping roof of a house (figure 4.1a) and imagine rain falling on it. The water will trickle down the steepest slope, that is, in the dip direction. The line at right angles to that direction will be parallel to the ridge of the roof, that is, horizontal. The ridge line of the roof therefore parallels the 'strike' direction. Thus if the house in figure 4.1a is 'south-facing', as estate agents say, in geological terms the front half of the roof is dipping S and striking E–W. If the slope of the roof were 45° we could express its orientation as 090/45° S. This expression conveys precisely and concisely the orientation of that part of the roof. Note that there would be ambiguity without the 'S' at the end. The northern half of the roof is oriented at 090/45° N.

Strike and dip, then, are ways of expressing the orientation of beds of rock (figure 4.1b). Any other geological plane can be treated in just the same way. The boundary surfaces of a map unit will have a strike and dip. If the formation comprises bedded sedimentary rocks, its boundaries are likely to be parallel to the beds within it.

The orientation of geological planes is commonly measured in the field during the map survey, and representative values added to the completed map by means of symbols. The map key will explain these. A variety of different symbols have evolved, both for bedding surfaces and for the various other structures to be discussed in later chapters. The guidelines of the governmental survey in Australia, the Bureau of Mineral Resources (1989; now AGSO), state that symbols

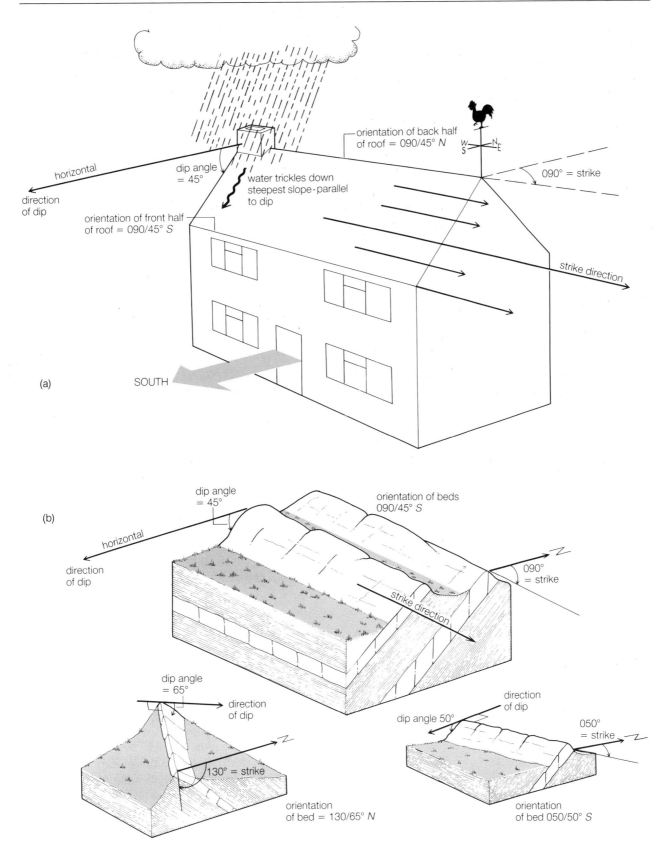

Figure 4.1 The concept of strike and dip. (a) Analogy with a house roof. (b) The strike and dip of inclined beds. The front edge of each diagram is oriented N–S. Only in the top diagram does this parallel the dip direction, at right angles to strike; in the bottom two diagrams the beds strike and dip obliquely to the edges of the figures.

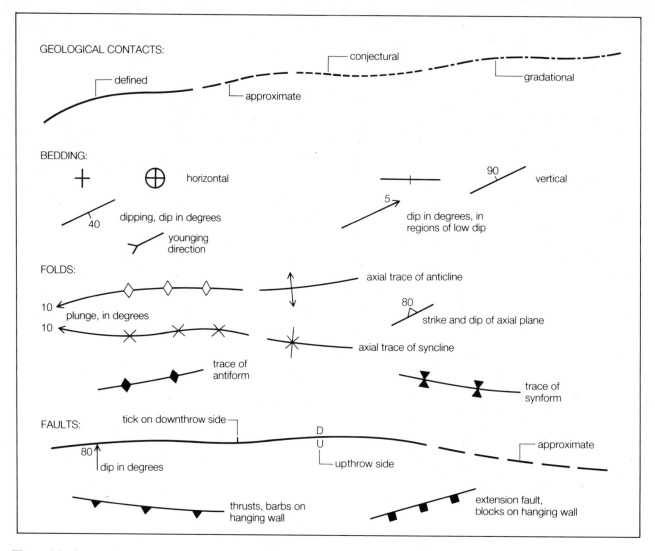

Figure 4.2 Some symbols commonly used on geological maps. The terms used are explained in the relevant chapters: chapter 8 for folds, and chapter 9 for faults. Numerous further symbols are given, for example, in Compton (1985, appendix 7), and the specifications issued by geological surveys and companies.

should be 'simple, easy to draw, clear, and reproducible on printed maps, preferably creating a mental image of the object or concept represented'. Some examples of commonly seen symbols are given in figure 4.2. If the strike and dip direction of a map unit is not known, an approximate orientation can be judged from the outcrop pattern of the formation (see chapter 6), and an accurate value can be derived by plotting some structure contours.

The strike direction of a formation is paralleled by structure contours. Because the contours are joining points of equal elevation, each contour line itself must be horizontal. Strike direction is horizontal, by definition. There can only be one horizontal direction on an inclined plane, therefore at any place along a structure contour, the course of the line represents the strike direction. Straight structure contours indicate a con-

sistent direction of strike; curving structure contours show that the strike direction is varying. At right angles to structure contours, decreasing in elevation, is the dip direction, and the spacing of the structure contours reflects the amount of dip. The dip value can be found from trigonometry or graphically (figure 4.3).

Because strike is horizontal, a horizontal bed cannot have a strike direction. Rather it will have an infinite number of horizontal directions. Any slight irregularities in near-horizontal beds will have a great effect on the strike direction. This is why the structure contour map of the Entre-Deux-Mers region (figure 3.3) is so complex looking. The Calcaire à Astéries has a very low dip, and so its strike direction is highly variable, as indicated by the structure contours. Conversely, steeply dipping beds tend to have well-defined strike directions, and, therefore, regular structure contours.

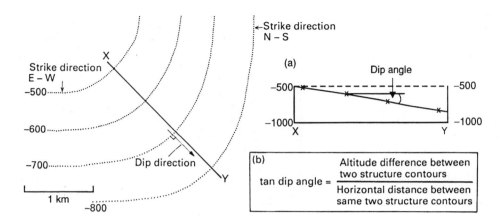

Figure 4.3 Deriving the direction and angle of dip from structure contours. (a) By graphical construction. Note that the section is drawn at right angles to the structure contours, i.e. in the true dip direction, and with a vertical scale equal to the horizontal scale. (b) By trigonometry.

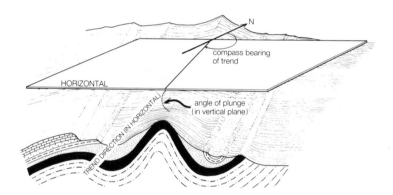

Figure 4.4 Diagram to illustrate the plunge and trend of a line.

The strike and dip concept will not work for *linear* geological features, for example, the intersections of unconformities or faults with the land surface. Here, plunge and trend are normally used (figure 4.4). **Plunge** is the inclination of the line from the horizontal, measured in a vertical plane. **Trend** is the direction of the line, measured in the horizontal as a compass bearing. The two systems of recording orientations exist because it is not possible to find the strike and dip of a line, nor the trend and plunge of a plane. The important practical difference between the two is that the angle of dip is measured at right angles to strike but the plunge is measured in the same direction as the trend.

4.3 Apparent Dip

If we draw a geological cross-section in the dip direction, the dip of the beds will be portrayed. On the other hand, a cross-section parallel to the direction of strike will show each bed as horizontal. It has to, be-

cause by definition the strike direction is a horizontal line running along the bed.

Therefore, a section along a line somewhere in between the dip direction and the strike direction will show the beds inclined somewhere between the dip angle and the horizontal. The closer the section line to the strike direction, the flatter the beds will look. The closer to the dip direction, the steeper they will appear (figure 4.5). In the actual dip direction the inclination reaches its maximum and is therefore the **true** dip. The intermediate angles of inclination are known as **apparent** dips (figure 4.6).

In mapwork, true dips are used wherever possible. Sometimes in practice, however, the surveyor can only measure an apparent dip, and it is this that is recorded on the map. An adjustment will be required to find the true dip. More commonly, the true dip is provided on the map, but for some reason the dip in another direction is required. It may be needed, for example, in engineering work where a road-cut or tunnel is being sited.

The conversion between true and apparent dips can be made in numerous ways, involving various

Figure 4.5 Sketch to illustrate true and
apparent dips. The formations are horizontal
at the front of the block diagram and show
maximum inclination on the right-hand side
of the diagram: the block is aligned parallel
with the strike and true dip directions. The
dip angle varies in magnitude in the quarry
walls because they occupy different
orientations. On the left wall of the quarry
the dip is steep, almost as the steep as the
true dip, because the wall is close to the true
dip direction. The back wall is parallel to the
strike direction, and so the formations appear
horizontal. Other walls show apparent dips
according to their orientation.

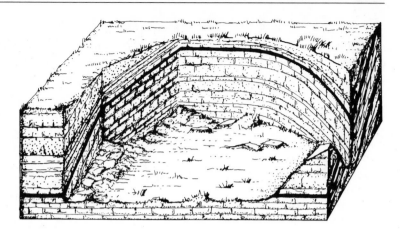

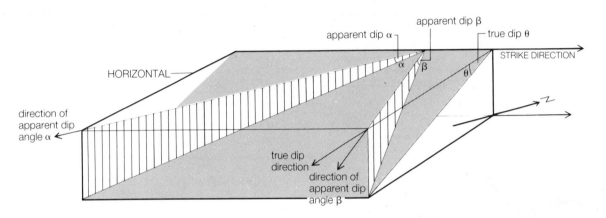

Figure 4.6 True and apparent dips. Any horizontal line along the inclined plane, i.e. the strike direction, parallels the front
and rear edges of the block. At right angles to this direction is the true dip direction, parallel to the side edges of the block, and
showing the true angle of dip. Dip angles in other directions, say α and β, are apparent angles of dip.

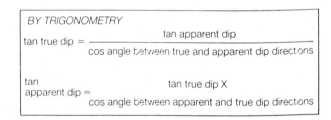

BY TRIGONOMETRY

$$\tan \text{true dip} = \frac{\tan \text{apparent dip}}{\cos \text{angle between true and apparent dip directions}}$$

$$\tan \text{apparent dip} = \frac{\tan \text{true dip} \times}{\cos \text{angle between apparent and true dip directions}}$$

Figure 4.7 Converting between true and apparent dips using trigonometry. Note that the directions of both the true and
apparent dips have to be known, and either the true or the apparent dip angle.

permutations of trigonometry, construction, and cun-
ning pre-determined devices. Travis and Lamar
(1987) listed over a dozen different methods. Explanations of the more common techniques are given by
Dennison (1968), Ragan (1985), and Rowland and
Duebendorfer (1994), and figure 4.7 shows the basic
trigonometric relations. The important thing, which-
ever technique you adopt, is to visualise and understand what you are doing.

4.4   Formation Thickness

Geologists often need to find from a map the thickness of a particular unit. It may be a matter of seeing
how the thickness varies from place to place in order
to build up a better picture of how the formation was
laid down, or the thickness may be required in order
to decide whether or not there is sufficient quantity of
a material of commercial interest for it to be worth

extracting. On published maps the thickness of units may be indicated on the key. If the thickness is variable, it may be given as a range. Some newer geological maps show the thickness to scale in a vertical column, sometimes incorporating a depiction of any lateral variations in thickness. Note that the vertical scale of such columns will be specified, and will not necessarily be the same as that of cross-sections or the horizontal scale of the map. Many maps, however, do not state thicknesses, and the geologist has to work them out.

The **true thickness** of a formation is the shortest distance between the top and the base of the unit. It is the distance measured *at right angles* to these bounding surfaces, assuming the thickness of the unit is constant. Measurements at any other angle will give a greater value, called the **apparent thickness**. When we look at the units appearing on a geological map, only in the case of the units being tilted to vertical will their outcrop width equal the true thickness (figure 4.8). In any other inclination of bedding, the outcrop width as seen on the geological map will be an apparent thickness only. Formation thicknesses encountered in a borehole or well, or derived from the difference in elevation between the top and bottom surfaces, will be vertical apparent values. Only in the case of horizontal beds will these equal the true thickness (figure 4.8).

As with apparent dips, there are various approaches to making the necessary conversions; simple trigonometric relationships are shown in figure 4.8. Note that the outcrop width/true thickness correction given there assumes that the land surface is horizontal. This is often a reasonable approximation on small-scale

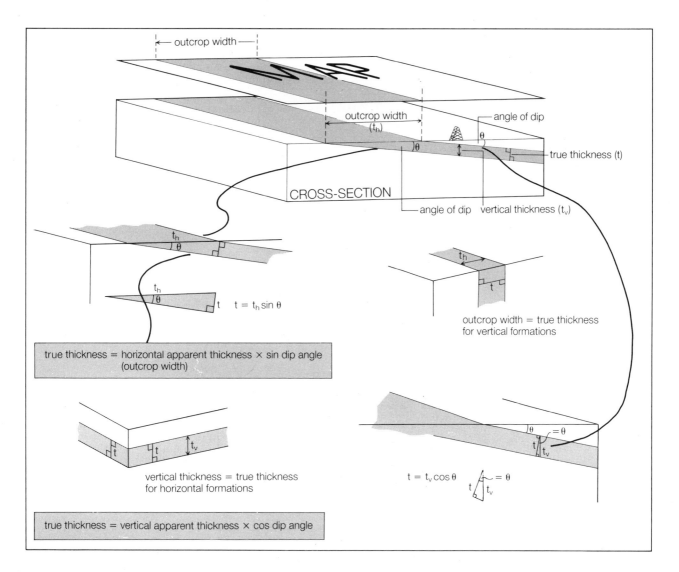

Figure 4.8 Formation thickness. The diagrams illustrate the trigonometric relationships between outcrop width and true thickness (assuming level ground), and vertical thickness (as observed in a borehole or well) and true thickness.

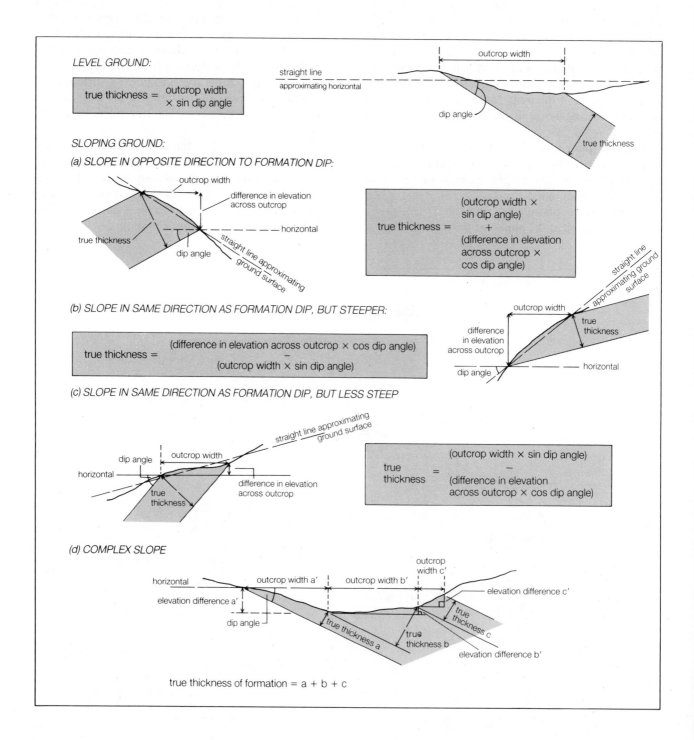

Figure 4.9   Diagrams to show trigonometric methods of deriving true thickness from outcrop width on various land surfaces.

maps, unless it is exceptionally rugged terrain, but a thickness value derived from a small-scale map is not going to be very accurate anyway. It is on large-scale maps that thickness determinations are most accurately made, and here the ground slope may have to be taken into account. If the slope is reasonably uniform and can be approximated in cross-section to a straight line, then the trigonometric methods shown in figure 4.9 can be applied. If the map scale is so large that considerable irregularities occur within the outcrop of the unit of interest, then the unit will have to be broken down into sub-units which adequately approximate to a straight line (figure 4.9d). An alternative approach, avoiding the problems of the land surface, is to construct an accurate scale drawing in the true dip direction of the unit, and simply to measure its scaled true thickness.

In some cases the thicknesses of units are portrayed directly on a published map, by a series of lines joining points of equal thickness. The lines are known as **isopachs**, from the Greek for 'equal' and 'thickness'. The device can only be used where the unit does actually vary in thickness across the map area, but this is always likely in natural deposits, at least to some extent. Such thickness information is especially useful for superficial deposits, where the values can vary very rapidly (section 10.5), with important practical repercussions (e.g. sections 12.2, 12.3, and 12.4).

Maps of thickness variations of subsurface formations are much used in the oil industry, as they can lead to ideas on geological histories (section 11.2) and likely hydrocarbon locations. The maps are usually referred to, from the lines plotted on them, as **isopach maps**. However, the name is an example – we will meet others later in the book – of how we have to be clear whether terms are being used rigorously or not. Properly, isopachs are lines joining points of equal *true* thickness, often called the **stratigraphic thickness** in the oil industry. But isopachs are usually derived by overlaying structure contour maps for the top and bottom surfaces of the formation – with the differences in altitude at points of intersection giving thickness values – and adding in information from boreholes. (This is not as intricate as it might sound because in much commercial practice all this kind of data is held in digital form and a computer rapidly carries out the exercise.) However, as we saw above and in figure 4.8, these thickness values will be in the vertical; they will not be *true* or *stratigraphic* thicknesses but vertical apparent thicknesses. Therefore such maps are not really isopach maps at all, even though they may well be called that. They are correctly termed **isochore** maps. The differences between a proper isopach map and an isochore map may be small with low formation dips but they will become increasingly significant with increasing dip angles. It may not matter with small-scale regional maps where it is overall patterns that are important, but if you are using an isopach map quantitatively, say to calculate volumes of hydrocarbon reserves, it is critical to know what kind of thickness is being portrayed – whether the lines really are isopachs or are just being loosely called that!

The vertical and true thickness values can be interchanged using the trigonometric formula given in figure 4.9 but, in practice, this will still leave errors. Boreholes/wells are rarely truly vertical and these days they are often drilled obliquely on purpose, so further trigonometric allowances will have to be made for that. Moreover, if the thickness of a formation varies – which is, after all, the whole point of isopach maps – although the true thickness will still be the shortest distance between the top and the base it cannot actually be at right angles to both surfaces. Further corrections have to be made (e.g. Bishop 1960; Tearpock and Bischke 1991), which are outside the scope of this book. Again, computers can manipulate such corrections easily, but the operator has to understand the problems that the machine is being asked to solve, as well as being able to assess the reasonableness of the solution.

4.5 Formation Depth

The **depth** of a formation is the vertical distance from ground level to the subsurface unit. If the formation has appreciable thickness, it will be necessary to specify whether the depth is being given to its upper or lower surface. Although depth is commonly measured to the top boundary, where the unit would be first encountered by vertical drilling, it is BGS practice to quote depths to the base of a formation. However, provided that the map trace of the upper or lower boundary is employed as appropriate, and that the dip angle as recorded at the land surface can be assumed to continue underground, either way it is a straightforward matter to calculate the depth of the unit, especially where the ground is reasonably flat (figure 4.10). A simple correction can be made if the ground is sloping. Of course, rarely in nature will the dip of the bed be constant. If there are indications on the map that the dip is variable, the method can only be approximate, and the error margin will grow with increasing distance of projection. In commercial applications where accuracy is imperative, it is

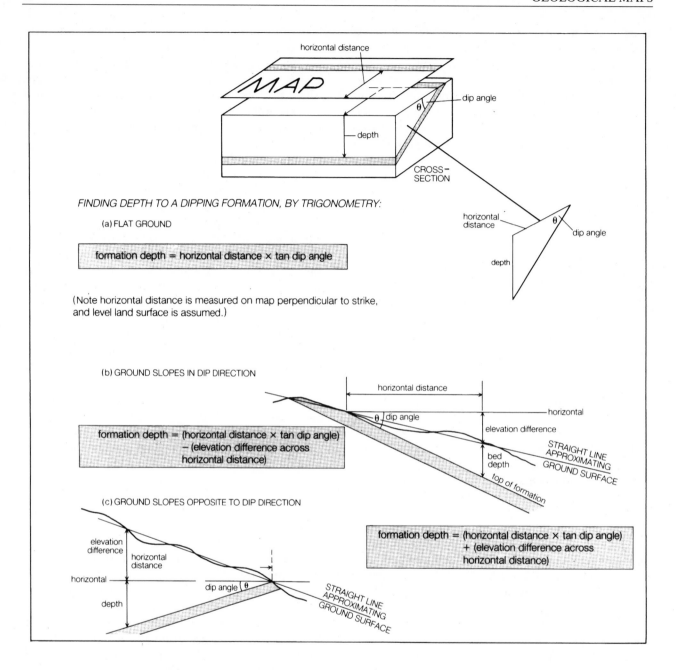

FINDING DEPTH TO A DIPPING FORMATION, BY TRIGONOMETRY:

(a) FLAT GROUND

formation depth = horizontal distance × tan dip angle

(Note horizontal distance is measured on map perpendicular to strike, and level land surface is assumed.)

(b) GROUND SLOPES IN DIP DIRECTION

formation depth = (horizontal distance × tan dip angle) − (elevation difference across horizontal distance)

(c) GROUND SLOPES OPPOSITE TO DIP DIRECTION

formation depth = (horizontal distance × tan dip angle) + (elevation difference across horizontal distance)

Figure 4.10 Formation depth. Diagrams to show trigonometric methods of finding the depth of a dipping formation.

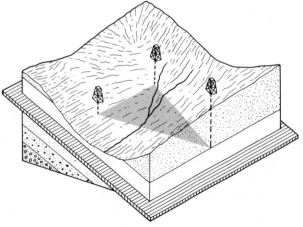

Figure 4.11 The 'three-point method'. Given a regularly dipping unit and its elevation at three points, various aspects of its orientation can be derived. The first step is to view the three points, here shown as boreholes to a buried unit, as corners of a triangle. The remaining procedures are given in figure 4.12. See also figure 3.4.

Knowing the elevation of a unit at three points – from outcrop at a known altitude, elevation in a drill core, etc., or some combination of these, the geologist can:

i   *FIND THE STRIKE DIRECTION*

① Connect the three known points to form a triangle. Indicate elevations of unit at corners for clarity.

② Divide side of triangle connecting biggest elevation difference to find point of same value as third corner of triangle, of intermediate elevation.
   Here, point halfway between 800m and 400m points will be 600m

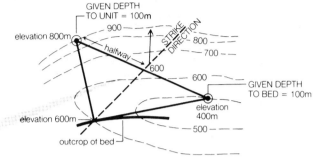

③ Join two points of same elevation. This line is the structure contour for this value, which indicates strike direction

ii   *CONSTRUCT STRUCTURE CONTOURS*

④ Make further subdivisions of triangle sides to find other intermediate elevations (if there are any). Connect with any other known elevation points.
   Here, point halfway between 600m and 400m will be 500m; can connect to point of outcrop at 500m topographic contour

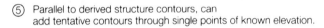

⑤ Parallel to derived structure contours, can add tentative contours through single points of known elevation.

iii   *FIND DIP DIRECTION AND ANGLE*

⑥ Dip direction is given by any line perpendicular to strike direction (structure contours) towards decreasing elevations.

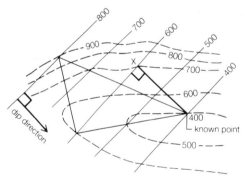

⑦ Draw line from either greatest or least of the three known elevation points to meet structure contour derived in i, above, at right angles. (These points are the most closely known – should give greatest accuracy). Label intersection with structure contour, say ×.

$$\tan \text{dip angle} = \frac{\text{elevation difference between} \times \text{and known point}}{\text{horizontal distance between} \times \text{and known point}}$$

iv   *FIND FORMATION DEPTH*

⑧ Derive strike direction (i, above) and dip angle (iii, above). Measure horizontal distance, at right angles to strike, between point at which depth required and any point where formation elevation is known (structure contour, outcrop, given point) From formation depth = horizontal distance × tan dip angle (Fig. 4.10), find depth difference from point of known formation elevation.

Figure 4.12   Some applications of the 'three-point method'.

common to obtain further information such as borehole/well data or seismic sections, with which the map predictions can be compared.

Bear in mind the matter of apparent dip (section 4.3). If the depth measurements on the map are not being made in the true dip direction, adjustments will have to be made. A further value of structure contours emerges here, for if they have been constructed on the map, the numerical difference between their elevation and that of the ground surface gives the bed depth directly, at any point of interest. Some maps involving superficial deposits show the depth to the underlying bedrock by a series of lines joining points of equal depth, sometimes referred to as **depth to bedrock** or **depth to rockhead** contours. The values are equivalent to the total vertical (apparent) thickness of the superficial deposits.

4.6   The 'Three-Point Method'

A common situation in applied geological work is the need for information about the underground arrangement of a unit from knowledge of its elevation at only a few places. Assuming the unit is reasonably planar, a minimum of three elevation points is required (figure 4.11). Each can be underground or at surface. Simple graphical constructions can provide different kinds of information on the dipping plane (figure 4.12). Of course, the greater the number of elevation points the greater the control, especially if the unit is thought to vary somewhat in dip. Actually, the technique has already been employed here, in section 3.4 and figures 3.4 and 3.5, for deriving structure contours from borehole data.

An alternative way to derive information from a map on bed dips, thicknesses, and depths, is to make a scale graphical construction in the vertical, a kind of cross-section. The accuracy of this approach for measurements depends a lot on drafting precision, but has the advantage of showing the situation pictorially. In fact, fully drawn cross-sections are the most visually effective way of representing the subsurface arrangement of rocks. We turn now to take a closer look at this extremely important facet of mapwork.

Summary of Chapter

1.   The orientation of a geological plane is specified by its strike – the compass direction of a horizontal line on the plane, and its dip – the maximum inclination from the horizontal.
2.   The direction of dip is at right angles to the strike direction.
3.   In other directions, the dip is a lesser value, called the apparent dip.
4.   The thickness of a formation is the shortest distance between its bounding surfaces. Except for vertical units, the outcrop width is an apparent thickness only. Except for horizontal units, the thickness encountered in a borehole or well is an apparent thickness only.
5.   The depth to a formation at a given point can be derived from the dip of the unit at a nearby outcrop.
6.   Given the elevation of a plane at three or more points, the orientation of the plane can be derived.

Selected Further Reading

Dennison, J.M., 1968. *Analysis of Geologic Structures.* Norton, New York, 209p.

Chapters 1, 2, and 3 cover true and apparent dips, thickness determinations, and depths, respectively. Derivations of the trigonometric equations are given. Chapter 5 includes three-point problems.

Rowland, S.M. and Duebendorfer, E.M., 1994. *Structural Analysis and Synthesis*, 2nd edition. Blackwell, Oxford, 279 p.

Pages 3–10 describe conversions between true and apparent dips using graphical projection, trigonometry and alignment diagrams (nomograms); pages 31–37 discuss the determination from maps of strike and dip and formation thickness.

Tearpock, D.J. and Bischke, R.E., 1991. *Applied Subsurface Geological Mapping.* Prentice-Hall, Englewood Cliffs, New Jersey, 648 p.

Chapter 10 of this book of advanced map methodsused in the oil industry deals with isopach maps.

Map 6   Boyd Volcanics, New South Wales, Australia

Part of the extreme southern tip of New South Wales appears on the map opposite, enlarged to about 1:14 500 and slightly simplified from Fergusson et al. (1979), by permission of the Australian Journal of Earth Sciences. The bedding symbol with 'facing determined' indicates that the beds become stratigraphically younger in the direction of the arrow in the normal way, i.e. in this map area there are no inverted successions.

What is the approximate orientation of the Banded rhyolite facies? Quote it as a strike and dip.

The main E–W road in the area crosses the Arkosic facies obliquely, for example at the point marked α. If there were road-cuts (parallel to the road) in these rocks, at what angle would they be seen to dip? How thick is this unit, at the north side of the road?

The base of the Basalt facies is shown to be dipping at 30°. At the point labelled β, at what depth are the phyllites and metaquartzites of the Mallacoota Beds? (The elevation of the stream at β is about 125 metres.)

Further examples of subsurface calculations

Referring to Map 1, of Raton, New Mexico, what is the approximate strike and dip of the Precambrian surface NE of Las Vegas? Calculate from the structure contours the dip angle of the Precambrian surface (a) ESE of Maxwell, and (b) SW of Cimarron.

Select three adjacent points on Map 2, say wells 115, 126, and 127, and find the dip direction and angle of the Neocomian in that part of the Lacq gas field. What is the strike direction and dip angle of the Neocomian surface in the vicinity of wells 122, 124, and 132?

Consult Map 3, of the Bear Hole area of Montana. Quote, in the conventional way, a representative strike and dip for the Mississippian. Calculate the thickness of the Middle and Upper Cambrian, bearing in mind the ground slope. At a point 1.5 km north of

Long Ridge, at the 8° dip symbol, find the depth of the Upper Devonian using structure contours, and by using trigonometry. Account for any discrepancy between results from the two methods. On the track NW from Commissary towards Spruce, at what angle would the beds be seen to dip?

On Map 4, of Maccoyella Ridge, New Zealand, what is the true thickness range of the Te Were Sandstone? At what depth is the Te Were Sandstone below the airstrip?

Vertical thicknesses of some of the units on Map 5 are readily derived from the depths given in the borehole logs. Convert these values to the true thicknesses for the Coal Measure fine sandstones in borehole G141, and for the whole of the Coal Measure sandstones as seen in boreholes G153 and G140.

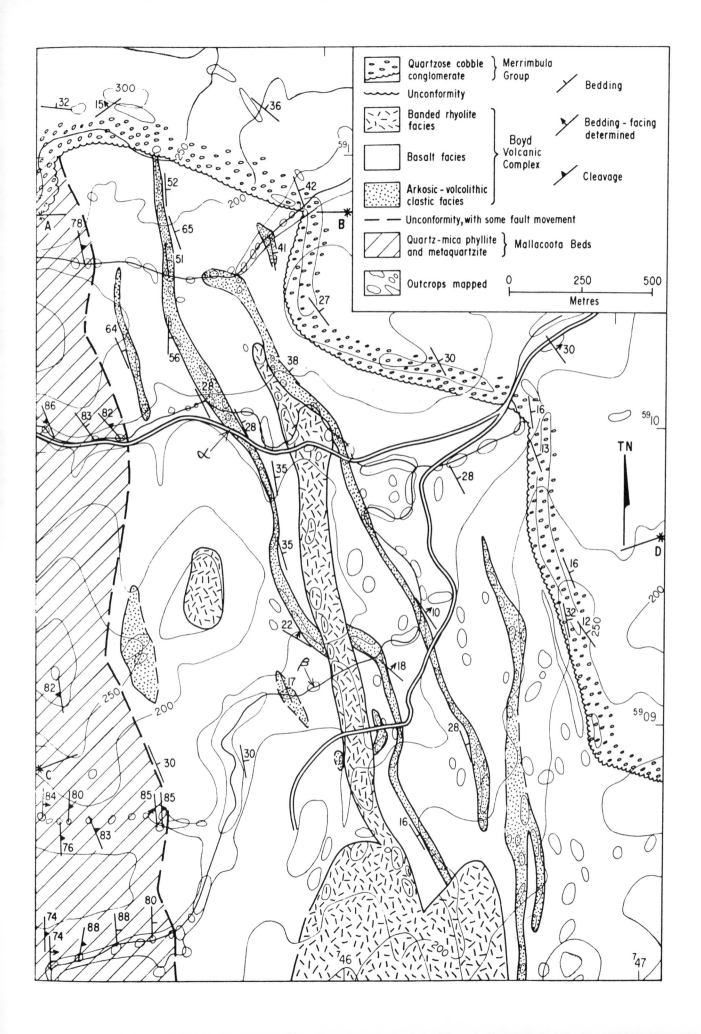

Map 7   'Northcrop' of the South Wales Coalfield, UK

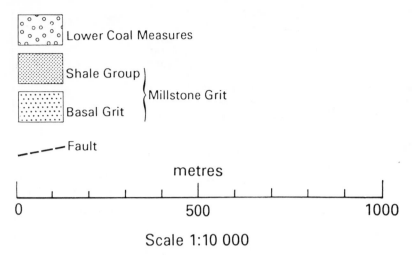

Scale 1:10 000

The map opposite is part of a series of 1:10 000 maps produced during a study of limestone subsidence commissioned by the Welsh Office. It is reproduced here by permission of the Controller of Her Majesty's Stationery Office. The map details part of the Carboniferous succession which comprises the northern flank, or 'northcrop', of the South Wales Coalfield. Part of this northcrop falls within Plate 1; the area enlarged here is around [SO0911]. The 200ft. 61m dashed contour line refers to the vertical thickness of the Basal Grit unit.

Draw structure contours for the top of the Shale Group of the Millstone Grit. What is the strike and dip of this surface? Comment on its structure.

What is the thickness of the Shale Group? (If you are solving this question by using outcrop width and dip angle, remember (a) to measure the outcrop width in the true dip direction, and (b) to allow for any topographic slope.)

If a borehole/well was sunk at Pen March, predict what rocks the core would contain. Draw a vertical section for Pen March, showing on the column the stratigraphic succession and, as far as possible, the thicknesses that would be observed in the borehole. (Remember that borehole thicknesses, being vertical, will not be the same as the true thickness of inclined beds.)

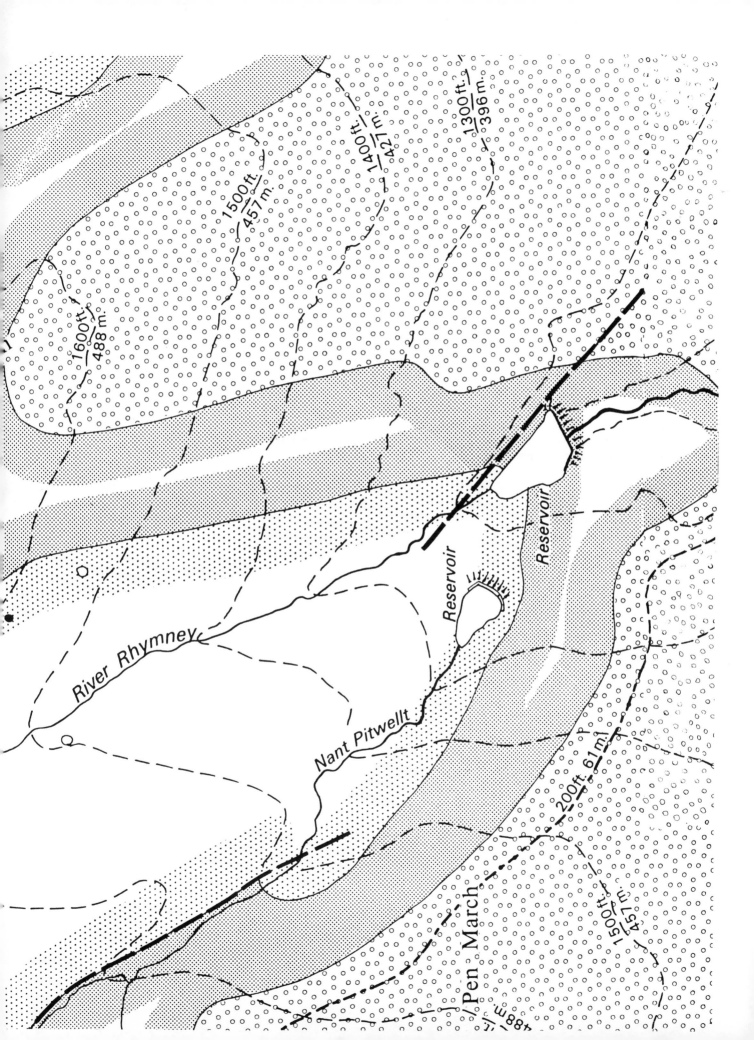

Map 8   Long Mountain, Powys, Wales

Key to letters on map
UL   Upper Ludlow
Ml   *Monograptus leintwardinensis*
Mt   *Monograptus tumescens*
Mn   *Monograptus nilssoni*
Mv   *Monograptus vulgaris*
Cl   *Cyrtograptus lundgrendi*

Opposite is a map (north to the top) at 1:10 000 scale of the area south of Long Mountain near Welshpool, in mid-Wales. The map units have been defined on the basis of the graptolite fossils they contain. The area can be seen on Plate 1 around [SJ2906]. What, from Plate 1, appears to be the overall structure of the rocks forming Long Mountain?

On the area covered by Map 8, the beds are of reasonably consistent orientation. Structure contours therefore need few control points for their construction. Note, from the above table which acts as a key, that the units on the map are mostly defined on the basis of the fossils they contain. The oldest unit, as always, is at the bottom of the list and the youngest at the top. In which stratigraphic system, according to the BGS 'Ten Mile' map (Plate 1), do these units occur?

Draw sufficient structure contours to be able to specify the strike and dip of the rocks. Quote the strike and dip in the conventional way (e.g. 110/30° N).

Plot the strike and dip on the map using a conventional symbol.

Using this value of strike and dip, calculate the depth to the top of the *Cyrtograptus* unit at Walton Hall and at Rowley Farm.

Draw some structure contours for this same surface and use them to predict its depth at the same two localities. What might the reasons be for any discrepancy with the depths obtained using trigonometry? Construct structure contours for other surfaces so that an accurate cross-section can be drawn (see next chapter).

If rocks are exposed along Binweston Lane, at what angle would they be seen to dip?

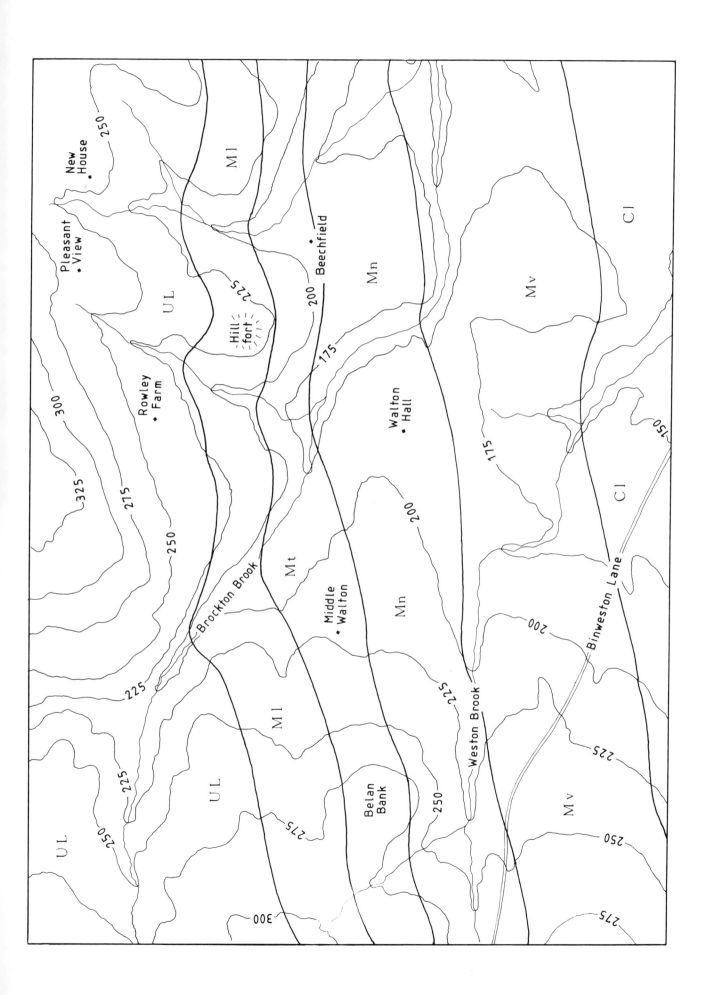

Map 9   Silverband orefield, North Pennines, England

On the western side of the Pennines in northern England is a complex of mineralised faults, a portion of which is shown on the large-scale accompanying map. Last century, areas such as this were the scene of much mining for lead, zinc, and silver.

Using trigonometry, what is the thickness of the Great Limestone? At what depth would this limestone be encountered below ground where the veins intersect at Swathbeck Vein?

The New Silverband Level was driven in 1838 roughly eastwards towards the mineralised vein at Rowpott's. Construct an accurate profile, including the land-surface topography, of the limestones that would be seen in the north wall of this level (i.e. construct the cross-section immediately to the north of the fault, taking the fault trace as a straight line). (Note, however, that the line of section is not in the true dip direction.)

The shaft at D would have to be driven what distance vertically before it encountered the Four Fathom Limestone?

Now use structure contours to find the following:
1. The vertical thickness of the Great Limestone.
2. The vertical thickness of the limestone *underlying* the Great Limestone.
3. The dip angle of the Great Limestone, where it underlies the Middle Tongue Beck.
4. The depth of the Four Fathom Limestone at Lord's Level.

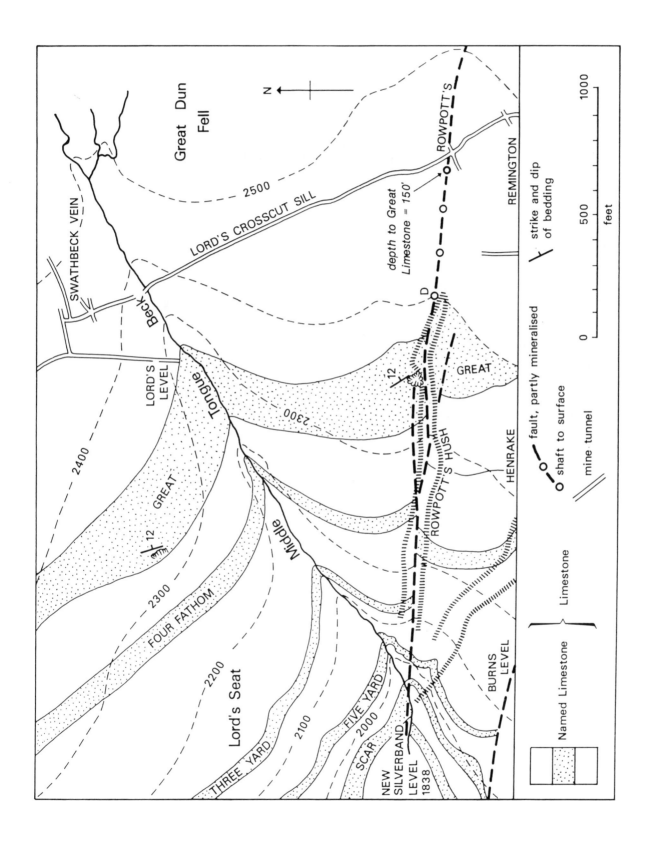

CHAPTER 5

Geological Cross-Sections

5.1 Introduction

Rock formations are readily observed in steep faces, such as cliffs, canyons, and mountainsides, in a kind of natural cross-section. It is perhaps for this reason that cross-sections seem more immediately familiar than geological maps, and give a more striking picture of the arrangement of the beds. This is borne out historically, as sections were being drawn long before geological maps. Cross-sections portray the arrangement of the rocks as seen in a vertical plane. They are extremely useful devices but, nevertheless, they are strictly two-dimensional. It is the combination of the vertical cross-sections with horizontal geological maps that forms such an effective means of working with three-dimensional geology on paper.

As with maps, cross-sections can be treated in a reconnaissance way, or they can be constructed accurately to enable measurements to be made. As mentioned in the last chapter, quantities such as bed dip, thickness, and depth can often be arrived at more conveniently from cross-sections than by using trigonometry. These days the more routine aspects of constructing and manipulating cross-sections are being done increasingly by computers. The present chapter is mainly concerned with the understanding of cross-sections, so that any available computer facilities can be used to best advantage.

The first half of the chapter discusses some fundamental aspects of cross-sections and their construction. The second half extends the concepts to three-dimensional devices such as block diagrams. In the present context we are largely concerned with constructing cross-sections from existing geological maps. However, the geological surveyor normally develops at least sketch cross-sections as part of his or her fieldwork, so as to keep in mind the spatial arrangement of the rocks. In some situations, particularly in the oil industry, it is common for maps to

be derived from the geological sections. Wells and seismic lines provide much of the subsurface data, which are compiled onto various kinds of cross-sections from which the geological maps are in turn produced. All this illustrates the importance of cross-sections and the way they complement geological maps.

5.2 Line of Section

Normally the line across a map along which the cross-section is drawn should be at right angles to the dominant strike of the rocks. This not only gives the best visual impression of structural relationships but, being in the dip direction, will allow the direct measurement of the true dip of the beds (provided there is no vertical exaggeration, see section 5.3). If the section is not perpendicular to strike, it will show apparent dip values only, and the trigonometric corrections given in section 4.3 will have to be applied. Strike-parallel sections will show nothing of the dip of the beds, although they do have their uses, such as illustrating thickness changes along strike.

If you are planning to draw a cross-section across an area of variable strike, say with two differently oriented rock sequences, you have several options. If one of the sequences seems more important, then the section line could be sited perpendicular to these rocks, with the understanding that the other sequence will have a somewhat illusory appearance on the section. If the section is being constructed by underground projections of dip readings at surface (section 5.4) rather than from structure contours, the dip amounts for the oblique beds will have to be adjusted to an apparent value (figure 4.7). For most purposes, however, corrections to section lines not more than about 15° away from the true dip direction are

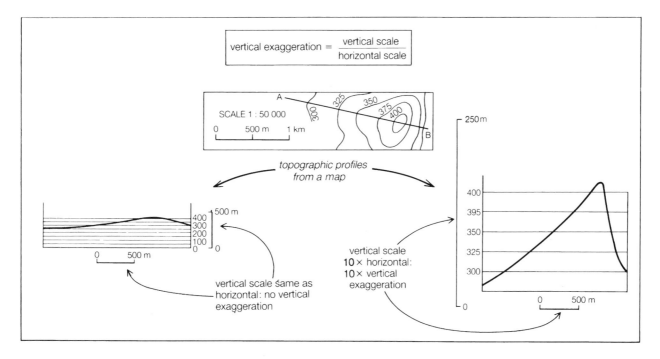

Figure 5.1 Vertical exaggeration and its effect on topographic slope.

negligible. An alternative procedure would be to draw two sections, each perpendicular to one of the sequences. In fact, a typical geological map needs several cross-sections to illustrate adequately the subsurface structure of the area. A third possibility is to bend the section line.

The problem of section line arises particularly where a section is being drawn from a map but is utilising some subsurface data. Such added information will probably greatly improve the reliability of the cross-section, but it is unlikely that seismic lines, boreholes, mine adits, etc., will lie exactly on the ideal line of section. Here it may be better to deflect the section to incorporate directly the subsurface information. Cross-sections used in oil exploration typically have highly zig-zag paths, passing through the greatest practicable number of wells. However, where the section is being drawn from a map but it is desired to incorporate some information not far from the strike-perpendicular section line, it may be best to project the data to the section rather than bend the line (see figure 5.6).

5.3 Scale and Vertical Exaggeration

In the great majority of cases in which a cross-section is being drawn from a map, the horizontal scales of map and section should be the same. It is with the vertical scale that an element of choice arises. Essen-

tially, using a vertical scale greater than the horizontal one allows greater detail to be included on the section but introduces distortion. In general it is best to use the same horizontal and vertical scales unless there is a strong reason for doing otherwise. If a larger vertical scale is employed it leads to **vertical exaggeration** (figure 5.1). This should be quoted on the section. Many published sections are clearly exaggerated, but leave it to the reader to divine by what amount.

If a cross-section is used for carrying out the kinds of measurements discussed in the previous chapter, then any vertical exaggeration has to be allowed for. An important effect arising from vertical exaggeration is the steepening of sloping lines on the section. By expanding the vertical scale on topographic sections subtle changes in landslopes can be magnified, say for geomorphological purposes (figure 5.1). But the disadvantage is obvious – realism is lost. Gently undulating land can look like alpine mountains!

In the same way, dips on geological cross-sections are increased by vertical exaggeration (figure 5.2). All dips, except perfectly horizontal and vertical features, are steepened. The shallower the dip the more it is affected (figure 5.3). As a result, two faults, for example, one of steeper inclination than the other, could appear misleadingly similar on a highly exaggerated section (figure 5.2). On the other hand, vertical exaggeration can be useful in bringing out small but potentially significant differences between shallow-dipping beds. This could be important, for example, in

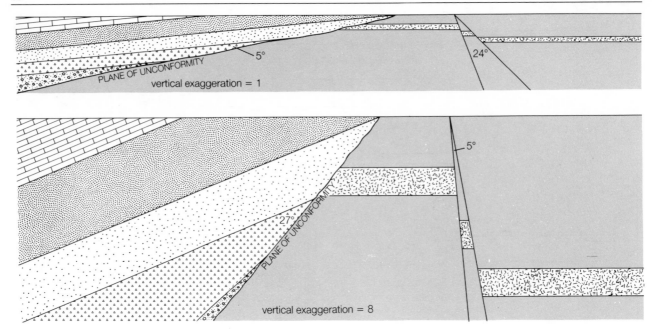

Figure 5.2 The effects of vertical exaggeration on geological surfaces. Note that, because vertical exaggeration affects shallow-dipping surfaces more than steep ones, angular differences *between* shallow-dipping surfaces are increased whereas angles between steep planes are reduced.

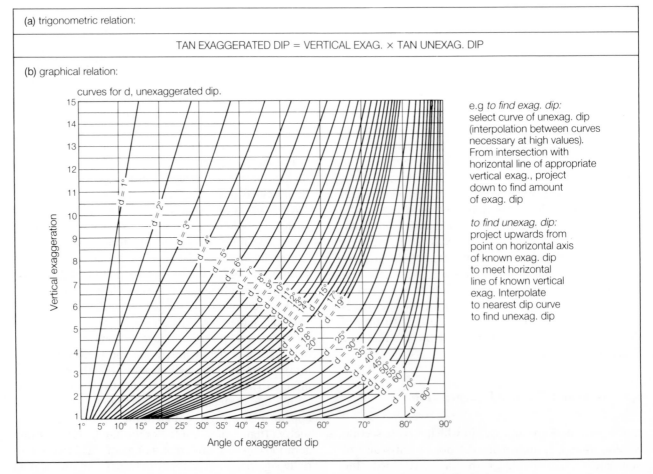

Figure 5.3 Trigonometric and graphical relations between vertical exaggerations and dip angles.

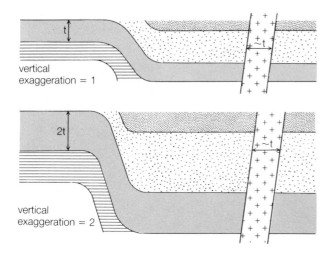

Figure 5.4 The effects of vertical exaggeration on formation thickness. Note the large effect on the thickness of horizontal units, the small effect on steep units, and the intermediate effect on moderately dipping beds. If the dip of a unit varies, an apparent attenuation effect can be introduced.

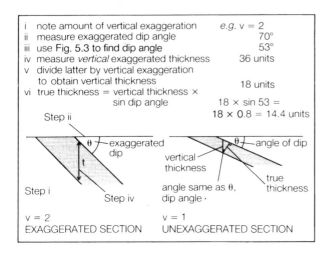

Figure 5.5 A simple graphical method for finding the true thickness of a unit shown in a cross-section with vertical exaggeration. Note that the method assumes the section to be drawn at right angles to strike, i.e. in the true dip direction.

stratigraphical and sedimentological studies. There may, therefore, be good reasons for expanding the vertical scale of the section, although the fact that the construction can be made with less care is not one of them! If vertical exaggeration is employed for some reason, the results have to be treated with care.

The thickness of beds is distorted on sections which employ vertical exaggeration (figure 5.4). The overall effect is to expand the thicknesses. Thickness *differences* will also be increased, so that exaggerated

sections can be used to show small variations more clearly. However, there is a complication to enhancing thickness differences in this way. The amount of thickness increase depends on the dip of the bed. The thickness of horizontal beds increases most, in the same ratio as the vertical exaggeration; vertical beds are not affected at all; and intermediate dips are influenced by various amounts. Dips in the range 1–20° are most susceptible. Two beds of the same real thickness but of different shallow dips will appear of differing thickness on an exaggerated section. A folded bed of constant thickness can take on an attenuated appearance (figure 5.4). Folded beds can look like this in nature, but here it is a purely artificial consequence of the vertical exaggeration. As long as you are aware of these effects, they can usually be allowed for when looking at exaggerated sections. The difficulties are that small real variations in thickness can be masked, and that cumbersome corrections will have to be applied to any thickness measurements (figure 5.5).

Where the exaggerated section is used to find depths to subsurface horizons, the necessary correction is less awkward than that for dips and thicknesses. The depth, by virtue of being measured vertically, is greater than the real depth by the same amount as the vertical exaggeration of the section. But to repeat, although such corrections can be applied, it is generally better to have the vertical scale of the section the same as the horizontal.

5.4 Manual Drawing of Cross-Sections

Cross-sections can either be sketched or accurately constructed, depending on the purpose for which they are being produced. When visualising the three-dimensional arrangement of rocks from a map, it is often useful to sketch cross-sections, both to clarify your own mental picture and to communicate it to others. The overall outcrop patterns are paramount in suggesting the arrangement to be sketched on the section. The principles of outcrop assessment were introduced in section 2.3 and are developed further in chapter 6. Any dip values provided on the map allow the subsurface lines to be sketched more realistically. With small-scale maps it is usually not possible to draw anything more precise than a sketch section, simply because more detailed information will not be shown.

On larger scale maps, say around 1:50 000 and larger, it should be possible to construct an accurate cross-section to the same scale as the map. Figure 5.6 shows the method. Good strike and dip information

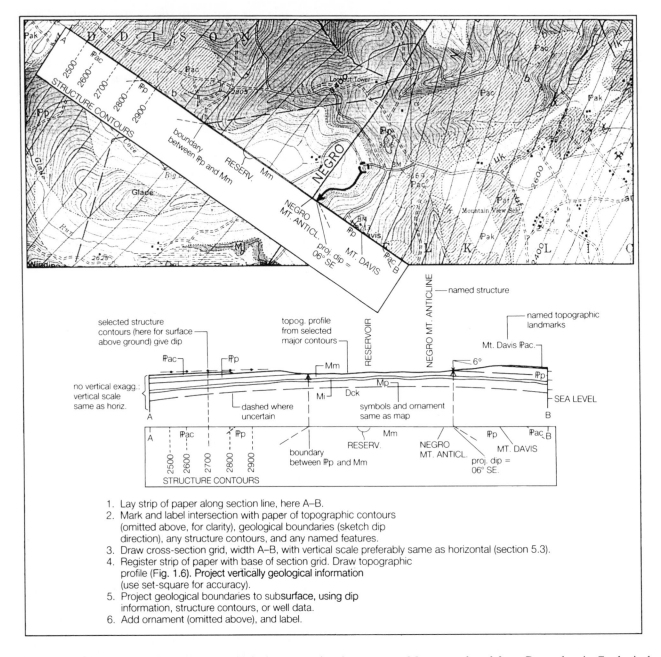

Figure 5.6 Instructions for drawing a geological cross-section from a map. Map reproduced from Pennsylvania Geological Survey Bulletin C56A, S Somerset County, by permission of the Geological Survey of Pennsylvania.

will be necessary, to allow judicious location of the section line and sound projection to depth. Ideally, it will be possible to draw some structure contours in the vicinity of the section line. They can then be plotted on the section in the same way as topographic contours.

It is usual to construct the topographic profile first and then project the subsurface geology from it. However, if structure contours are being used to position the geological surfaces, it is more accurate to draw the topography last, as the elevations of the out-

crops of those surfaces can be used in addition to the topographic contours to control the course of the topographic profile. Good information on the subsurface form of one formation will help constrain the projections of adjacent surfaces. The greater the control on the cross-section, the more reliable any measurements from it are likely to be. The completed section should provide a basis for confident advice about subsurface conditions.

If the section is drawn in a direction other than that of true dip, dip values given on the map may have to be adjusted (section 5.2). Formations are assumed to retain constant thickness at depth, that is, the lines representing the top and bottom of a unit remain parallel, unless there is evidence on the map of thickness change. Folded beds are extrapolated at depth with smooth curves, although for more advanced work there is a range of techniques for more refined projection of folded beds (e.g. see Ramsay and Huber 1987). Many faults dip at about 70°, but it is conventional to show faults as vertical, unless there is direct evidence otherwise. Faults may flatten to give listric geometry (section 9.7.1), but additional subsurface information is usually necessary to support projections beyond shallow levels. Indicate the sense of fault displacement (section 9.3) if it is known.

The same ornament as used on the map should be added to the section; the datum, horizontal and vertical scales are indicated; and any named structures in the rocks and important topographic landmarks are labelled. Make sure that the line of section is indicated on the map. It is not possible to lay down rules on how rocks should be projected underground. Each situation has to be considered on its own merit. In general, aim for the simplest arrangement which is compatible with the known information. Study examples of published cross-sections; a selection is reproduced in figure 5.7.

5.5 Structure and Stratigraphic Sections

All cross-sections are drawn with reference to a particular datum plane, which appears as a horizontal line on the section. Each formation or feature on the section is shown at its scaled elevation with respect to this datum. By far the most common datum is sea level. Some other marker could be used, and especially on large-scale maps a local datum may be preferred, perhaps a nearby benchmark if the area is in high terrain or deep in a mine. Irrespective of the elevation of the datum used, the effect of the section is to depict the *structure* of the rock units. Therefore these cross-sections are properly called **structure sections**. They show the configuration of the rocks, *as they are now.*

For some purposes, however, the present-day arrangement of the rocks is of less interest than the relationships at some past geological time. In this case, a cross-section can be drawn for which the datum is a stratigraphic boundary that formed at the time of interest. Such a construction is called a **stratigraphic section**. The datum horizon normally has to be a reasonably prominent, persistent, stratigraphic unit. It appears on the stratigraphic section as a straight line and the positions of all the other surfaces are referred

to it. Rocks below the datum will show the arrangement they had at the time of formation of the datum surface. For example, any intertonguing between units and thickness variations will be highlighted, together with any sedimentary features such as reefs or channels. Any tilting, folding, etc., subsequent to the datum will not appear, otherwise the datum surface would not be horizontal. Such later effects cannot therefore obscure aspects of stratigraphic interest around the time of the datum.

A straightforward way of manually constructing stratigraphic from structure sections is shown in figure 5.8. It is only approximate because of the implication that the originally horizontal beds got to their present-day form by purely vertical movements – which this construction method simply reverses – and this will not usually be realistic. Any obvious non-vertical displacements, such as low-angle faults (chapter 9), will have to be restored separately.

Stratigraphic sections are much used in the oil industry, partly because the stratigraphic information is important in exploration, and partly because this kind of section is readily constructed from well data. Elaborations of the device are employed too, so it becomes important where different kinds of cross-sections are being employed to specify their nature. This is by no means always done, leaving the reader to work out which datum has been used. However, in most fields of geology, constructions which are called by some combination of 'section', 'cross-section', and 'geological section', will almost certainly be structure sections.

5.6 Three-Dimensional Diagrams

Although the usual way of representing the geometrical arrangement of rocks is to use sections and maps in conjunction, the role of cross-sections can be extended by giving them a three-dimensional aspect. The two main ways of doing this are to construct either a fence diagram or a block diagram. Each takes two or more conventional cross-sections and converts them into diagrams which look three-dimensional. They are still drawn on a flat surface, but are more strikingly three-dimensional to the eye, and are a highly effective way of visually conveying the spatial arrangement of rocks. The following section avoids detail on manual methods of construction, but outlines the principles behind them.

5.6.1 Fence diagrams

Fence diagrams, sometimes called panel diagrams, are a three-dimensional network of cross-sections drawn in two dimensions (figure 5.9). They have the

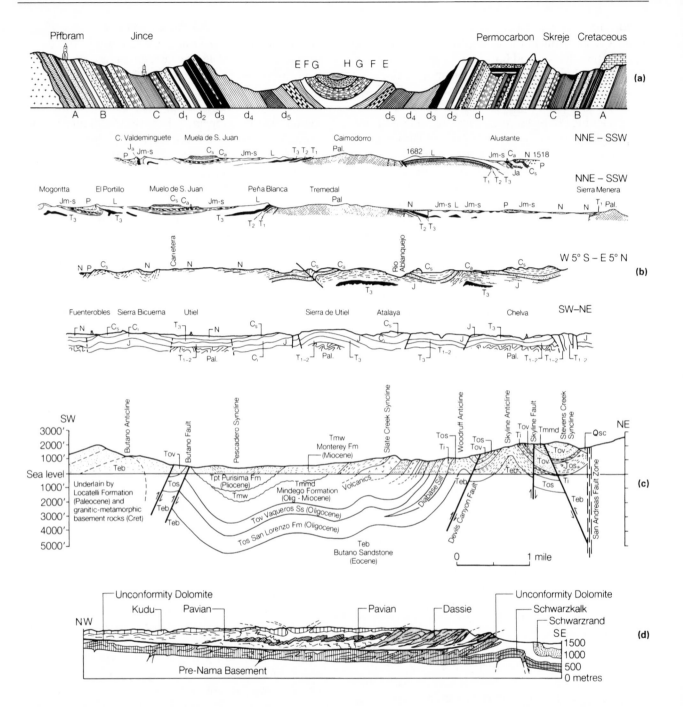

Figure 5.7 Some examples of published geological cross-sections. (a) An early cross-section, now looking very diagrammatic, across the Bohemian basin, Czechoslovakia, published in 1852 by Joachim Barrande. (b) Four cross-sections of the Iberian fold belt of western Spain. Reproduced from Lemoine (1978), by permission of Elsevier BV. (c) Cross-section of the Coast Ranges, California, USA. Reproduced from Page (1966), by permission of the California Division of Mines and Geology. (d) Cross-section of the Naukluft Mountains, SW Africa. Reproduced from Holmes (1965) after reduction from Korn and Martin (1959), by permission of Thomas Nelson and Sons.

appearance of interlocking fences, rather like those used in a cattle market. It is useful to draw the diagram on a simplified base map, to show the locations of the sections. The viewing direction is chosen to show the sections to the best advantage. Fences at

right angles to the viewer will appear as conventional cross-sections; any sections close to parallel with the viewing direction will show little information.

Deriving fence diagrams from maps requires the construction of several cross-sections, which will

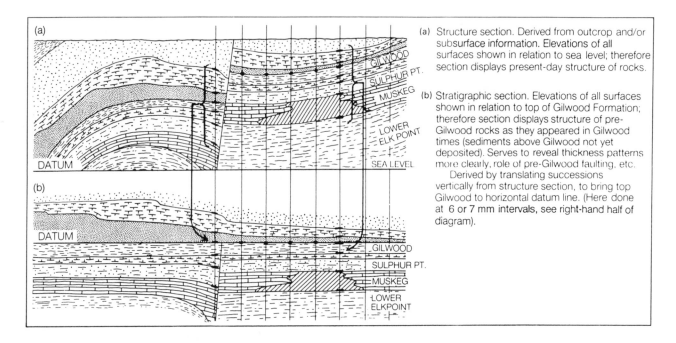

(a) Structure section. Derived from outcrop and/or subsurface information. Elevations of all surfaces shown in relation to sea level; therefore section displays present-day structure of rocks.

(b) Stratigraphic section. Elevations of all surfaces shown in relation to top of Gilwood Formation; therefore section displays structure of pre-Gilwood rocks as they appeared in Gilwood times (sediments above Gilwood not yet deposited). Serves to reveal thickness patterns more clearly, role of pre-Gilwood faulting, etc.
    Derived by translating successions vertically from structure section, to bring top Gilwood to horizontal datum line. (Here done at 6 or 7 mm intervals, see right-hand half of diagram).

Figure 5.8   Illustrations of structural and stratigraphic sections, and a simple graphical method for converting between them.

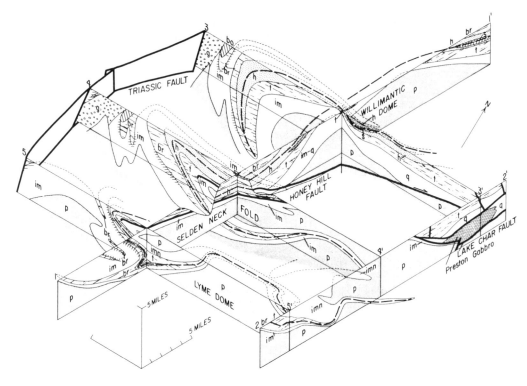

Figure 5.9
An example of a fence diagram, used here to illustrate complex fold structures in Connecticut, USA. From Dixon and Lawrence (1968).

5.6.1 Fence diagrams

eventually become the panels of the diagram. The various corners of the fences can then be positioned on the base map of the diagram, and the information from the sections transferred to vertical scales extending upwards from the corners. You can then fill in the panels from the cross-sections, if necessary erecting vertical scales within the panels to aid the accurate transfer of the lines from the sections. Lines on rear panels are dashed if they are obscured by part of a forward panel.

If the cross-sections had vertical exaggeration then this will be translated to the fence diagram. In this case, and if the panel spacing is close, there may be an unacceptable obscuring of the rear panels. A different

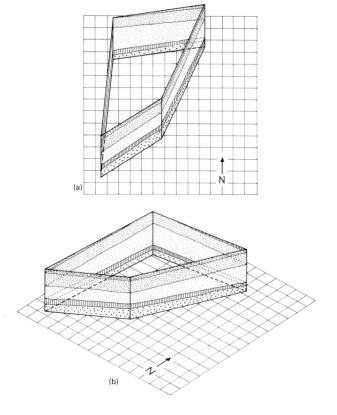

(a)

N

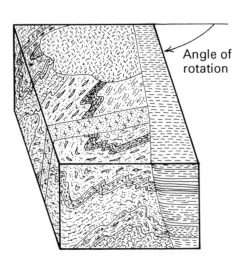

(b)

N

Figure 5.10 Some aspects of fence diagrams. (a) The orientation of some of the panels, particularly in the west, suppresses the visibility of the information. (b) Greater visibility is gained by transforming (a) into an isometric projection. The gridded base allows this to be done manually. Note, however, that only the vertical lines and those parallel to the N–S and E–W axes preserve their length; in other orientations some distortion is introduced.

line of sight may help, but it may be useful to transform the fence diagram and its map base to an isometric projection (figure 5.10). The rectangular map base becomes a parallelogram, the sides of which are paralleled by lines that were N–S or E–W on the map. Vertical lines remain vertical. All these lines, which were originally at right angles, retain the same linear scale in the isometric projection. However, any lines at intermediate angles will change in length, so a grid system such as that shown on the map base in figure 5.10 will aid in transferring their locations.

5.6.2 Block diagrams

Block diagrams are two-dimensional drawings of a rectangular block (e.g. Lobeck 1958). The top of the block is the map plane, and the sides are cross-sections. It is a fairly intuitive way of visualising three dimensions on paper, and has been used frequently already in this book (e.g. figures 2.1 and 2.2). However, for quantitative work there are snags, and the kind of projection used to draw the block diagram has to be considered.

Isometric block diagrams are both easy to draw and amenable to direct vertical measurements, because the vertical scales are the same as on cross-sections. However, they have an awkward visual effect (figure 5.11). They are constructed in a similar way to isometric fence diagrams, so a horizontal grid is helpful here, too, in manually transferring the information from the starting cross-section to the block diagram.

Perspective block diagrams, on the other hand, give a much more realistic appearance to the block, but introduce further distortion (figure 5.12). Block diagrams can be drawn in **one-point perspective**, where the front of the block faces the observer squarely, and all lines not in the plane of the paper meet at one

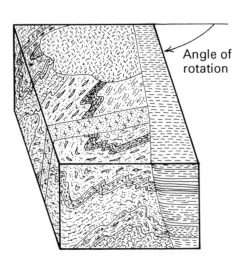

Angle of rotation

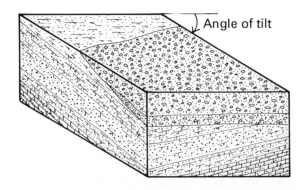

Angle of tilt

Figure 5.11 Isometric block diagrams. Note that opposing faces of the blocks are parallel. The left-hand block has a high rotation angle (measured in the horizontal plane and indicated by the arrow) giving a good view of the front of the block but not of the sides. It also has a high tilt angle (measured from the horizontal in the vertical plane) to emphasise the top of the block. The right-hand block has a medium tilt, indicated by the arrow, and a medium rotation, giving adequate views of the top and sides of the block.

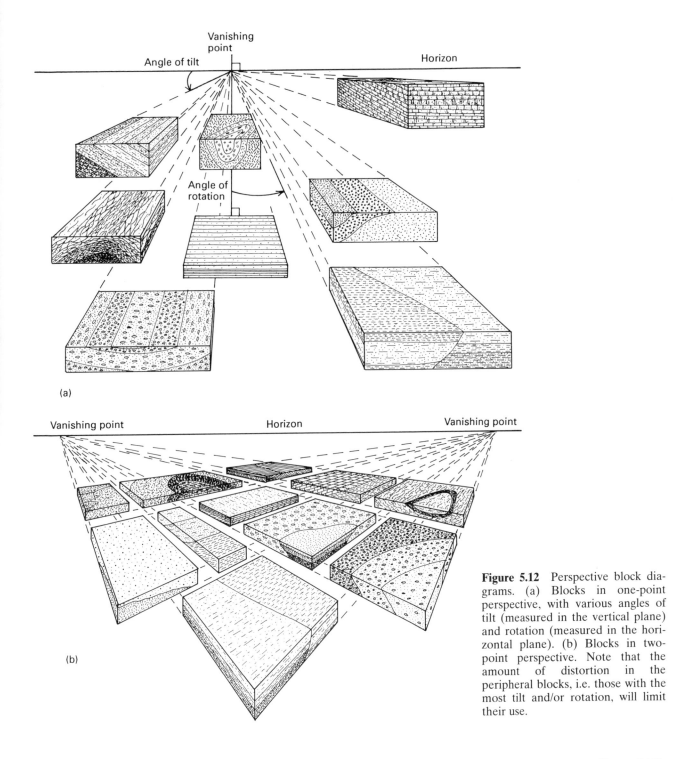

(a)

(b)

Figure 5.12 Perspective block diagrams. (a) Blocks in one-point perspective, with various angles of tilt (measured in the vertical plane) and rotation (measured in the horizontal plane). (b) Blocks in two-point perspective. Note that the amount of distortion in the peripheral blocks, i.e. those with the most tilt and/or rotation, will limit their use.

point on the horizon. Sketch versions can achieve the effect by simply making the back of the block narrower than the front.

Block diagrams in **two-point perspective** have their front corner pointing towards the observer, with all non-vertical lines merging to one of two points on the horizon. With both kinds of perspective, the amounts of rotation and tilt of the block can be selected in order to display features to the greatest advantage (figure 5.12). However, all the changes in dimensions made in drawing block diagrams interfere with any measurements that are to be made. The manual construction of properly scaled drawings is laborious, and corrections to measurements tedious. Hence, block diagrams are used less for quantitative purposes than for the rapid visual communication of three-dimensional relationships.

Computers are ideally suited to the construction and manipulation of cross-sections and three-dimensional diagrams (section 15.3). Trying various lines of section and different amounts of vertical exaggeration; transferring structure sections to stratigraphic sections, and sections to fence and block diagrams; transforming between isometric and perspective projections; varying the amount of rotation and tilt: all these tedious procedures are executed swiftly and effectively by the kinds of computers and graphic capabilities available today. Undoubtedly, these applications will grow. However, the software continues to have built-in limitations, varying according to its complexity and cost. Mistakes can be made in the processing, and any section or diagram can only be as reliable as the geological information on which it is based. Pity the geologist who does not understand what it is that the computer is doing!

Summary of Chapter

1. Cross-sections are drawn in a vertical plane.
2. They therefore closely complement maps, with which they make a powerful way of communicat-ing the three-dimensional arrangement of rock formations.
3. Sections should be drawn at right angles to the dominant strike, as far as possible, and should not involve vertical exaggeration unless for good reason.
4. Structure sections are drawn with an altitude such as sea level as the datum surface; stratigraphic sections have some stratigraphically significant surface as the datum.
5. Accurate fence diagrams, and isometric and perspective block diagrams, are laborious to construct manually. Sketched diagrams are effective at rapidly conveying the overall structure of the rocks.

Selected Further Reading

Langstaff, C.S. and Morrill, D., 1981. *Geologic Cross Sections*. International Human Resources Development Corporation, Boston, 108 p.

A straightforward introductory account of the use of sections and three-dimensional diagrams in the oil industry.

Map 10 Zambian copper belt

Opposite is a 1:100 000 reconnaissance map of the region containing the Zambian copper belt. The geological units recognised during this preliminary survey are listed in the key, but *not necessarily in their correct stratigraphic order*.

Make a cross-section to display what appears to be the regional structure. (To do this, it will first be necessary to utilise the principles of cross-cutting relationships and younging directions introduced in chapter 2 to establish the stratigraphic sequence.) List the units in stratigraphic order. Indicate any units which are of unclear stratigraphic age.

Further examples of cross-section work

Construct a cross-section at right angles to strike across the Bear Hole, Montana, area in Map 3, and from C–D on Map 6 of the Boyd Volcanics, Australia.

Cross-section work is involved with the questions that accompany Plates 3 and 4, of Epernay, France, and Root River, Canada.

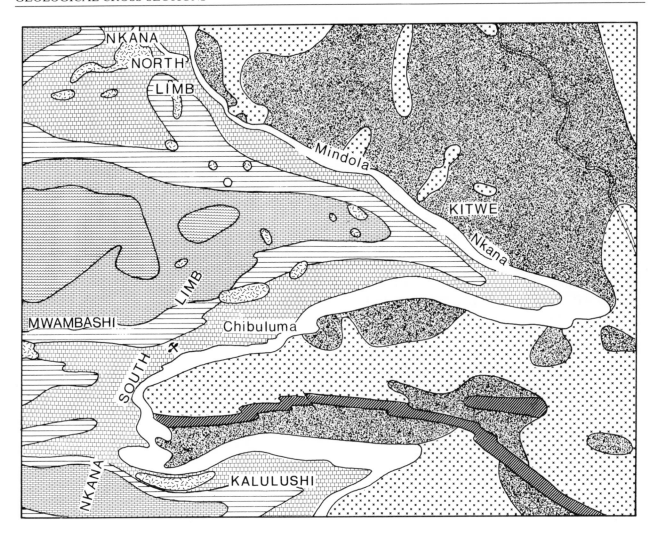

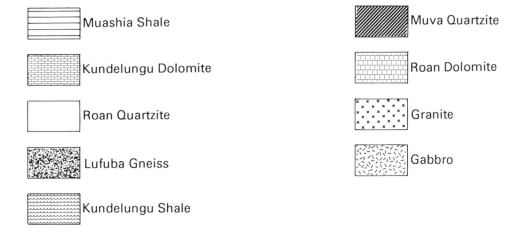

Muashia Shale

Kundelungu Dolomite

Roan Quartzite

Lufuba Gneiss

Kundelungu Shale

Muva Quartzite

Roan Dolomite

Granite

Gabbro

Map 11   Builth Wells, Powys, Wales

The facing map is of a small area near Builth Wells in mid-Wales. The map units are largely defined on the basis of fossils. The area appears on Plate 1, centred on [SO0950]. According to Plate 1, what is the stratigraphic age of the units?

Construct, as accurately as possible, a geological cross-section. Seiect carefully the appropriate scales and line of section. (Even if structure contours are not used throughout for the section construction, a few will be needed to establish representative strikes and dips of the units.)

List the true thicknesses of the units, as derived from the cross-section, as closely as you can (i.e. quote minimum thicknesses if the bases or tops are not available).

List the vertical thicknesses, as they would be seen in a borehole or well.

Draw an isometric block diagram to illustrate the geological structure of the area. (Make the sides parallel to the dip direction and the front and back parallel to the strike. Establish the four corners of the diagram, at right angles on the map, and trace their positions on to a separate sheet. Erect vertical columns at each corner, draw the skeleton of the block, add altitude scales, and transfer the vertical stratigraphic successions to the three visible corners and a few intermediate locations on the two visible panels. Connect the subsurface units. Add topographic profiles to all four panels, and sketch in the surface geology and drainage.) Comment on the angles of rotation and tilt of the block diagram.

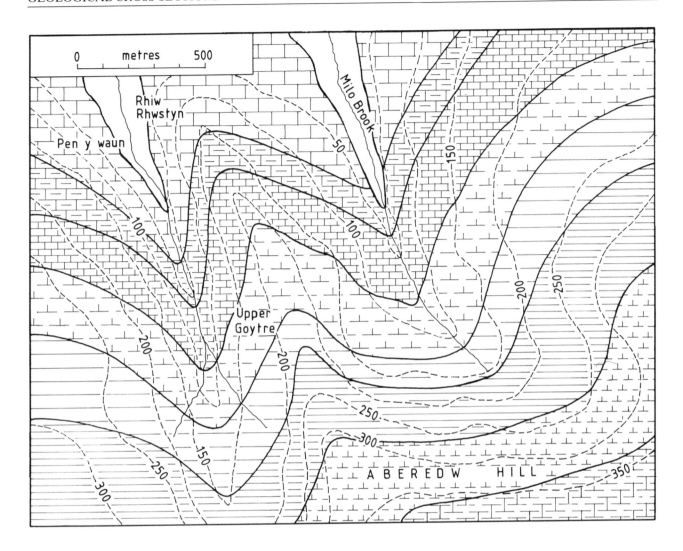

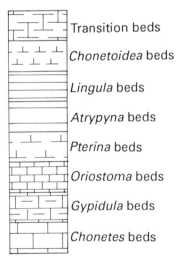

Transition beds

Chonetoidea beds

Lingula beds

Atrypyna beds

Pterina beds

Oriostoma beds

Gypidula beds

Chonetes beds

CHAPTER 6

Visual Assessment of Outcrop Patterns

6.1 Introduction

The previous three chapters dealt with using maps in a quantitative way. Methods were discussed of making precise measurements, where in commercial applications small errors could be costly. Now we return to the other approach to geological maps, that of making rapid visual assessments of the geology of an area. Central to this use is an understanding of the outcrop patterns which appear on maps. Strictly, it is the junctions of outcrops, the trace showing where adjacent units meet, that we look at. The idea was introduced in chapter 2, and even at that beginning stage we deduced from the maps the presence of features such as variably dipping beds, folds, unconformities, and faults.

However, we have assumed so far that the land surface is fairly level. This is reasonable on the small-scale maps we worked with in chapter 2, but on large-scale maps the relief can become significant. This can greatly complicate the outcrop patterns and produce misleading effects. On the other hand, much can be deduced by looking at how the map units interact with the land surface. Here we develop some general points to aid the visual appraisal of outcrop patterns on large-scale maps. We start by looking at horizontal beds, and then the effects produced by successively steeper beds.

6.2 Horizontal Formations

Horizontal beds make horizontal outcrops on the ground, irrespective of the gradients and irregularities of the land surface. Steep cliff or slight incline: the outcrops are still horizontal. And because topographic contour lines are horizontal (they must be, connecting as they do points of equal altitude), the outcrops of horizontal units always parallel the topographic contours. In terrains of intricate relief the outcrop patterns can therefore be highly irregular.

A famous example is the Grand Canyon. The deeply dissected canyon walls give very jagged topographic contours and hence a distinctive outcrop style (figure 6.1). The outcrops may look complex at first glance but they are simply following the topography. A look at the canyon walls leaves no doubt that the beds are virtually horizontal.

6.3 Dipping Formations

6.3.1 Recognition

Section 2.3.1 introduced the appearance of dipping formations on small-scale maps. It was mentioned there that if the relative ages of the rocks are known, the dip direction can be inferred from the direction in which the units become successively younger (providing they are not inverted). On large-scale maps, however, the effect of topography will be more marked, and can be a complicating factor. In fact, the outcrop shapes made by gently dipping beds intersecting with an irregular land surface can be very awkward to assess visually.

Outcrops of units dipping gently in the same direction as the land slope are not too difficult to work with. They *tend* to follow topographic contours in the same way as horizontal beds. The tendency decreases

Figure 6.1 (opposite) Outcrop patterns of horizontal beds: the Grand Canyon, Arizona, USA. The area shown here is Marble Canyon, about 40 km north of the main canyon. Reproduced from the 1:62 500 Preliminary Geologic Map of the Grand Canyon and vicinity, Arizona, by permission of the Grand Canyon Natural History Association. The inset engraving is reproduced from J.W. Powell's account of the surveying of the Grand Canyon (Powell 1895), described briefly in section 14.3.5.

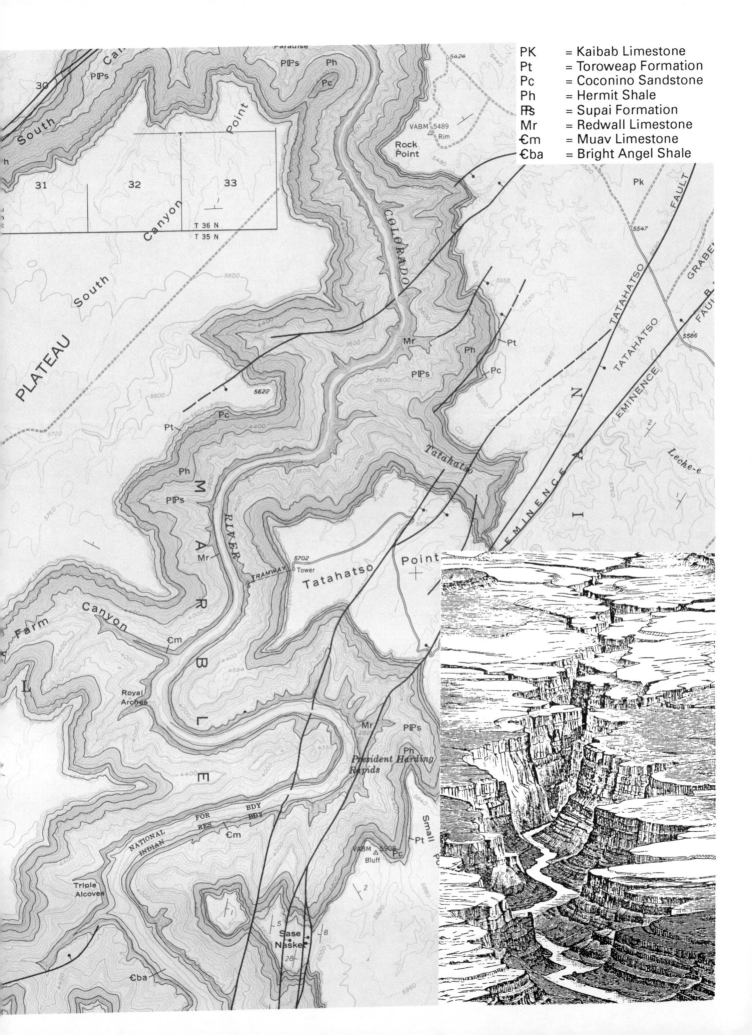

PK	= Kaibab Limestone
Pt	= Toroweap Formation
Pc	= Coconino Sandstone
Ph	= Hermit Shale
PⱣs	= Supai Formation
Mr	= Redwall Limestone
€m	= Muav Limestone
€ba	= Bright Angel Shale

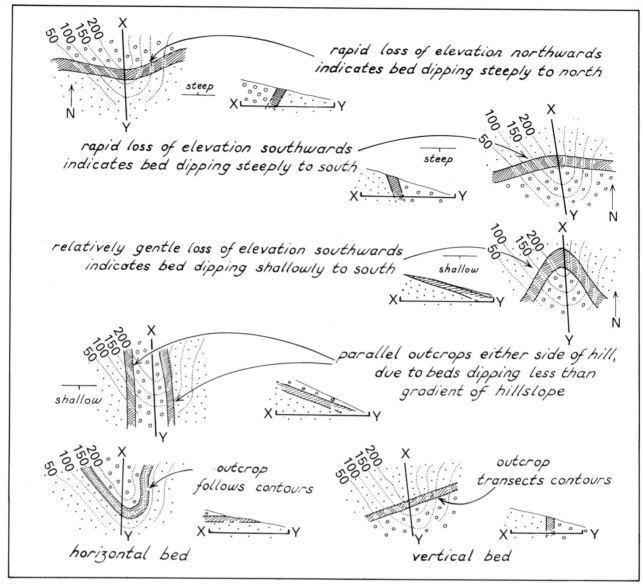

Figure 6.2 Some examples of the visual assessment of outcrops crossing hills. In each example, the formation is regularly dipping and the hillside of roughly uniform gradient. In practice, less regular outcrops are likely.

as the dip of the formations increases. It is not realistic to give a dip angle at which the mimicking effect ceases, partly because topographic slopes can be so intricately variable and partly because the beds themselves may be influencing the relief. For example, the tendency for durable beds to form escarpments is well-known (e.g. see the bottom illustration of figure 6.6), and the topographic contours of the escarpment tend to parallel the outcrop of the dipping bed. The parallelism does not extend to any irregularities in the escarpment in the way that would happen with horizontal beds, but it does mean that the interpretation of dips from roughly parallel topography and outcrop has to be treated with some care.

Formations dipping gently in the opposite direction to land slopes can, depending on the irregularity of the topography, produce very complex looking outcrop patterns. The safest approach to determining the direction and amount of dip is to look for where the outcrops cross a series of topographic contours, and see in which direction the beds consistently lose overall altitude (figure 6.2). Rapid crossing of successive topographic contours implies a steep dip.

Outcrops of formations with moderate dips, in general, cross rounded hills with an arcuate pattern (figure 6.3). If the arc curves in the same direction as the topographic contours, the dip direction is opposite to the downslope direction of the hillslope. If the

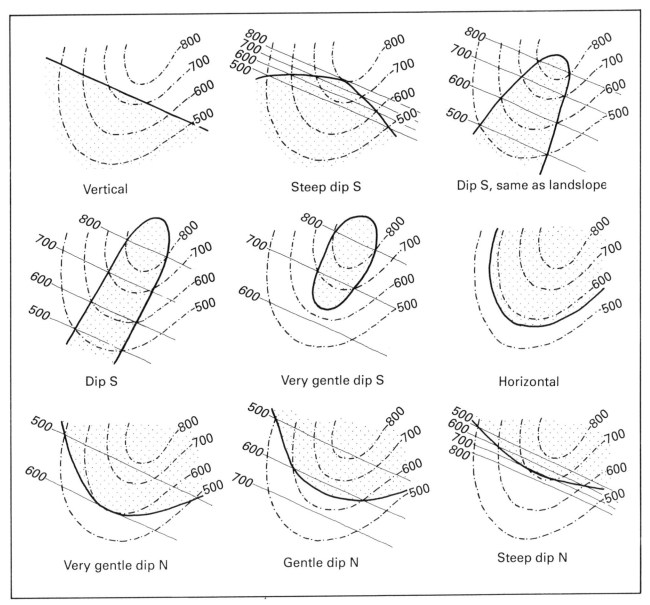

Figure 6.3 Outcrop shapes made by a uniformly dipping surface crossing a rounded hill. The trace of the boundary between two formations is shown in a heavy line and topographic contours in a dot–dash line. Some structure contours, in light continuous lines, have been added to indicate the differing amounts and directions of dip that produce the different outcrop shapes. North is to the top of the page; south and north below the sketches are approximate directions.

outcrop arc is in the opposite direction to the curve of the topographic contours, the dip direction is the same as the downslope direction of the hillslope. In both cases, the openness of the outcrop arc increases as the dip steepens. But hills are so diverse in form that these ideas can be precarious in practice. Visualising the structure contours is a better guide: their direction will be at right angles to the dip direction and their spacing reflects the amount of dip. In desperation you could sketch in a few structure contours, temporarily!

6.3.2 Assessment of formation dip in valleys

The one situation in which rules for visual assessment can be applied with some confidence is where formations cross a valley. Drainage patterns or topographic contours enable any valleys to be identified on a map; the so-called 'V-rules' refer to the *outcrop* shape typically made by formations crossing the valley (figures 6.4 and 6.5):

● the apex of the 'V' points in the direction of bed dip;

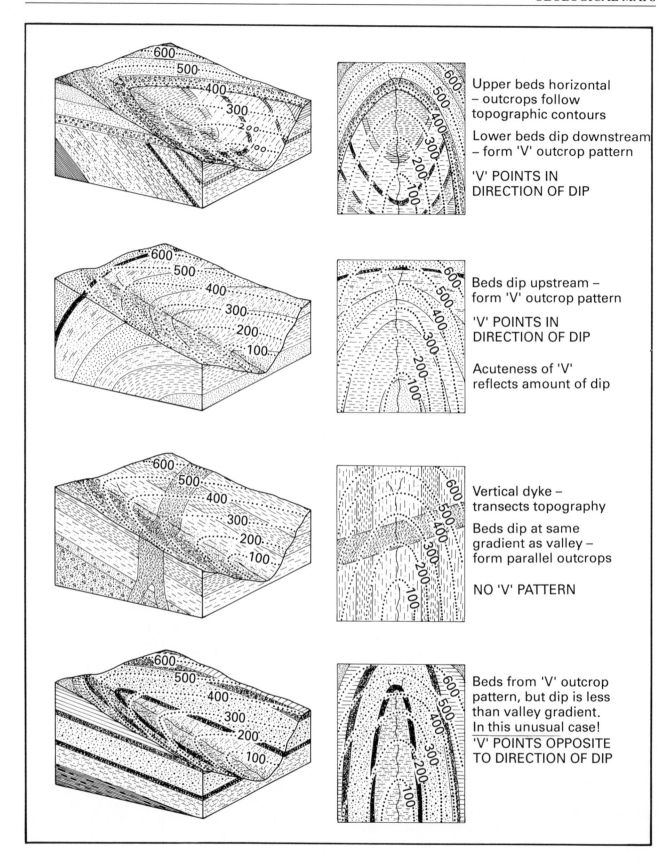

Figure 6.4 Block diagrams and maps to illustrate the 'V-rules' of outcrops crossing valleys.

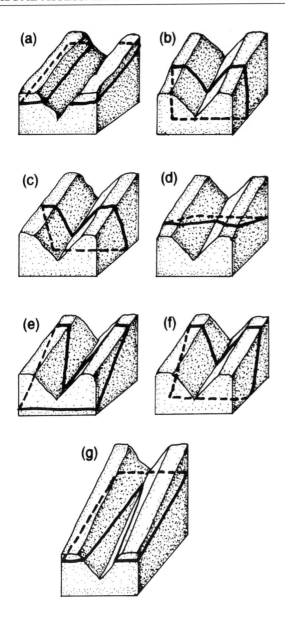

Figure 6.5 Conceptual diagrams to illustrate how the 'V-rules' work. (a) Horizontal formation: its outcrop following the valley sides, paralleling topographic contours. (b) Vertical formation: cross-cutting the valley making a straight trace. (c) Steeply dipping formation: making open V-shape pointing in direction of dip. (d) Formation dipping gently, in same direction as in (c): making a tighter V-shape, pointing in same direction as in (c), the direction of dip. (e) Formation makes a moderate dip, towards front of diagram: V-shape points in same direction, opposite to (c) and (d). (f) Very steeply dipping formation, in same direction as (e): makes an almost opened-up V-shape (approaches the straight trace seen in (b), but does point towards front of diagram, the direction of dip. (g) The exceptional situation: the floor of the valley slopes more steeply than the dip of the formation. The V-shape points in the *opposite* direction to its direction of dip.

- acute ('tight') 'V'-shapes reflect shallow dips;
- open 'V'-shapes indicate steep beds.

These are useful guides for swift outcrop appraisal. However, beware of the exception that proves the rule:

- beds dipping less steeply than the valley floor give 'V's pointing in the direction *opposite* to bed dip! This uncommon circumstance can be recognised by the outcrops having an even more acute 'V'-shape than the topographic contours.

Regarding very steeply inclined units, their outcrop patterns are normally straightforward to analyse. The 'V-rules' work well. The complicating effect of topography progressively lessens as beds become steeper.

6.4 Vertical Formations

Vertical beds are characterised by straight-looking outcrops. They completely ignore topography (figures 6.3, 6.4 and 6.5). With horizontal beds they form two ends of a spectrum of the influence topography has on outcrop shape. Any twists in the outcrops of a vertical unit represent a real change of the strike of the formation.

This behaviour applies to any vertical geological features, not just beds of rock. Igneous dykes commonly appear on maps with straight traces, because they are typically vertical. Faults are usually depicted as fairly straight lines. This is no intrinsic property of faults, but simply follows from many faults being oriented very steeply or vertically.

6.5 Assessment of Formation Thickness

It was mentioned in section 2.3.1 that knowing something about the thickness of a map unit allows us to judge on a small-scale map the amount of dip, from the outcrop width of the unit (figure 6.6). Steeper beds make narrower outcrops than shallower beds of the same thickness. Conversely, if we know something about the dip, we should be able to estimate from the outcrop width the approximate thickness of the unit. Thicker beds make wider outcrops than thinner beds of the same dip. However, there is a further factor – the relief of the land surface. On a small-scale map its effect will be diminutive and can be neglected. The ground can be treated as being flat.

On large-scale maps, especially of rugged areas, the effect of topography can be very significant. In general, increasingly steep slopes progressively suppress outcrop width (figure 6.7). With horizontal beds, the effect of topography will be at its greatest and the land gradient will have a profound influence on outcrop width. The effect is well illustrated by the Grand Canyon (figure 6.1). Shallow-dipping beds, too, give outcrop widths which are highly dependent on topographic slope, but the effect diminishes as beds become steeper. Vertical beds form the other end of the spectrum – their outcrop width equals their thickness irrespective of the topography. So, on large-scale maps the visual assessment of formation thickness from the approximate dip, or the estimation of

dip knowing the thickness, has to be made with a careful eye on the topographic slope.

Summary of Chapter

1. Outcrops of horizontal formations parallel the topographic contours.
2. The tendency for dipping formations to parallel topographic contours decreases as the angle of dip increases.
3. It is important to visualise mentally in three dimensions how dipping formations cross

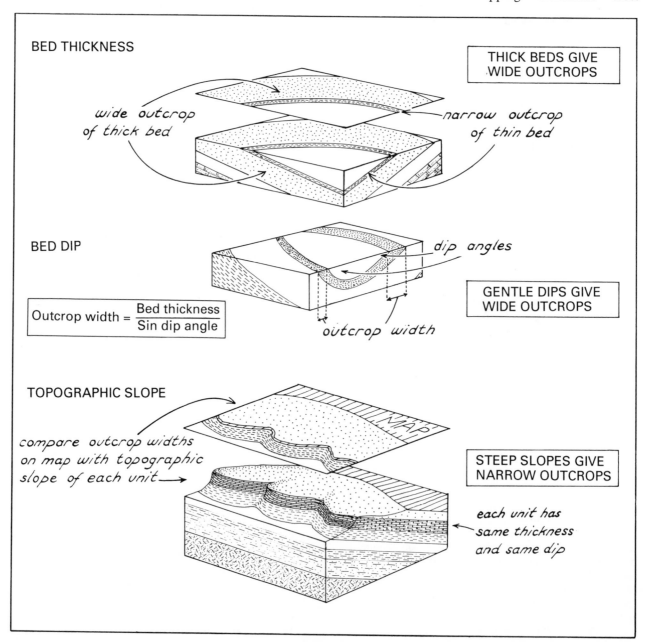

Figure 6.6 Block diagrams to illustrate the three factors which influence the outcrop width of a formation. The equation relating outcrop width to bed thickness assumes level ground, and is a rearrangement of the equation given in figure 4.8.

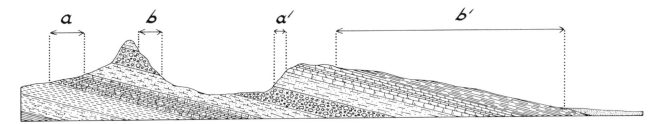

Figure 6.7 The relationship between outcrop width and topography. All the formations shown have the same thickness and dip angle. Note that steeper land slopes suppress the outcrop width: compare a with a', both slopes inclined in the same direction, and b with b'. Slopes inclined in the opposite direction to the formation dip tend to suppress outcrop width relative to slopes in the same direction: compare a with b', both with similar slope angles, and a' with b.

landforms, but some generalisations can be made, particularly regarding the 'V'-shapes of outcrops crossing valleys.

4. Outcrops of vertical formations ignore topography and tend to be straight.

5. Outcrop width depends not only on formation thickness and dip, but on the land slope. Increasingly steep slopes progressively suppress outcrop width.

Examples on maps

Find areas on Map 3, of Bear Hole, Montana, where the outcrop patterns do not follow the normal 'V-rules'. Indicate an area where the 'V-rules' are obeyed.

On the basis of visual assessment, what is the approximate orientation of the units appearing on Map 8, Long Mountain, and on Map 11, near Builth Wells, Wales?

Using visual assessment alone, deduce the overall orientation of the beds on Plate 3, Epernay, France, and on the eastern half of Plate 8, of Malmesbury, England. Comment on the dip of the andesite and basalt (Ac1) on Plate 5 (Sanquar, Scotland). What is the likely dip of the Southern Uplands Fault, which appears on Plate 5?

On Plate 6, of the Heart Mountain district, Wyoming, USA, explain why the Carboniferous and Permo-Trias units in the SE of the area have more pronounced 'V'-shaped outcrops than the Cretaceous formations in the east, yet both sequences have approximately similar dips. What is the approximate orientation of the Carboniferous rocks in Paint Creek (lat. 44° 43', long. 109° 20'), and of the Cambrian Pilgrim Limestone in Dead Indian Creek (lat. 44° 42', long. 109° 25')?

Colour Plates

Plate 1   Part of the 'Ten Mile' map of the UK

Reproduced opposite with a simplified key is a portion of the 1:625 000 map discussed in chapter 2. Some background on how the map was made is mentioned in section 14.3.7. The numbers shown in red on the map refer to BGS one-inch and 1:50 000 sheets; the heavy dot-dash lines are faults.

Where on the map area do the oldest rocks occur?

Where are the youngest rocks?

Where are the oldest sedimentary rocks?
Why might the ground making the Longmynd [SO4293] be higher than the land around it?

Following the principles introduced in chapter 2, it should be possible to make interpretations of the outcrop patterns on the map. A journey southwards from Llangynidr [SO1619] would cross rocks dipping in what direction?

Why might the outcrop width of the Carboniferous rocks around Brynamman [SN7214] be greater than the same rocks at Abersychan [SO2604]?

Between Cemmaes [SH8406] and Llanelltyd [SH7219], the overall dip of the rocks is in which direction?

What is (a) the strike direction, and (b) the dip direction of the Silurian rocks around Wenlock Edge [SO5090]?

Comment on the nature of the junction at the base of the Silurian rocks around [SO2694].

What is the nature of the junction of the Permo-Triassic sandstones (unit 89) with older rocks?

Explain the junction of the base of the Lower Old Red Sandstone (unit 75) with older rocks at [SO3505].

Why do the outcrops of Silurian rocks around Leintwardine [SO4173] have a curving pattern?

What geological evidence is there on the map for the existence of a fault running NE–SW from Church Stretton [SO4693]? What can you say about the stratigraphic age of this fault?

If a borehole (well) were sunk at Michaelchurch Escley [SO3332], what is likely to be the first unit to be encountered below the Lower Old Red Sandstone?

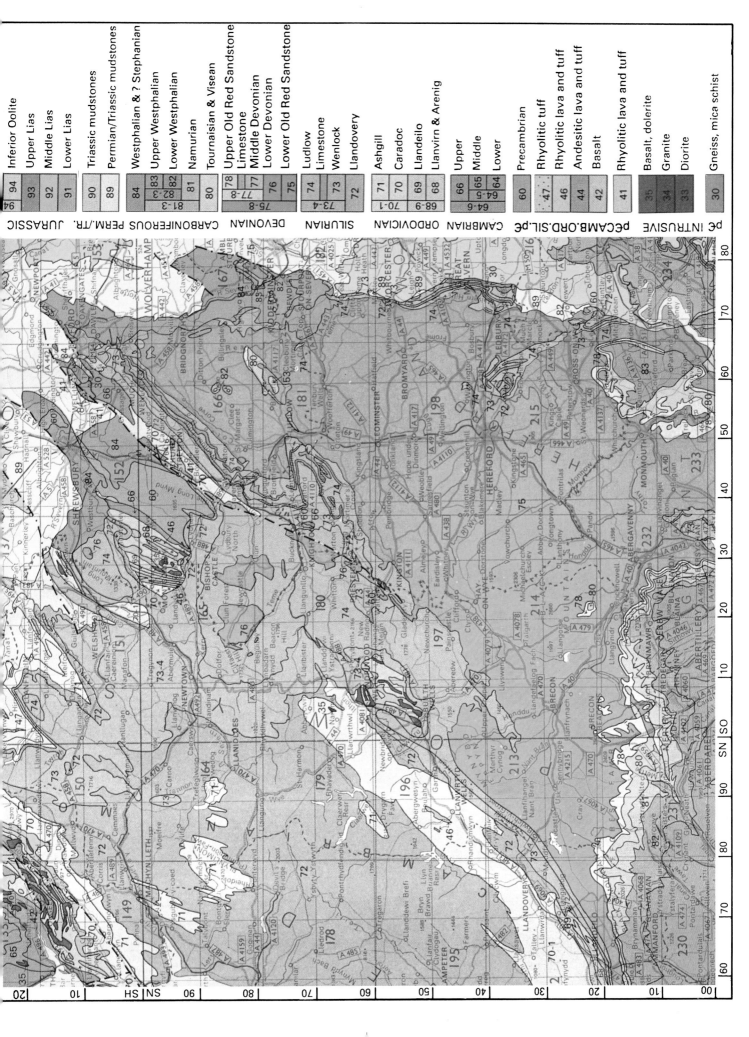

JURASSIC	94	Inferior Oolite
	93	Upper Lias
	92	Middle Lias
	91	Lower Lias
PERM./TR.	90	Triassic mudstones
	89	Permian/Triassic mudstones
CARBONIFEROUS	84	Westphalian & ? Stephanian
	83 / 82·3 / 81·3	Upper Westphalian
	82	Lower Westphalian
	81	Namurian
	80	Tournaisian & Visean
DEVONIAN	78	Upper Old Red Sandstone
	77 / 75·8 / 77·8	Limestone
	76	Middle Devonian
	75	Lower Devonian
		Lower Old Red Sandstone
SILURIAN	74 / 73·4	Ludlow
		Limestone
	73	Wenlock
	72	Llandovery
ORDOVICIAN	71	Ashgill
	70 / 70·1	Caradoc
	69	Llandeilo
	68 / 68·9	Llanvirn & Arenig
CAMBRIAN	66 / 64·6	Upper
	65 / 64·5	Middle
	64	Lower
pЄCAMB.ORD.SIL.pЄ	60	Precambrian
	47	Rhyolitic tuff
	46	Rhyolitic lava and tuff
	44	Andesitic lava and tuff
	42	Basalt
	41	Rhyolitic lava and tuff
INTRUSIVE	35	Basalt, dolerite
	34	Granite
	33	Diorite
	30	Gneiss, mica schist

Plate 2 Part of the 1:2 500 000 map of the US

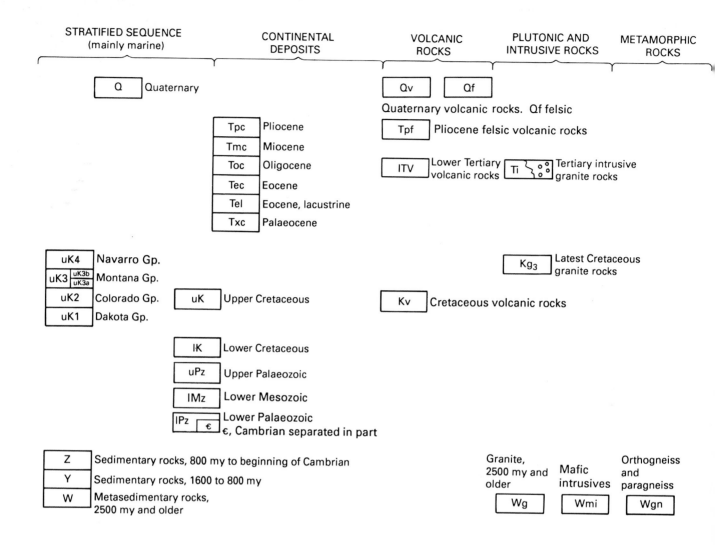

STRATIFIED SEQUENCE (mainly marine)

CONTINENTAL DEPOSITS

VOLCANIC ROCKS

PLUTONIC AND INTRUSIVE ROCKS

METAMORPHIC ROCKS

Q   Quaternary

Qv   Qf
Quaternary volcanic rocks. Qf felsic

Tpc   Pliocene
Tmc   Miocene
Toc   Oligocene
Tec   Eocene
Tel   Eocene, lacustrine
Txc   Palaeocene

Tpf   Pliocene felsic volcanic rocks

ITV   Lower Tertiary volcanic rocks

Ti   Tertiary intrusive granite rocks

uK4   Navarro Gp.
uK3   uK3b uK3a   Montana Gp.
uK2   Colorado Gp.
uK1   Dakota Gp.

uK   Upper Cretaceous

Kg₃   Latest Cretaceous granite rocks

Kv   Cretaceous volcanic rocks

lK   Lower Cretaceous

uPz   Upper Palaeozoic

lMz   Lower Mesozoic

lPz   Lower Palaeozoic
€   €, Cambrian separated in part

Z   Sedimentary rocks, 800 my to beginning of Cambrian
Y   Sedimentary rocks, 1600 to 800 my
W   Metasedimentary rocks, 2500 my and older

Granite, 2500 my and older   Mafic intrusives   Orthogneiss and paragneiss
Wg   Wmi   Wgn

Reproduced opposite, by permission of the US Geological Survey, is a portion of the map discussed in chapter 2. The legend given above is abstracted from the comprehensive explanation that accompanies the original map.

Towards which direction are the beds around lat. 45° 30′, long. 110° 20′ dipping?

Comment on the nature of the junction at the base of the Miocene continental deposits (Tmc) around lat. 45° 50′, long. 111° 30′.

Describe the structure west of Cody, centred on lat. 44° 30′, long. 109° 15′.

What can be inferred about the fault in Madison County, Montana, centred on lat. 45° 0′, long. 111° 30′? (An earthquake in this vicinity in 1959 caused serious damage and numerous fatalities.)

Plate 3 Epernay, France

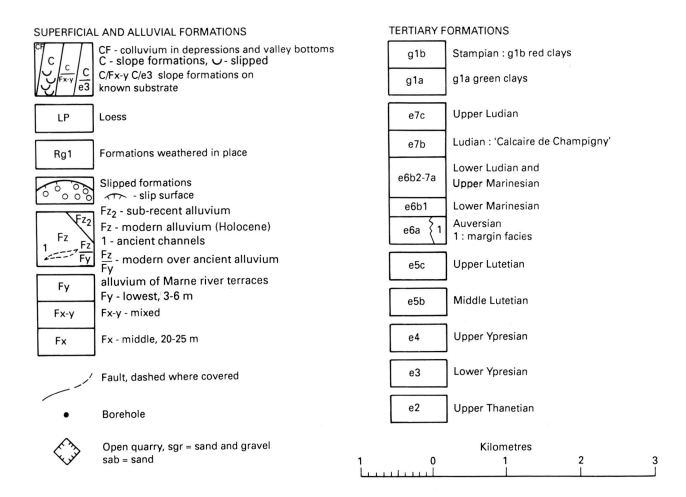

SUPERFICIAL AND ALLUVIAL FORMATIONS

CF, C, Fx-y, C/e3	CF - colluvium in depressions and valley bottoms C - slope formations, ᴗ - slipped C/Fx-y C/e3 slope formations on known substrate
LP	Loess
Rg1	Formations weathered in place
	Slipped formations ⌢ - slip surface
Fz₂, Fz, 1, Fz/Fy	Fz₂ - sub-recent alluvium Fz - modern alluvium (Holocene) 1 - ancient channels Fz/Fy - modern over ancient alluvium
Fy	alluvium of Marne river terraces Fy - lowest, 3-6 m
Fx-y	Fx-y - mixed
Fx	Fx - middle, 20-25 m

Fault, dashed where covered

● Borehole

◇ Open quarry, sgr = sand and gravel
sab = sand

TERTIARY FORMATIONS

g1b	Stampian : g1b red clays
g1a	g1a green clays
e7c	Upper Ludian
e7b	Ludian : 'Calcaire de Champigny'
e6b2-7a	Lower Ludian and Upper Marinesian
e6b1	Lower Marinesian
e6a {1	Auversian 1 : margin facies
e5c	Upper Lutetian
e5b	Middle Lutetian
e4	Upper Ypresian
e3	Lower Ypresian
e2	Upper Thanetian

Kilometres

1   0   1   2   3

Reproduced opposite is part of the French BRGM 1:50 000 map 157, 'Epernay', published in 1977. It is of the area around the River Marne, in the Champagne region of north-central France. The key given above is translated from the French of the original, and various details have been omitted.

Looking at the topographic contours, printed in brown, and the outcrop patterns, what is the overall orientation of the Tertiary beds? Which of the superficial formations tend to be arranged horizontally and which do not? Where do the Tertiary beds have the steepest dips?

What is the dominant control on which bedrock unit outcrops at a particular place? Which bedrock unit tends to make the steepest topographic slopes? Which units are most prone to landslips? Draw a cross-section to show the relation between relief and bedrock geology. What appears erroneous about the portrayal of the geological succession west of Courthiezy, around [6938 1510]?

The pink lines are structure contours for the top of the chalk (Cretaceous). What do they indicate about the overall arrangement of the chalk? Construct a cross-section to show the structure. In the centre of the village of Chassins [6943 1541], at what depth is the top of the chalk?

(Map reproduced by permission of the Bureau de Recherches Gèologiques et Miniéres, France.)

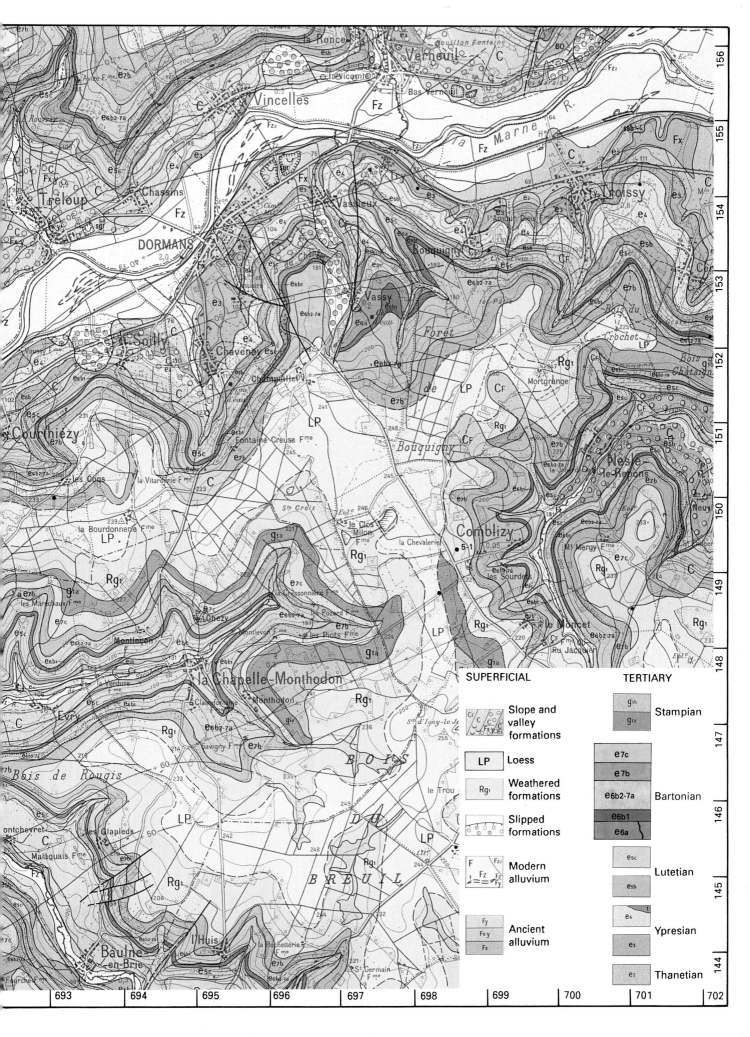

SUPERFICIAL

Slope and valley formations	
LP	Loess
Rg₁	Weathered formations
	Slipped formations
F · Fz	Modern alluvium
Fy · Fxy · Fx	Ancient alluvium

TERTIARY

g₁b · g₁a	Stampian
e7c · e7b · e6b2-7a · e6b1 · e6a	Bartonian
e5c · esb	Lutetian
e4 · e3	Ypresian
e2	Thanetian

Plate 4 Root River, District of Mackenzie, Canada

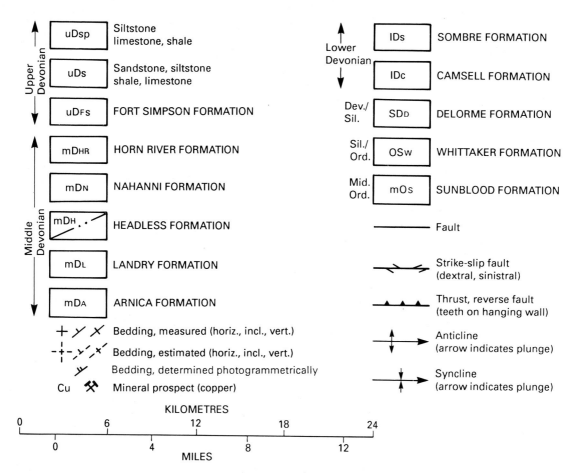

Opposite is part of the 1:250 000 Geological Survey of Canada map of the Root River district, in the Northwest Territories of Canada. The area falls within an orogenic belt, the Mackenzie Fold Belt, and so the rocks are deformed, particularly by folding.

Are there any folds present for which the axial traces are not marked? Find examples of periclinal synclines and periclinal anticlines. Comment on the plunge direction of the Delorme Syncline. Describe the orientation, attitude, and style of the Delorme Syncline and the Whittaker Anticline. (Assess the limb dips by noting, for example, the

'V'-shape of the outcrops of the Headless Formation in the syncline and the Delorme Formation in the anticline.)

Account for the irregular shape of the outcrops of Camsell Formation in the Painted Mountains Anticline.

Draw a cross-section across the map, along the line indicated between long. 125° 40' and long. 124° 40'.

(Map reproduced by permission of the Geological Survey of Canada.)

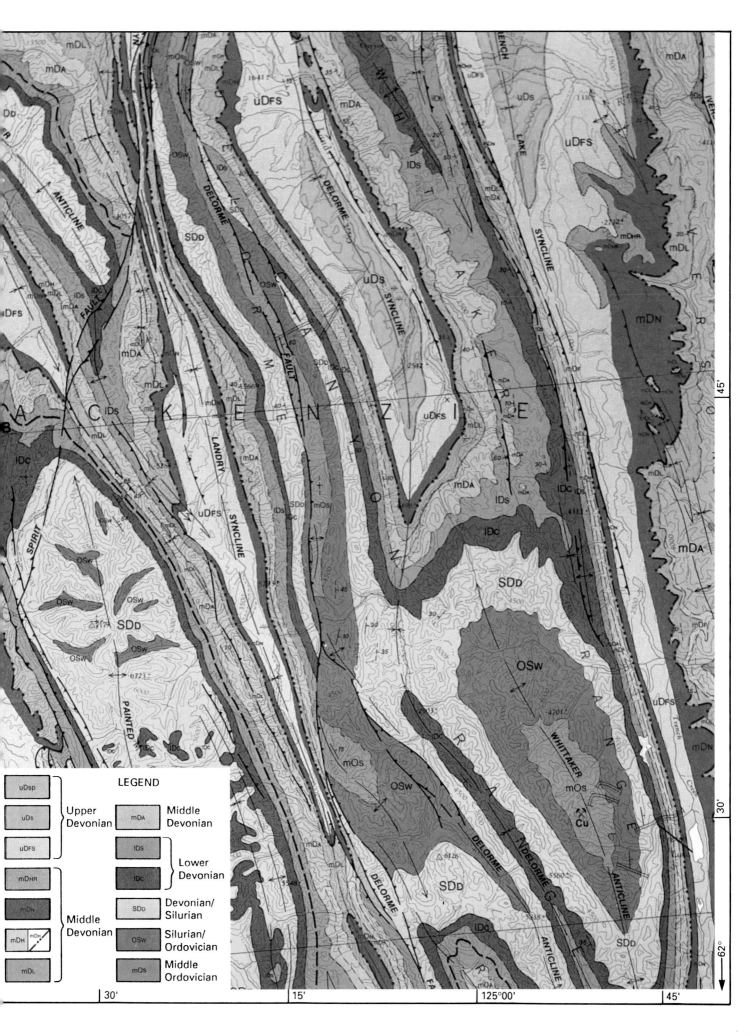

LEGEND

Upper Devonian: uDsp, uDs, uDFS

Middle Devonian: mDHR, mDN, mDH, mDL

Middle Devonian: mDA

Lower Devonian: lDS, lDC

Devonian/Silurian: SDd

Silurian/Ordovician: OSw

Middle Ordovician: mOs

Plate 5 Sanquar, Southern Scotland

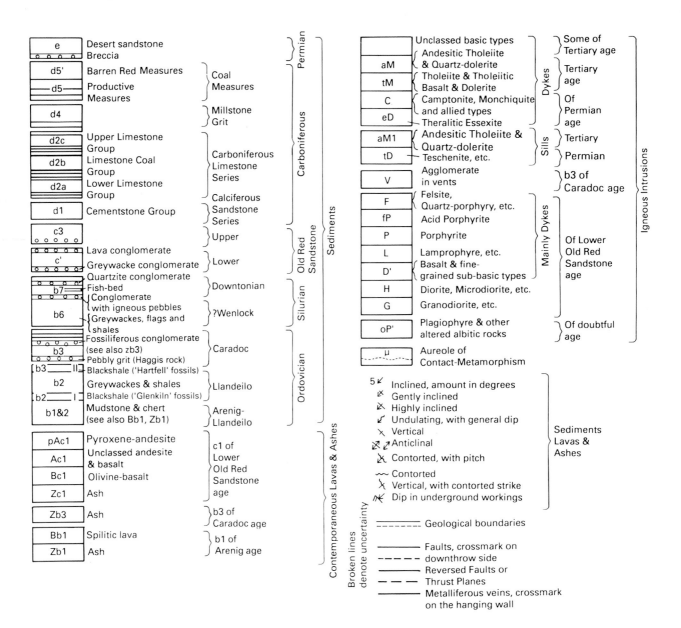

Reproduced opposite is a part of the BGS Sheet 15, Sanquar. It was published in 1937 after a series of revisions, some involving famous BGS geologists such as Ben Peach, John Horne and Sir Edward Bailey (section 14.3.7). It is reproduced here at its original 'one-inch' scale (1:63 360), although it is now available as two 1:50 000 sheets. A remarkable amount of the geology of southern Scotland is encapsulated in the portion of the map given here.

Discuss the role of the Southern Upland Fault (the major NE-SW fault in the centre of the area) in influencing sedimentation, volcanism, and structure. Define its age. What else may have influenced the distribution of volcanism?

Judging from the lithologies, how might the environments of deposition have changed drom the Silurian to the Old Red Sandstone, from the Old Red Sandstone to the Carboniferous, and from the Carboniferous to the Permian?

What are the main geological trends? Summarise the structural styles seen in the Lower Palaeozoic, the Old Red Sandstone, and the Carboniferous rocks. What evidence is there from the stratigraphic succession that earth movements were active during this period?

(Map reproduced by permission of the Director, British Geological Survey: Crown/NERC copyright reserved.)

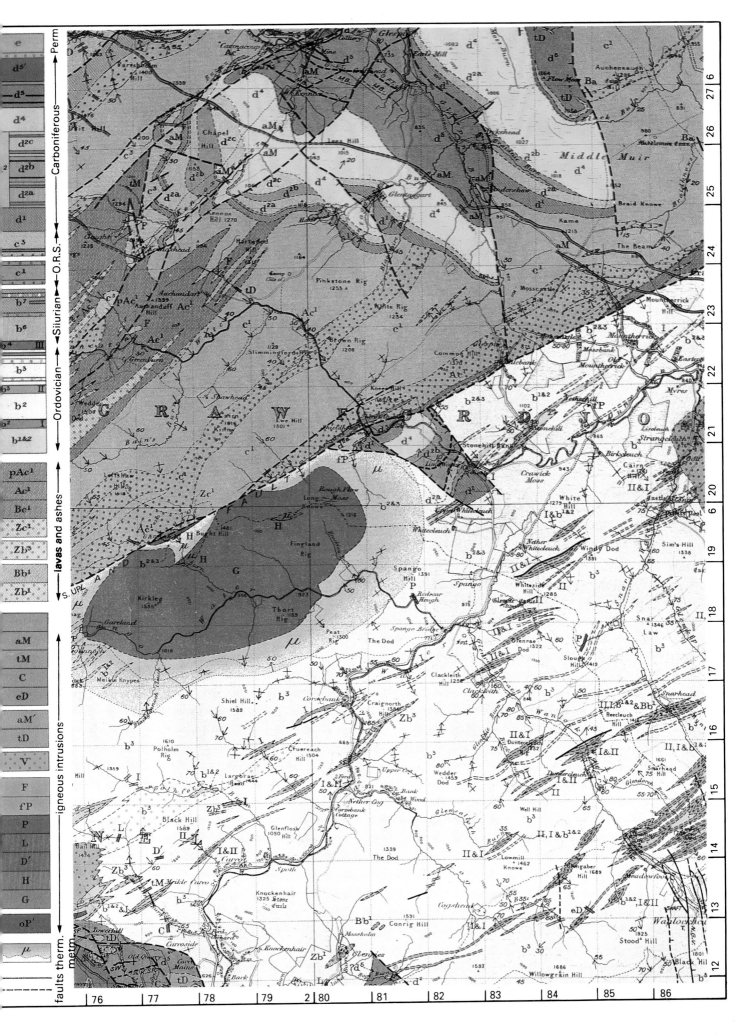

Plate 6 Heart Mountain, Wyoming

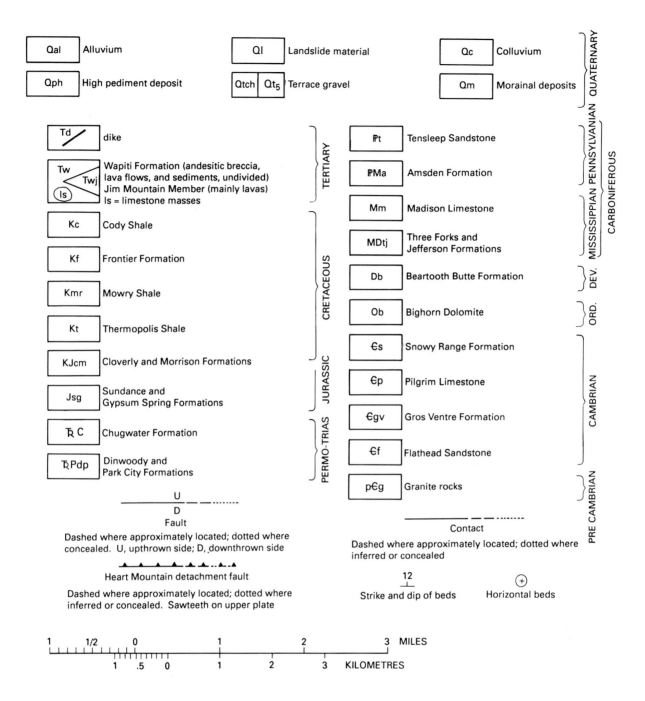

Qal — Alluvium

Qph — High pediment deposit

Ql — Landslide material

Qtch Qt₅ — Terrace gravel

Qc — Colluvium

Qm — Morainal deposits

QUATERNARY

Td — dike

Tw Twj ls — Wapiti Formation (andesitic breccia, lava flows, and sediments, undivided) Jim Mountain Member (mainly lavas) ls = limestone masses

TERTIARY

Kc — Cody Shale

Kf — Frontier Formation

Kmr — Mowry Shale

Kt — Thermopolis Shale

KJcm — Cloverly and Morrison Formations

Jsg — Sundance and Gypsum Spring Formations

ŘC — Chugwater Formation

ŘPdp — Dinwoody and Park City Formations

CRETACEOUS

JURASSIC

PERMO-TRIAS

Pt — Tensleep Sandstone

PMa — Amsden Formation

Mm — Madison Limestone

MDtj — Three Forks and Jefferson Formations

Db — Beartooth Butte Formation

Ob — Bighorn Dolomite

Єs — Snowy Range Formation

Єp — Pilgrim Limestone

Єgv — Gros Ventre Formation

Єf — Flathead Sandstone

pЄg — Granite rocks

PENNSYLVANIAN MISSISSIPPIAN CARBONIFEROUS

DEV.

ORD.

CAMBRIAN

PRE CAMBRIAN

U
D
Fault

Dashed where approximately located; dotted where concealed. U, upthrown side; D, downthrown side

▲ ▲ ▲ ▲ ▲ ▲ ▲ ··▲·▲
Heart Mountain detachment fault

Dashed where approximately located; dotted where inferred or concealed. Sawteeth on upper plate

Contact

Dashed where approximately located; dotted where inferred or concealed

12
⊥
Strike and dip of beds

⊕
Horizontal beds

1   1/2   0   1   2   3   MILES

1   .5   0   1   2   3   KILOMETRES

The map opposite is reproduced from parts of two USGS 15′ quadrangles in NW Wyoming (Pat O'Hara Mountain, GC-755, and Deep Lake, GQ-478), by permission of the US Geological Survey. It includes a famous structure called the Heart Mountain Fault, described as a 'showpiece' of North American tectonics. Although the geological history of the area was remarkably stable throughout much of the Phanerozoic, at one point huge masses of rock, themselves intact, were dispersed over a vast area along a detachment fault with a puzzlingly low dip angle. How it happened is still the subject of debate.

Discuss the geological history of the area shown, paying particular attention to the nature of the Cambrian – Cretaceous succession, any tectonism before or after the Heart Mountain Fault, and the fault itself, for example, its attitude, any stratigraphic influence on its location, and its timing.

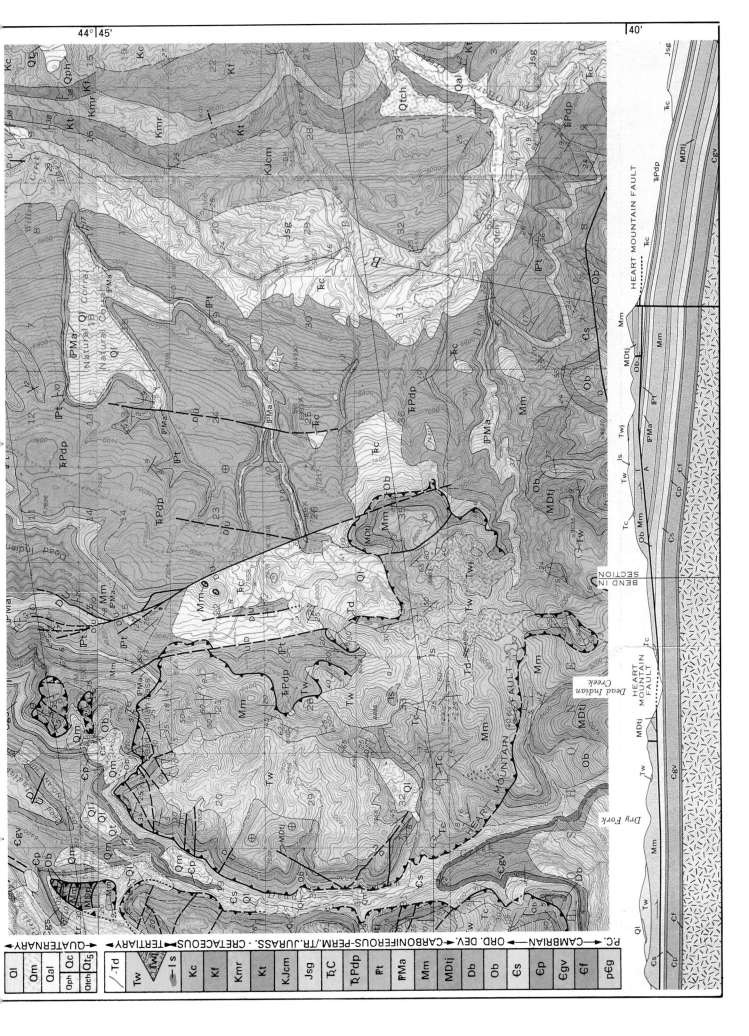

Plate 7 Marraba, Queensland, Australia

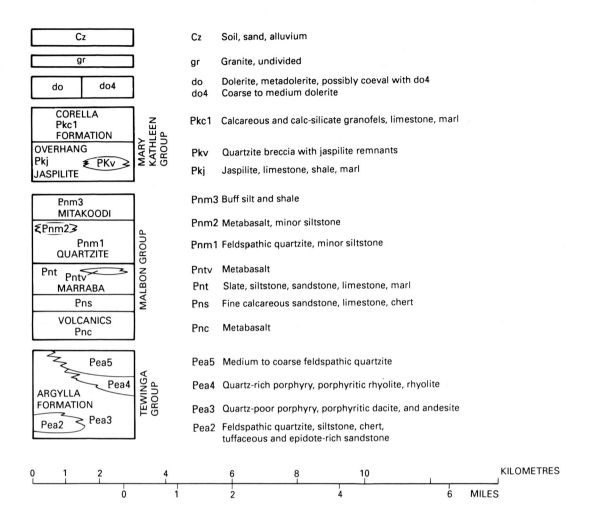

Cz	Soil, sand, alluvium
gr	Granite, undivided
do	Dolerite, metadolerite, possibly coeval with do4
do4	Coarse to medium dolerite
Pkc1	Calcareous and calc-silicate granofels, limestone, marl
Pkv	Quartzite breccia with jaspilite remnants
Pkj	Jaspilite, limestone, shale, marl
Pnm3	Buff silt and shale
Pnm2	Metabasalt, minor siltstone
Pnm1	Feldspathic quartzite, minor siltstone
Pntv	Metabasalt
Pnt	Slate, siltstone, sandstone, limestone, marl
Pns	Fine calcareous sandstone, limestone, chert
Pnc	Metabasalt
Pea5	Medium to coarse feldspathic quartzite
Pea4	Quartz-rich porphyry, porphyritic rhyolite, rhyolite
Pea3	Quartz-poor porphyry, porphyritic dacite, and andesite
Pea2	Feldspathic quartzite, siltstone, chert, tuffaceous and epidote-rich sandstone

Facing is part of the 1:100 000 Marraba Sheet 6956 produced jointly by the Bureau of Mineral Resources in Australia and the Geological Survey of Queensland. The area is immediately west of the town of Cloncurry in western Queensland, one of the hottest parts of Australia, having reached 53° in the shade!

The area reproduced here is just part of the extensive and intricate complex of Proterozoic igneous and metamorphic rocks that is finely portrayed on the full sheet. For simplicity, the numerous structural and other symbols have been omitted from the key above. Also, mining details (the area falls in the important Mount Isa Pb-Zn-Ag region) are here omitted. Note that the cross-section reproduced here does not extend across the width of the map because the section line bends.

On the basis of the map patterns and the cross-section, describe the geological structure of the area, and interpret the successive stages in its geological history.

(Map and section reproduced by permission of the Bureau of Mineral Resources, Canberra, and the Department of Mines, Queensland, Australia.)

Plate 8 Malmesbury, England

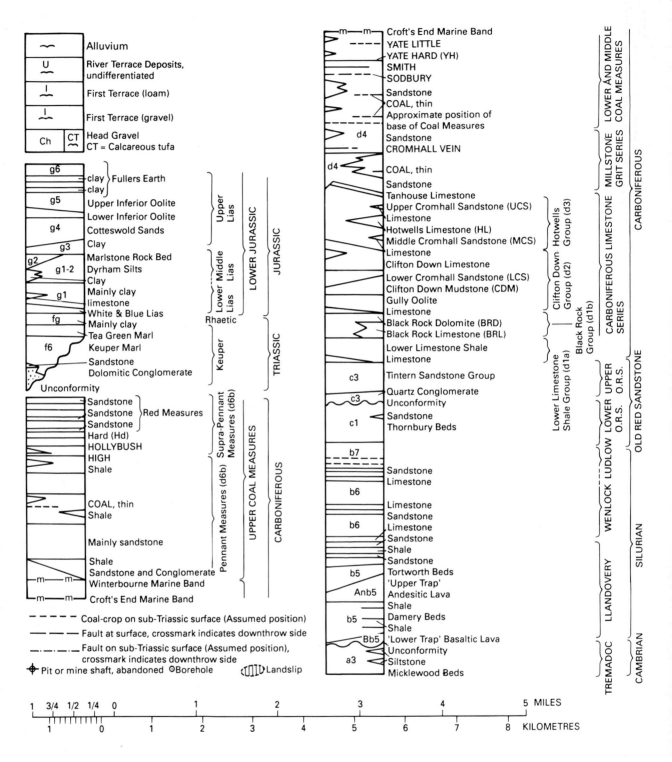

Reproduced opposite is part of the BGS 'one-inch' map (1:63 360) of the Malmesbury district, England. The key given above is simplified from the scaled vertical sections which accompany the original map.

(Map and section reproduced by permission of the Director, British Geological Survey: Crown/NERC copyright reserved.)

Interpret the geological history of the area shown.

CHAPTER 7

Unconformities

7.1 Introduction

In the spring of 1788, James Hutton peered from his small boat at the cliffs of Siccar Point in SE Scotland, and, seeing horizontal red sandstones lying directly upon vertical beds, became the first to perceive the long time interval implied by this kind of junction. The underlying material had to have been buried, lithified, rotated, and uplifted before the red sandstones were deposited on it.

The concepts of geological time have become central to an understanding of this kind of junction, known as an unconformity. Use of the term has grown in a somewhat complex and confused way, though the details do not concern us here. The first part of this chapter summarises current terminology. The important thing in the present context is the identification and manipulation of unconformities on maps, and the interpretation of what they mean. These aspects are developed in the second part of the chapter.

7.2 Terminology

An **unconformity** is a significant time-break in the stratigraphic succession of an area. It is usually visualised as the junction where the formations of differing stratigraphic age come together, properly called the **surface of unconformity**. The time gap implied by the junction may be relatively small and the unconformity developed only locally, or entire stratigraphic systems may be missing and the unconformity of regional importance.

Most adjacent formations in a stratigraphic succession are probably separated by some time gap, to account for the production of the different rock types which define the formations. However, this alone would not be considered enough to call the junction an unconformity. There has to be an indication that part of the known geological record is absent. If the units are discordant at the junction, as at Hutton's Siccar Point locality, then an unconformity is clearly indicated, because there has to have been an interval sufficient to allow tilting and erosion of the older rocks before the younger ones were formed. However, if the two units are arranged parallel to each other, their stratigraphic age will have to be known in order to gauge whether or not there is sufficient stratigraphy missing for the junction to rank as an unconformity.

The main kinds of unconformity are depicted in figure 7.1. In this scheme, the discordant arrangement originally described by James Hutton would merely be one kind of unconformable relationship, namely an **angular unconformity**. In contrast to the original use of the term, the time gap is now central to the concept, rather than merely following by implication from the geometrical discordance between the adjacent units.

Geologists also talk about formations resting 'unconformably' upon older rocks, or having an 'unconformable junction'. The surface of unconformity represents the surface of the lower rocks upon which the upper material was deposited. With most unconformities, the younger materials are sedimentary rocks, but they could be volcaniclastic or extrusive igneous rocks, such as the lavas in the example of section 2.3.2. The boundaries of igneous intrusions are not regarded as unconformities.

7.3 Recognition on Maps

Unconformities are surfaces and therefore appear on maps as linear traces (figure 7.2). The lines are not normally shown with any particular symbolism; it is up to the map reader to recognise them for what they are.

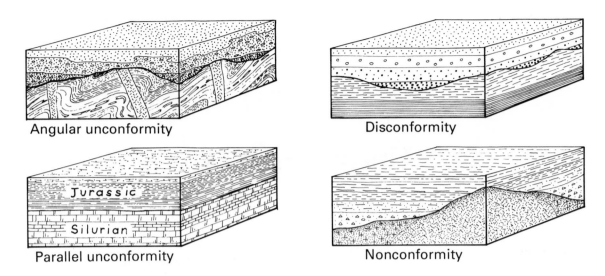

Figure 7.1   Diagrams to show the four main kinds of unconformity.

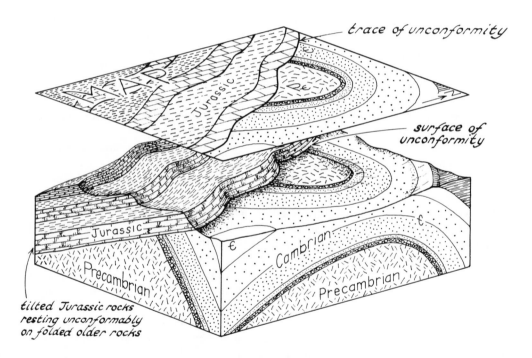

Figure 7.2   The appearance of an angular unconformity, in block diagram and map view.

The map key may give some indication of any unconformities that are present.

Some kinds of unconformity are easier to recognise than others. Angular unconformities are among the most dramatic features that can be seen in rocks, and on maps, too, they can have a conspicuous appearance. The unconformable units meet at the line of unconformity with discordance (figure 7.3a), which, if the angularity of the unconformity is large, will be pronounced. The discordance may have to be searched for if the angularity is small, and it may be masked if the outcrop patterns are made irregular by the relief of the land surface (figure 7.3b).

Parallel unconformities cannot produce a discordant pattern on maps. They have to be recognised by the detection of missing units, by reference to the map key (figure 7.3c). Absence of some unit that may only have been deposited locally is insufficient; the stratigraphic break has to be significant at the scale of the map.

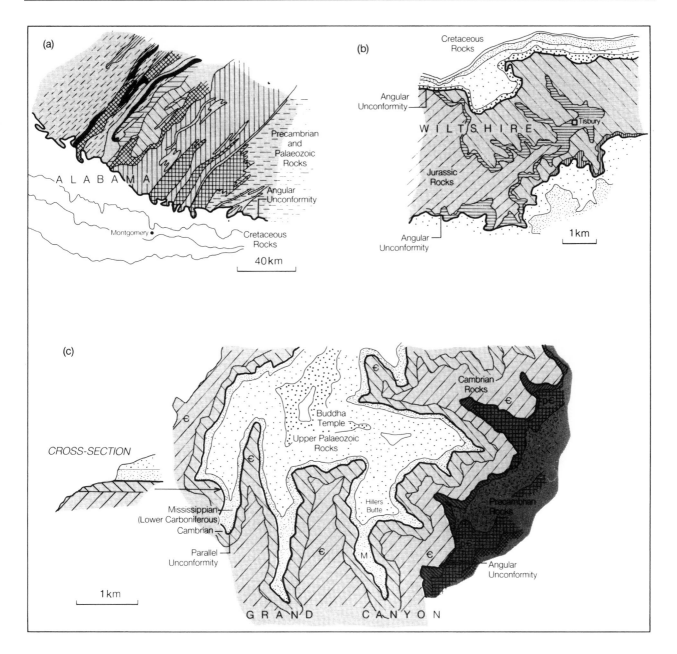

Figure 7.3 Some examples of the appearance on maps of unconformities. (a) SE USA. (b) Shaftesbury district, southern England. (c) Grand Canyon, Arizona, USA.

Unconformities can be difficult to distinguish from faults, both in nature if the junction is not exposed and on maps that do not distinguish faulted contacts. In most cases, however, the surveyor will have made a decision in the field, and indicated on the completed map any junctions that he or she thinks are faults by some particular line symbol. Unconformable junctions, on the other hand, are normally drawn with the same line as for ordinary stratigraphic boundaries, and their nature left for the reader to deduce. That the geological surveyor can have this difficulty is illus-

trated by the 'Ten Mile' map of the UK. Part of the *third edition* (1979) of this map is reproduced here as Plate 1. About 10 km NW of Shrewsbury [SJ4521] rocks of Triassic age (unit 90) are shown to be faulted against Permian/Triassic mudstones (unit 89). However, on the *second edition* (1957) of this same map, the junction is portrayed as an unconformity. Presumably in the years between the two editions of the map, new evidence came to light which prompted the surveying team to reassess the nature of this particular junction.

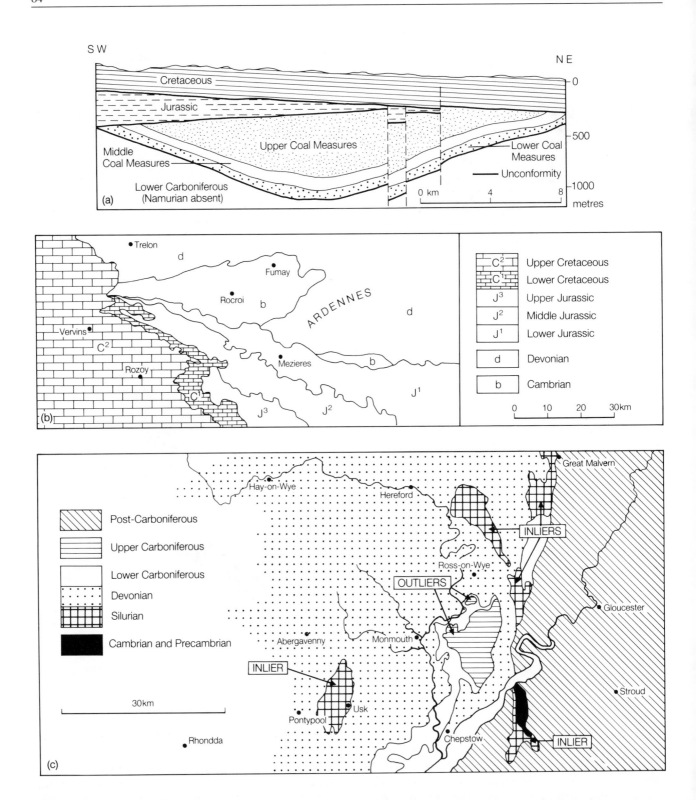

Figure 7.4 Some features associated with unconformities. (a) The concealed coalfield of Kent, SE England, in cross-section, showing three unconformities. The unconformity at the base of the Cretaceous shows overstep, the Cretaceous rocks overstepping the Jurassic onto Coal Measures. (b) A map, simplified from the 1:1 000 000 map of France, showing Devonian unconformably overlying Cambrian rocks (because Ordovician–Silurian strata are absent), and both systems unconformably overlain by Mesozoic rocks, J^2 overlaps J^1, and C^2 overlaps C^1. (c) A map, simplified from the 'Ten Mile' map of the UK, showing examples of inliers and outliers. (The northern part of this area is included in Plate 1, and the SE part enlarged in Plate 8.)

The analogous problem confronts the reader of a map which does not distinguish between faulted and unconformable contacts. There may be clues. For example, unconformities tend to be horizontal, unless they have been tilted later, whereas faults tend to be steep. Some unconformities have an irregular form, reflecting an uneven surface of deposition, so their map trace is irregular, unlike the smooth trace of faults. Although the rocks either side of an unconformity may be influencing relief, the surface of unconformity itself rarely has any direct topographic effect. Faults, in contrast, because of the weakening effect they produce on the rocks, commonly induce negative topographic features such as valleys. Nevertheless, in the absence of further knowledge of the rocks in the area, it may remain impossible to distinguish on the map between unconformities and faults.

7.4 Associated Features

We can consider the relationship with the surface of unconformity of the rocks below and those above. **Overstep** is concerned with differences in the rocks *below* the unconformity (figure 7.4a). It refers to the way in which they are crossed by the overlying unit and depends largely on the way the lower rocks happened to be disposed at the time the material above the unconformity was being laid down. Geologists speak, for example, of the upper unit 'overstepping older rocks northwards', 'overstepping highly folded rocks in the south', or 'overstepping limestones east of the river'. Overstep can only arise with an angular unconformity.

Other terms are concerned with the arrangement of rocks *above* the unconformity. If progressively younger formations come into contact with the unconformity as it is traced laterally, they are said to **overlap** (figure 7.4b). The situation commonly arises as a result of a marine transgression (section 11.2). An example is sketched in figure 11.2b. Often the overlapping sedimentary units that accumulated at a particular place become progressively finer-grained rock types upwards, reflecting deposition in gradually deepening water.

Offlap is the reverse situation (figure 11.2c). The oldest formation of the upper rocks is everywhere forming the unconformable contact with the lower rocks, and is overlain by successively younger units which have a less and less extensive distribution. In general, fine-grained rocks such as shales and fine limestones are replaced upwards by coarser rocks such as sandstones and conglomerates.

It may not be possible to see these relationships on a map, either because they never existed – uniform older rocks were evenly blanketed by younger material – or the map is not sufficiently detailed to provide the information. If, however, they can be discerned, they should be noted. They can be of significance if commercial deposits are involved, because the quantities of material are affected, and they can be of great value in interpreting the geological history of the region (section 11.2).

Two features that commonly arise in association with unconformities, though they are produced in other ways as well, are inliers and outliers (figure 7.4c). An area of older rocks that appears in map view to be completely surrounded by younger rocks is called, particularly in the UK, an **inlier**. The converse situation, of younger rocks surrounded by older, is an **outlier**. Both effects can be produced simply by differential erosion, but more significant inliers and outliers are produced where the junction between the younger and older rocks is an unconformity or set of faults, or a combination of both (figure 7.5). Inliers provide 'windows' through the covering rocks to what is below. Outliers can give a glimpse of the younger rocks that were once more areally extensive.

7.5 Use on Maps

Probably the most usual application of unconformities as they appear on maps is their great value in helping decipher the geological history of an area. The gap in the stratigraphic record will prompt palaeoenvironmental explanations for why beds were not deposited during the unrepresented interval, or why they were removed by erosion. Moreover, the three-dimensional shape of the unconformity surface provides a record of the land surface at the time the upper beds were deposited. The pattern of covering of the lower rocks – whether they are 'blanketed' by the upper material or overlapped by successively younger beds – gives information on the depositional environments of the younger sediments. Interpretations of geological histories in these ways are developed in section 11.2.

Unconformities can have practical significance, too. For example, impermeable beds overlying porous reservoir rocks can form important groundwater accumulations and petroleum traps. A number of the oil and gas fields in the North Sea involve unconformities. Here the precise shape of the unconformity can influence the location and method of hydrocarbon production and the economics. In situations like this, it is necessary to treat unconformities in a more quantitative way. Because unconformities are surfaces, structure contours can be drawn to quantify their shape. The contours are drawn in exactly the same way as for other geological boundaries. If the

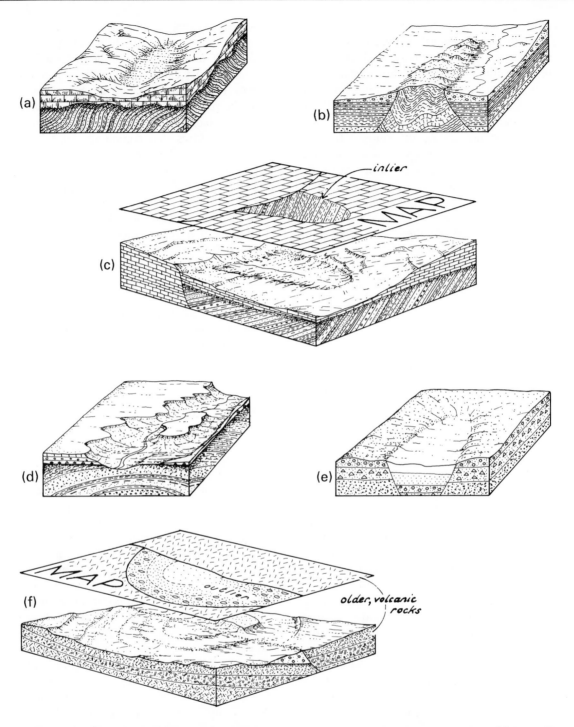

Figure 7.5 Block diagrams of inliers and outliers, to show some ways in which they are formed. Inliers: (a) formed by erosion down to the older rocks below an unconformity; (b) by upfaulting; (c) as they might appear on a map, with a combination of erosion below an unconformity and faulting. Outliers: (d) formed by differential erosion leaving an isolated patch of younger rocks; (e) by downfaulting; (f) as they may appear on a map, resulting from the faulting of a plunging synform.

unconformity outcrops at various known altitudes (section 3.5) it may be possible to draw the structure contours from a geological map alone, although usually supplemental borehole information will be required, especially if the surface of unconformity is irregular.

In the diagrammatic example shown in figure 7.6, the surface is fairly smooth and so its structure contours are reasonably straight and evenly spaced. In contrast, the irregular unconformity at the base of the Triassic in the Cheddar area of

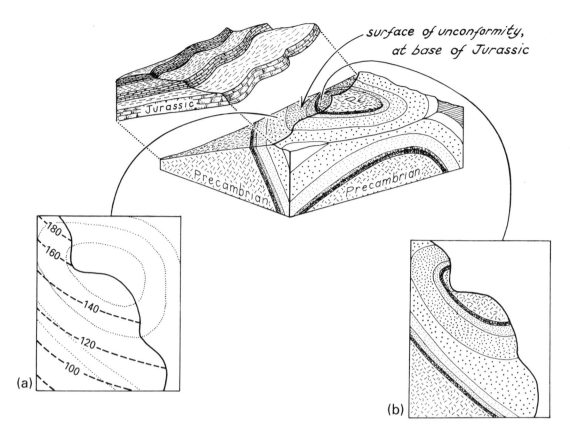

surface of unconformity, at base of Jurassic

Figure 7.6 Block diagram of the unconformity shown in figure 7.2, 'exploded' at the surface of unconformity to reveal its form and the sub-crop of the older units. (a) Structure contour map of the surface of unconformity. Note that this is the configuration as it is seen *today*, presumably after post-Jurassic tilting had affected the area. Dotted lines indicate the distribution of the Precambrian rocks. (b) Sub-crop map, at the base of the Jurassic.

western England produces the contour pattern shown in figure 7.7.

7.6 Palaeogeological Maps

The structure contours for the unconformity in figure 7.7 can be thought of in a different way. They could equally be regarded as topographic contours for the land surface as it was in Triassic times, when the Triassic sediments were first being deposited. In other words, drawing structure contours for an unconformity can provide information on a past landscape. In the example of figure 7.6, the structure contours reflect the land surface when the first Jurassic sediments were being deposited, except that here, judging by the dip which the Jurassic rocks now show, there has been subsequent tilting.

Extending this concept further, it is possible to draw the actual geology of this land surface, as well as its relief, as it was in Triassic times. The younger rocks are mentally stripped away (figure 7.6) to reveal the

intersections of the older rocks with the Jurassic land surface – what is now the surface of unconformity. Such a map would show the rock outcrops at the land surface in the normal way, but here it is the *Jurassic* land surface. Because those outcrops are now subsurface, buried by the younger deposits, this kind of construction is called a **sub-crop** map.

Maps such as those mentioned above, which attempt to reconstruct some aspect of the geology of a bygone time, are called **palaeogeological maps** (e.g. Levorsen 1960). Every palaeogeological map has to specify the stratigraphic time for which it is constructed. Unconformities are amenable to this kind of treatment, but in principle any kind of evidence can be used to help assemble palaeogeological maps, hence they take on very varied forms.

In some instances, a whole series of palaeogeological maps has been devised for a particular area, thus mapping its evolution through geological time. Reconstruction of past geological conditions is another major attribute of geological maps, and the ideas will be developed later (chapter 11). Before doing that, however, we have to develop further the three-

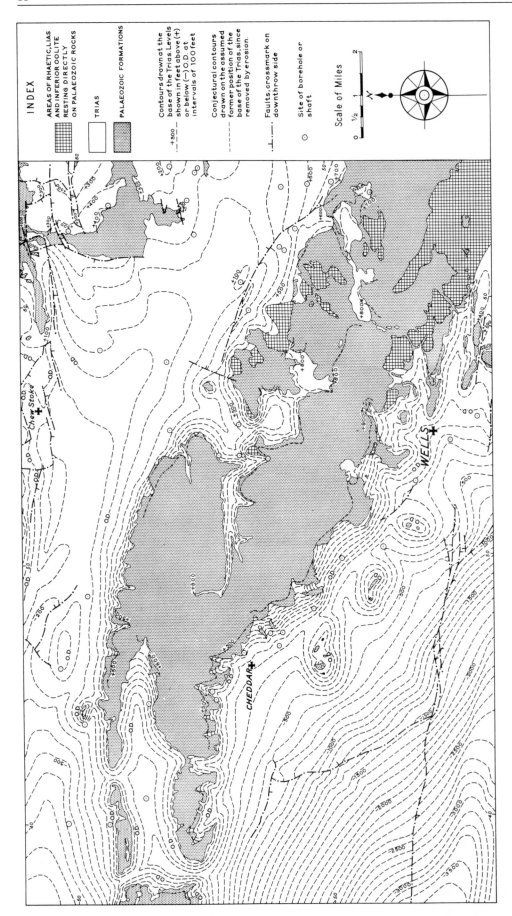

Figure 7.7 An example of a structure contour map for a surface of unconformity, Wells district, England. The structure contours shown here are for the unconformity at the base of the Triassic rocks, which were deposited on an irregular surface of Palaeozoic rocks. Reproduced, at reduced size, from Green and Welch (1965), by permission of the Director, British Geological Survey. Copyright NERC. All rights reserved.

dimensional considerations of maps. For so far we have taken map formations to be more or less planar, in the form in which they were deposited. There are profound repercussions for mapwork if stresses in the earth have deformed the beds from their original shape.

Summary of Chapter

1. An unconformity is a significant time-break in the stratigraphic succession of an area.
2. It is usually visualised as the contact between adjacent units that have differing stratigraphic ages. This surface of unconformity represents the time-span that separates the formations.
3. Angular unconformities are usually readily recognisable on maps from the discordance between units, but the identification of parallel unconformities requires stratigraphic information from the map key.
4. The unconformity may overstep the older rocks in different ways, and the upper rocks may be arranged with overlap or with offlap.
5. Unconformities are valuable in reconstructing the geological histories of regions. Sub-crop and palaeogeological maps can sometimes be constructed from them.

Selected Further Reading

Roberts, J.L., 1982. *Introduction to Geological Maps and Structures*. Pergamon Press, Oxford, 332 p.
   Chapter 7, Unconformities and the geological record, contains a detailed discussion of the use of unconformity and related terms.

Map 12   The Helderberg, South Africa

The map opposite is reproduced from de Villiers (1983), by permission of the Geological Society of South Africa. It is of an area known as the Helderberg, near Stellenbosch, in the Cape Province of South Africa. Outside the NE corner of the area shown, the Table Mountain Group, predominantly a tough sandstone, connects with the rocks that make the famous Table Mountain at Cape Town. The term foliation, mentioned in the key, refers to planar structures developed in the metamorphic and igneous rocks.

What is the nature of the junction at the base of the Table Mountain Group?

What is its approximate orientation?

What term can be applied to the outcrop of the Table Mountain Group around the Dome?

Why does the Table Mountain Group not appear elsewhere in the area?

Draw a NE–SW cross-section across the map area to illustrate the geological arrangement of the Dome.

Further examples of unconformities on maps

Say what you can about the unconformity at the base of the Coalport Group on Map 5.

Discuss the nature of the unconformities that appear on Map 6, of the Boyd area, Australia.

Locate and identify the various unconformities that appear on Plate 5, of Sanquar, Scotland; on Plate 6, of the Heart Mountain district, Wyoming, USA; and on Plate 8, of Malmesbury, England.

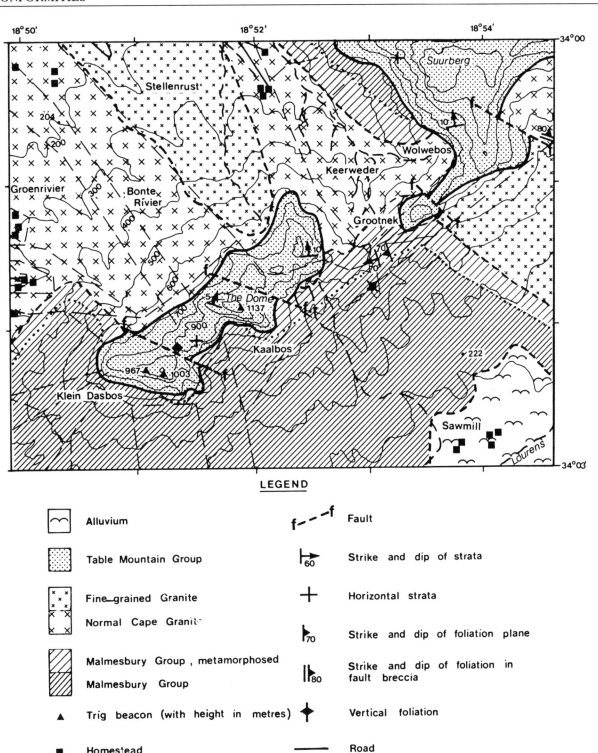

LEGEND

Alluvium	Fault
Table Mountain Group	Strike and dip of strata
Fine-grained Granite / Normal Cape Granite	Horizontal strata
Malmesbury Group, metamorphosed / Malmesbury Group	Strike and dip of foliation plane
▲ Trig beacon (with height in metres)	Strike and dip of foliation in fault breccia
■ Homestead	Vertical foliation
1 : 50 000	Road
0 ——— 1km	River
	Farm boundaries

Map 13   A sub-Permian unconformity and inlier

The map opposite is of the kind of geology found in the north of England. The conspicuous unconformity is at the base of the Permian, which in northern England includes units known locally as the Magnesian Limestone and the Brockram.

What evidence is there on the map for an unconformity at the base of the Carboniferous?

Comment on any structure shown by the Lower Palaeozoic, and by the Carboniferous rocks.

Trace with a coloured pencil the course of the unconformity at the base of the Permian. Notice that the surface of unconformity is highly uneven, so that structure contours drawn for it will be irregular. The locations of a number of boreholes are indicated on the map, together with the *depth* in feet at which they encountered the sub-Permian rocks. Using this borehole data together with the topographic information, carefully construct structure contours for the sub-Permian unconformity.

Describe, from the structure contours, the form of the land surface as it was in early Permian times.

Can you detect any structure in the pre-Permian rocks that may help explain the form of this landscape?

Is there any other junction on the map area, besides the sub-Permian and sub-Carboniferous boundaries, that could be referred to as an unconformity?

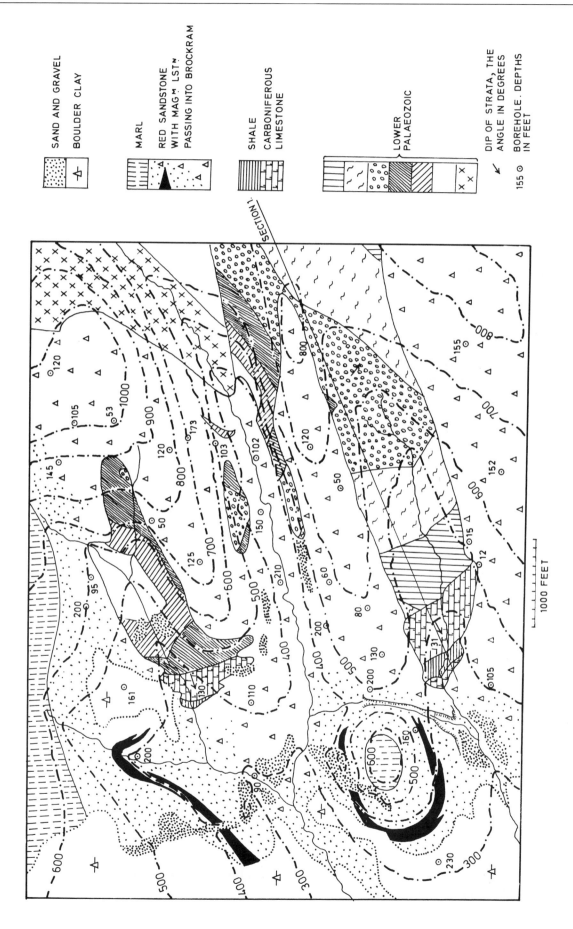

SAND AND GRAVEL

BOULDER CLAY

MARL

RED SANDSTONE WITH MAGᴺ LSTᴺ PASSING INTO BROCKRAM

SHALE

CARBONIFEROUS LIMESTONE

LOWER PALAEOZOIC

DIP OF STRATA, THE ANGLE IN DEGREES

BOREHOLE. DEPTHS IN FEET

155 ⊙

SECTION 1.

1000 FEET

CHAPTER 8

Folds

8.1 Introduction

Geologists are still discovering just how mobile the rocks of the earth can be. The movements that are constantly taking place within the earth generate stresses, and these are capable of deforming rocks, that is, changing their overall shape. This deformation can occur in a brittle way, where the rocks change shape by fracturing, or in a ductile way. In the latter case, because of factors that come about at depth in the earth such as increased pressure and temperature and longer times of deformation, the rocks respond to any stress by flow, and there is no fracture. The common way in which bedded sedimentary rocks deform in a ductile way is by warping into the wave-like shapes known as **folds**.

We saw something in chapter 2 of the way folded units are detected on maps. Any folds will certainly have to be taken into account in our interpretations of the underground arrangement of the beds. Folds can be of great economic significance – in section 3.3 we had a glimpse of the importance of certain kinds of fold to the oil industry – and we now take a closer look at how to deal with folded formations as they appear on geological maps and sections. We begin by defining a few basic parts of folds, in order to be able to report the nature of any folds we see. The chapter goes on to outline the great variety of appearances which natural folds can have, and how they are recognised and measured on maps.

8.2 Description from Maps

8.2.1 The parts of a fold

Most folds do not look like half-cylinders. In cross-section, a fold typically has a narrow zone of sharper curvature, called the **hinge zone** (figure 8.1). It occurs

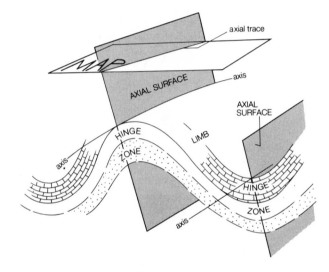

Figure 8.1 The parts of a fold.

between two broad, less curved zones called the **limbs**. We can imagine the greatest curvature in three dimensions making a line running along the folded bed. For introductory purposes, this is the **fold axis**. Advanced work requires a more precise definition (e.g. see Fleuty 1964; Ramsay and Huber 1987). Similarly the definition of the fold **axial surface** has to be rigorous if it is to cover all circumstances, but here it is sufficient to regard it as the surface which passes through the axes of the successive beds in a fold (figure 8.1). Commonly, although by no means always, the axial surface has the appearance of dividing the fold into two fairly symmetrical halves.

The intersection of the axial surface with another surface, such as a map, cross-section, or the ground, is called the fold **axial trace** (figure 8.1). It is often useful to draw it on maps where the rocks have been folded, if it is not already shown. It will emerge later that the axial trace on a map does not normally coincide with the axis of the fold.

8.2.2 Fold orientation

The value of the terms fold axis and axial surface lies in their ability to report succinctly the three-dimensional orientation of a fold. However, as we shall see in sections 8.3 and 8.4, it is much more difficult than you might at first think to work out the orientation of a fold from a geological map. Hence the geological surveyor usually makes a point of measuring fold axes and axial surfaces *in the field*, where it is a relatively straightforward operation. His or her measurements are then reported on the map by symbols such as those shown in figure 4.2. Even if the surveyor cannot actually see a fold in its entirety there are routine methods for finding its orientation from indirect field observations (e.g. see Ramsay and Huber 1987).

Many geological maps, then, have marked on them symbols to indicate the orientation of any folds in the area. A fold axial surface is reported as a strike and dip (figure 8.2), exactly analogous to bedding or any other geological plane (section 4.2), and a fold axis or an axial trace, both being lines, are given as a plunge and trend (section 4.2). Remember, from section 4.2, that the trend is expressed as a compass bearing from north. Thus geologists commonly speak loosely of folds 'trending northeastwards' and the like, though it

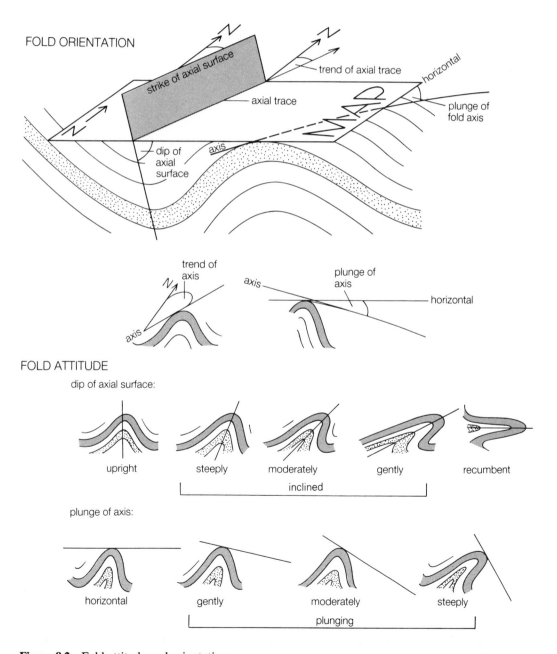

Figure 8.2 Fold attitude and orientation.

FOLD SHAPE

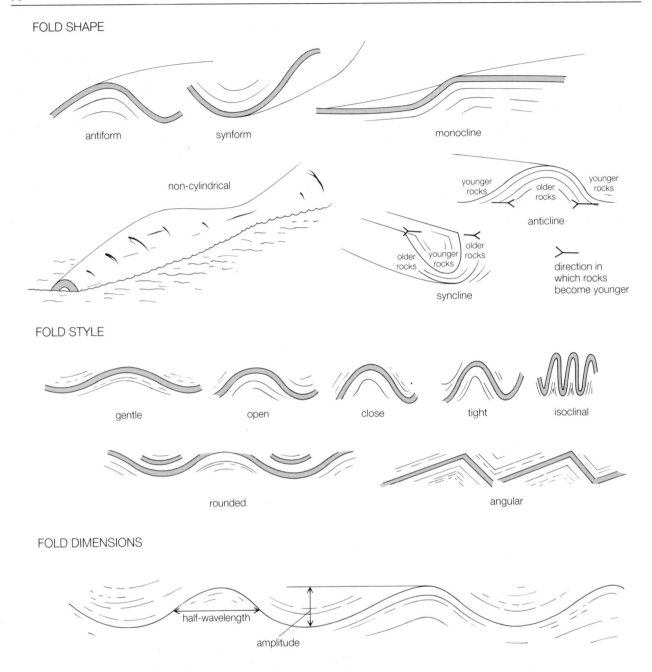

Figure 8.3 Fold shape, style, and dimensions.

really should be specified whether the axial trace or the axis is meant. The plunge of the axial trace will normally average out as horizontal, being in the horizontal map plane, but it may fluctuate locally as the axial surface crosses various topographic slopes.

8.2.3   Fold attitude

In addition to specifying the spatial orientation of a fold, geologists often record in words how much the fold itself is inclined on its side or tipped on end, what is called the fold's **attitude**. The terms commonly used

for this are given in figure 8.2. For example, it is sometimes useful to speak of a map area being dominated by steeply plunging folds, without specifying in degrees the exact range of plunge values.

8.2.4   Fold shape

We saw in section 2.3.3 how map units can be folded into basins and arches, synclines and anticlines. Although these terms are useful on the regional scale being discussed there, for smaller structures the terms **antiform** and **synform**, respectively, are used. They are illustrated in figure 8.3. As introduced in section 2.3.3, a fold with the oldest rocks in its inner part is called an anticline, and with the youngest rocks in the middle, a syncline (figure 8.3). It follows from the discussion of the age sequences seen in tilted beds (section 2.3.1 and figure 2.1) that ordinarily antiforms will be anticlines, and synforms will be synclines. This is not necessarily so, however, because in highly deformed areas, outside the scope of this book, folds can become entirely inverted, such that synforms have the oldest rocks in the middle, etc.

Figure 8.3 illustrates the single-limbed deflection of beds known as a **monocline**, and also the term **non-cylindrical**. This means that the axis varies in plunge along its length. You should make sure you can visualise what this signifies for the three-dimensional appearance of a fold. A situation commonly seen on maps is a non-cylindrical fold with steepest plunges at the ends of its axial trace and least plunge in the centre. Such folds are referred to as being **periclinal**. It is in effect a small-scale dome or basin. Thus those folds discussed in section 2.3.3 with a roughly circular outcrop pattern must be non-cylindrical.

8.2.5 Fold style

In addition to describing the basic shape of a fold using the above terms, geologists find it useful to mention what is called the **style** of a fold. The adjectives commonly used for this are given in figure 8.3. Describing the fold style is not just a point of detail, because it has major implications for structural geologists interested in the mechanics of how folds form, for more advanced graphical constructions involving folded beds, and for quantitative estimates of the volumes of industrial materials involved in folded rocks (e.g. Badgley 1959).

8.2.6 Fold dimensions

Finally, to communicate usefully what a particular fold or collection of folds looks like, the size needs to be indicated. The simplest approach to this is to treat the folds as waves, and to measure the fold amplitude and half-wave length (figure 8.3).

8.3 Visual Assessment on Maps

Because many folds are too large to be observable in the field, geologists commonly detect their presence by compiling relevant information on a map. Dip directions may systematically reverse, representing the different limbs of folds, or outcrop patterns may emerge which indicate folded units. Figure 8.4 shows

the kinds of outcrop patterns produced on maps by folds which are horizontal or of very shallow plunge. They are essentially symmetrical repetitions of the same beds outwards from the hinge zone of the fold. Note that this pattern of repetition differs from that produced by faults (section 9.5; figure 9.5). With increasing plunge of the fold axis there is a growing tendency for the beds on the two limbs of the fold to converge. If the plunge is sufficient, the beds meet and the hinge zone of the fold then appears on the map (figure 8.4). Open folds give arcuate patterns like those seen in figure 2.2c whereas tighter folds give more rapidly convergent shapes. Note, however, that on large-scale maps all these appearances can be complicated by an irregular relief.

Unless we have some other information, it is not possible to deduce anything further about the nature of the fold. In fact, at first glance the converging appearance of the beds can be confused with the effects of topography discussed in chapter 6, such as the 'V- shape' made by dipping beds crossing a valley. On a large-scale map without the topography marked it may not be possible to recognise folded beds with certainty, but normally topographic effects tend to be small and local in comparison with the appearance of folds. If symbols are provided on the map or there is information on the topography, it should be possible to assess the dip of the beds and, hence, to ascertain whether the fold is a synform or an antiform. It may help to sketch in a few structure contours on each limb to find which way the limbs are inclined (figure 8.4).

It is important to understand that the outcrop pattern of the folded beds does not necessarily indicate the shape or style of the fold. The converging pattern of the beds on the map does not show directly whether the fold is an antiform or synform. The map view is just a horizontal section through the folds. These may well be plunging, and – especially if there is marked topographic relief – producing illusory effects. Similarly, a cross-section can be misleading. Only if the fold is horizontal will the cross-section give a true portrayal of the appearance of the fold. There will be distortion in a cross-section through a plunging fold. To see plunging folds properly we should construct a section through the fold *at right angles to the fold axis*, a plane which is called the fold **profile**.

A handy device for the visual assessment of fold profile is 'down-plunge' viewing of the map. By inclining your line of sight to the map at an angle roughly equal to the fold plunge, the outcrop pattern you see will approximate the fold profile (figure 8.5). Sketching where you think the axial surface runs in this profile view gives a better estimate of its course than normal 'head-on' viewing of the map. Figure 8.5 gives an example of how the axial trace drawn this way can differ

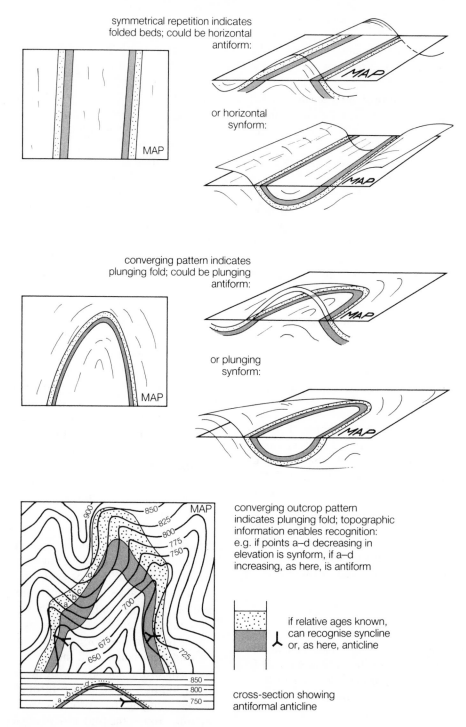

Figure 8.4   Some guides to the visual assessment of folds on maps.

from where you might draw it without looking down the fold plunge. It is necessary, though, in order to employ this useful technique for visually assessing folds, to have knowledge of the fold plunge. If this is indicated on the map by a symbol there is no problem; otherwise you will first have to visualise or sketch the structure contours in the hinge zone of the fold (see next section).

Figure 8.6 shows some of the outcrop patterns typically seen on geological maps of folded areas. Almost all of them have some element of the converging aspect so common with folds. But because folds can be highly non-cylindrical, beware of assuming that all the folds which look alike at first glance are of the same kind. Notice also that to tell whether the folds are

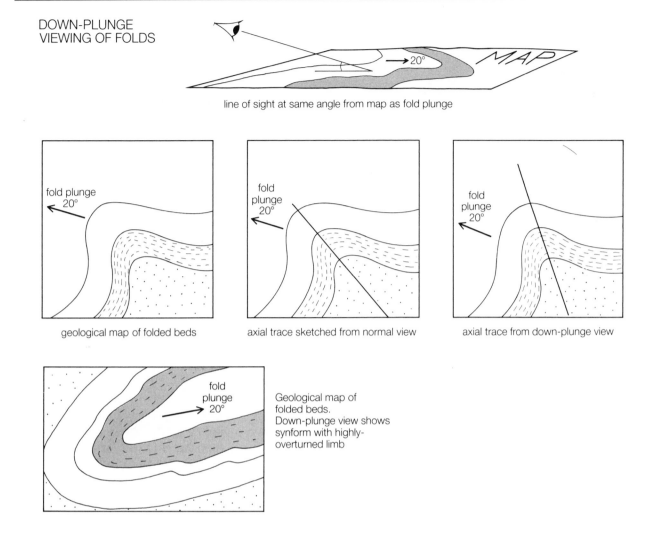

DOWN-PLUNGE
VIEWING OF FOLDS

→ 20°

line of sight at same angle from map as fold plunge

fold plunge
20°

geological map of folded beds

fold
plunge
20°

axial trace sketched from normal view

fold
plunge
20°

axial trace from down-plunge view

fold
plunge
20°
→

Geological map of
folded beds.
Down-plunge view shows
synform with highly-
overturned limb

Figure 8.5   Illustrations of the down-plunge method of viewing plunging folds on maps.

anticlines or synclines, we must have information on the relative ages of the units. This may well be provided in the key; otherwise it could be impossible to tell. In weakly deformed areas it is likely that the beds become younger in the dip direction, that antiforms are anticlines, etc., but if there is suspicion that beds may be inverted, this assumption could be invalid.

8.4   Measurements on Maps

If the locations of fold axes are known on a map and they are horizontal then true cross-sections can be constructed in the normal way, at right angles to the axes. Being a fold profile, the attitude, shape, style and scale of the fold can be determined directly. Difficulties arise where the folds are not horizontal. If the plunge is less than 30° or so then the usual vertical cross-section may well be sufficient for most purposes,

but otherwise the information from a cross-section may be misleading, especially on the large-scale maps likely to be used for measurement purposes. It would be necessary to establish the fold profile, a time-consuming procedure to carry out by hand.

The amount of fold plunge may be given on the map by a symbol, otherwise a few structure contours will have to be drawn in the hinge zone of the fold. From their spacing the plunge can be derived either graphically or by trigonometry (figure 8.7). To obtain the axial trace on a map, it will be necessary to construct at least two sections across the fold, and add vertical axial traces to each section in order to derive points where the axial surface intersects the ground surface. These can then be connected to route the axial trace in map view. If the axial trace is suspected to curve, several sections may be needed. Bear in mind, though, that all this is only approximate if the folds plunge significantly, where fold profiles should really be employed.

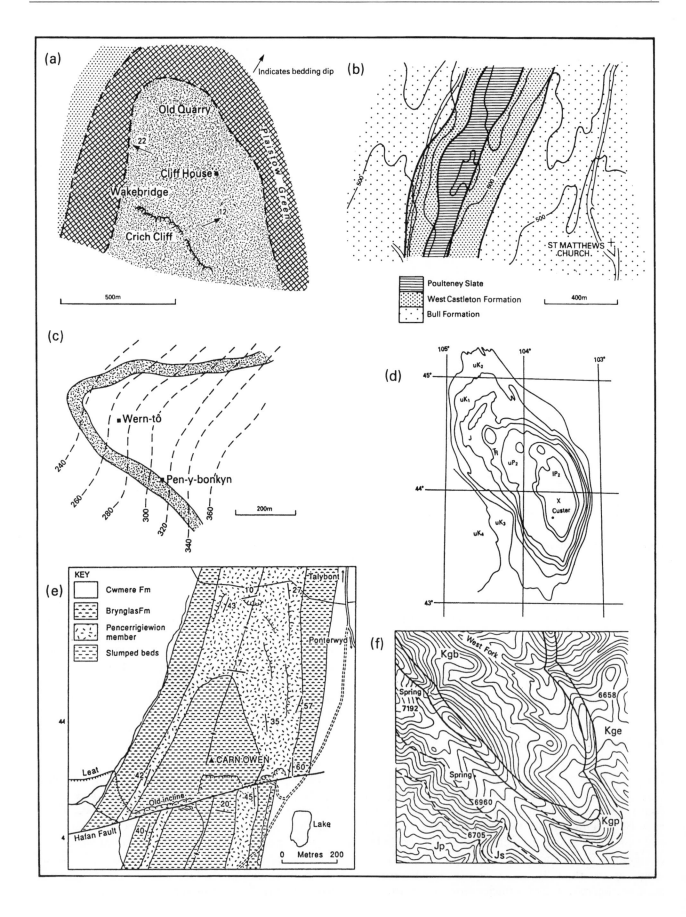

(a)

Indicates bedding dip

Old Quarry

Cliff House

Wakebridge

Crich Cliff

Plaistow Green

22

12

500m

(b)

ST MATTHEWS
CHURCH

Poulteney Slate

West Castleton Formation

Bull Formation

400m

(c)

Wern-tô

Pen-y-bonkyn

240
260
280
300
320
340
360

200m

(d)

105°   104°   103°

45°   uK₂

uK₁

J
R
uP₂
IP₂

44°   X
Custer

uK₃

uK₄

43°

(e)

KEY

Cwmere Fm

Brynglas Fm

Pencerrigiewion member

Slumped beds

Talybont

Ponterwyd

CARN OWEN

Old incline

Leat

Hafan Fault

Lake

44

4

10   27
43
17
57
35
42
60
45
20
40

0   Metres   200

(f)

West Fork

Kgb

Spring
7192

Spring

6658

Kge

6960

6705

Jp   Js

Kgp

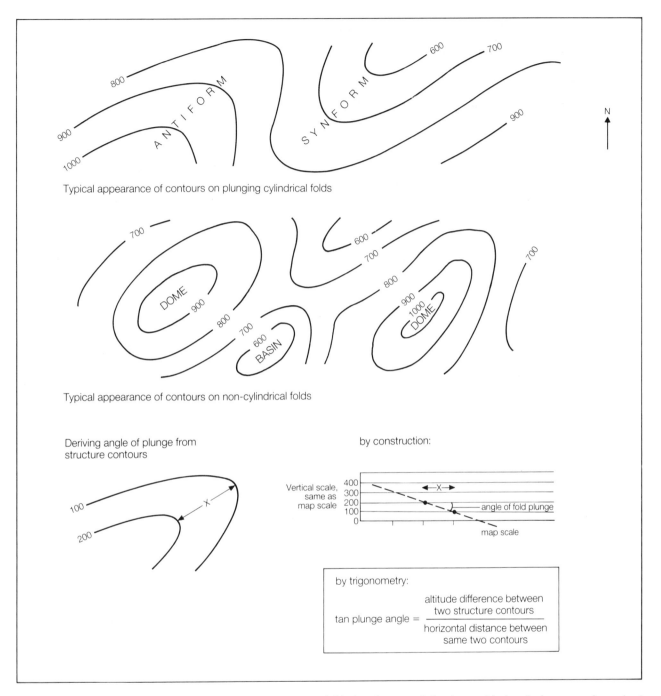

Typical appearance of contours on plunging cylindrical folds

Typical appearance of contours on non-cylindrical folds

Deriving angle of plunge from structure contours

by construction:

Vertical scale, same as map scale

angle of fold plunge

map scale

by trigonometry:

$$\tan \text{plunge angle} = \frac{\text{altitude difference between two structure contours}}{\text{horizontal distance between same two contours}}$$

Figure 8.7 Illustrations of how structure contours represent folded surfaces, and simple graphical and trigonometric methods for deriving fold plunge.

Figure 8.6 (opposite) Examples of interpreting outcrop patterns of folds. (a) Arcuate outcrop pattern suggests a plunging fold; lack of relation with cliff suggests a non-topographic effect. Dip symbols confirm presence of antiform. May be an anticline, but relative ages of rocks not known. Near Crich, Derbyshire, England. (b) Symmetrical repetition of formations suggests a fold; relative ages of formations suggest a syncline. Near Lake Bomoseen, Vermont, USA. (c) Without further information, arcuate outcrop could represent a fold, but is adequately explained by a uniformly dipping unit crossing a hillside. Compare with figure 6.2. Near Llanidloes, Wales. (d) Symmetrical repetition of units suggests a fold; unlikely to be a topographic effect at this scale (note degrees of latitude and longitude). Relative ages (K = Cretaceous, J = Jurassic, T = Triassic, etc.) indicate an anticline. Black Hills, South Dakota, USA. (e) Symmetrical repetition of formations suggests a fold; relative ages of formations suggest an anticline, dip symbols confirm an antiform. Shape of Slumped beds outcrop suggests a plunge towards the NNE. Carn Owen, near Aberystwyth, mid-Wales. Redrawn from Dobson (1995), by permission of the Geologists' Association. (f) Symmetrical repetition of formations suggests a fold; V-shaped outcrops suggest a synform. J = Jurassic and K = Cretaceous, therefore a syncline, plunging towards the SE. Based on Jenkins (1981) with permission.

We are now in a position to consider the strike and dip of the axial surface. If the fold axial trace is available on the map and crosses topographic contours, it may be possible to sketch in some structure contours in order to derive the orientation of the fold axial surface. Its strike will be the same as the trend of the axial trace, if the latter is horizontal and not crossing topographic slopes. A vertical cross-section, drawn perpendicular to the horizontal axial trace, will reveal the true dip of the axial surface.

There are more sophisticated ways of dealing with folded beds, such as the use of stereographic projections (e.g. Ramsay and Huber 1987). These methods provide a much more direct treatment of plunging folds but can be difficult for beginners. All this underlines the usefulness of obtaining as much information as possible on folds while in the field. This is why most completed geological maps, especially those concerned with folded units, have much of the orientation data already marked on the map. Although folds on maps commonly produce visually attractive outcrop patterns, they have to be interpreted with care.

Summary of Chapter

1. The axis and axial surface are useful concepts in specifying the orientation of a fold – as plunge and trend, and strike and dip, respectively.
2. This orientation information is awkward to interpret from outcrop patterns; it is commonly provided on the map by symbols.
3. Folds are recognised on maps by the symmetrical repetition of units in a horizontal fold, and by the converging outcrops of a plunging fold.
4. It may be possible to interpret from a map something about the attitude, shape, style, and dimensions of folds, but the appearance of plunging folds on maps and sections can be illusory.
5. Most useful are fold profiles, but these can be tedious to construct manually.

Selected Further Reading

Powell, D., 1992. *Interpretation of Geological Structures through Maps*. Longman Group, London, 176 p. Pages 65–93 give further illustrations of folds in the context of maps.

Map 14   Llandovery, Dyfed, Wales

The map opposite provides detail of the district appearing on Plate 1 around [SN8338], NE of Llandovery. It is based on the work of O.T. Jones (1949), by permission of the Geological Society. The rocks at Llandovery were discussed in section 2.3.1 as examples of dipping beds. However, in a number of places on Plate 1 the outcrops curve, suggesting that, seeing as there is no indication of pronounced topography, the units are folded. Notice, incidentally, that these rocks belong to the lowest series of the Silurian, known in the UK and elsewhere as the 'Llandovery'. They are therefore here in their 'type' area.

Based on visual assessment, what is the structure of the area shown?

Draw lightly on the map where you gauge the axial traces to run.

Carefully construct structure contours for the three stratigraphic boundaries shown. (Draw the tentative contours lightly in pencil; when 'firming up' use coloured pencils or different line symbols to distinguish the different sets of contours.)

By what amount does the fold plunge in the Cwm-Crychan–Bryn-ffoi area? And by what in the Cwm Clyd–Cefn-y-gareg area?

Is this fold showing any tendency to be periclinal?

How does its plunge compare with that of the adjacent anticline?

At a point between Pen-y-rhiw and Cwm Coed-Oeron, draw a NW–SE cross-section (note the orientation of N on this map) across the Middle Llandovery outcrop. How does the dip on the NW limb of the fold compare with that on the SE limb?

On the cross-section draw a line representing the fold axial surface. The intersection of the line with the land surface gives the position of the axial trace on the map. How does its position derived in this way from a cross-section compare with your earlier location from the map alone?

(Note that the axial trace has still not been properly located. This would require a fold *profile*, i.e. a section at right angles to the fold axis, which the vertical section you have drawn is not.)

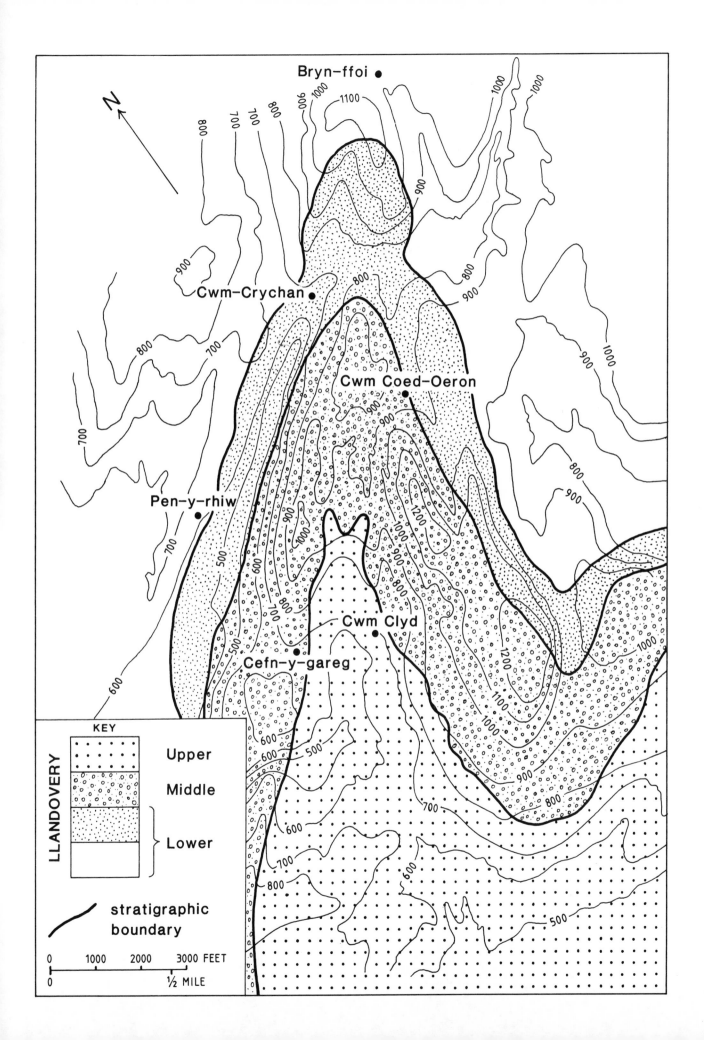

Bryn-ffoi •

Cwm-Crychan •

Cwm Coed-Oeron •

Pen-y-rhiw •

Cwm Clyd •

Cefn-y-gareg •

KEY

LLANDOVERY

Upper

Middle

Lower

stratigraphic
boundary

0   1000   2000   3000 FEET

0   ½ MILE

N

Map 15   Millstone Grit of northern England

The facing map, at a scale of 1:20 000, is of the kind of geology found in parts of the Pennines of northern England. The 'Millstone Grit' is a time-honoured name for the coarse sandstones, once much used for millstones, which dominate the Namurian series of the Carboniferous in this region. Many parts are uniformly shallow-dipping, but in some places the beds have been folded by Variscan earth movements.

Describe, from visual assessment, the kind of fold present in this area.

One of the Millstone Grit bands has a much wider outcrop in the NW of the area than in the E. Why might this be?

Using the dip information provided, draw a cross-section from E–W in the vicinity of Black Gill.

In order to illustrate the plunge of the fold, draw a cross-section from N–S across the hinge zone outcropping in the south of the area. Because the values of dip indicated are so variable, a more accurate section will result from drawing structure contours for some of the surfaces.

Is there any relationship in the area between the geology and the relief of the land surface?

Further examples of folds on maps

On Map 4, of Maccoyella Ridge, New Zealand, having constructed structure contours for the base of the Te Were Sandstone, calculate the plunge of the major synform and describe its form.

Consider the rocks lying below the sub-Permian unconformity in Map 13. Interpret visually the structure in each of the Lower Palaeozoic inliers. Construct structure contours in the southern inlier for the top of the unit with an open circle ornament, and in the northern inlier for the junction of the two units with diagonal line ornaments. Describe the form of the structures revealed by the contours.

Note that the questions accompanying Plate 4, Root River, Canada, involve the interpretation of folded units, and that folds appear on Plate 7, of Marraba, Australia, and Plate 8, of Malmesbury, England.

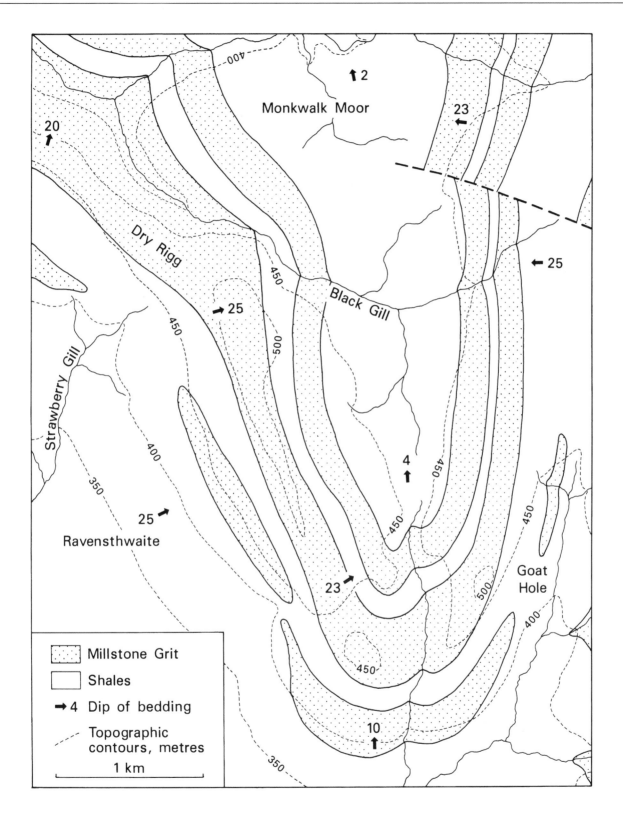

Millstone Grit

Shales

➔4 Dip of bedding

Topographic
contours, metres

1 km

CHAPTER 9

Faults

9.1 Introduction

Faults are perhaps the most frustrating structures to deal with on maps. On the one hand, they are usually conspicuous features on maps and they are tremendously important in both academic and industrial work. They can, for example, interfere with the predictions of the mining geologist and cause special problems for the civil engineer. Even non-geologists may well know something about faults, if only the devastating effects of earthquakes that can arise from earth movements along them. Yet on the other hand, it can be very difficult to deduce much from a map about a fault. The amount by which the rocks either side of the fault have moved is commonly extremely elusive; even the direction of movement is often unclear. Moreover, faults are very awkward to classify and describe in a rigorous way. Much of the terminology that surrounds them is very inexact, but deeply entrenched. We will have to be careful in this chapter to distinguish between those terms that are used in a time-honoured but loose way, and those that have to be employed exactly.

We saw in section 8.1 that stresses in the earth's crust sometimes cause rocks to deform in a ductile way, producing folds. In other circumstances, in general where the stresses are operating near the earth's surface, rocks can respond in a brittle way. This behaviour is characterised by fracture of the rocks. Faults result from brittle behaviour. These geological terms brittle and fracture mean exactly what they do in everyday speech, say about someone's bones.

By far the most common fractures in rocks are **joints**, along which there has been no movement of the rocks. Virtually every rock exposure you see has joints in it, and they can be very important in contexts such as engineering geology. Normally, joints are only represented on large-scale geological maps, and only then where they are of special significance. Being planar surfaces, their orientation is shown by a special strike and dip symbol, which will be explained in the key. It is those fractures along which there has been displacement of the rocks – **faults** – which are important for mapwork, especially if the amount of movement has been large enough to affect the outcrop patterns on the map.

Let us begin by looking at faults in the same order as we considered folds: by first outlining some basic descriptive features so that we can report on faults more efficiently, and then by looking at the principles of dealing with faults on maps. Finally we will take into account some of the situations in which faults occur in nature.

9.2 Fault Parts, Orientation, and Dimensions

Faults commonly have an undulating form, but are often loosely referred to as planes. For most mapwork, it is realistic to think of a fault as a single plane, although on the ground there may be a broad zone of broken rocks, or a network of smaller fractures. Treated as a single surface the fault appears on a map or cross-section as a **fault trace** (figure 9.1). It is dealt with exactly like any other line that arises from a plane intersecting with the map surface. Hence if the fault trace crosses topographic contours it should be possible to work out something of the three-dimensional orientation of the fault plane. Like other geological surfaces, its orientation can be specified as a strike and dip. The dip of the fault as seen in cross-section will be an apparent value, unless the section is drawn at right angles to the true dip direction of the fault (cf. section 5.2).

In most cases the map units will appear to have moved either side of the fault trace. Either or both sides may have actually carried out the movement – it

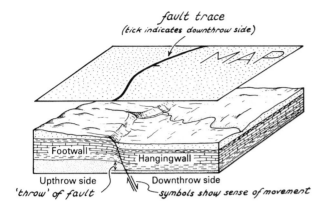

fault trace
(tick indicates downthrow side)

Footwall

Hangingwall

Upthrow side
'throw' of fault

Downthrow side
symbols show sense of movement

Figure 9.1 Block diagram to show the main elements of faults relevant to mapwork.

is normally impossible to tell which. We can refer to the result of this movement, irrespective of the direction in which we are looking, as the **displacement**. This is a very useful but loose term, and by no means necessarily represents the *actual* movement along the fault. The tricky matter of attempting to deduce the real three-dimensional movement is deferred until section 9.6.

Displacement in a steep direction – the 'up and down' part of the movement – is often called the **throw** of the fault. However, this is an example of a term that is sometimes used loosely and other times with an exact meaning. Throw is often used in the loose sense just mentioned, but it also has the precise definition given in section 9.6. The relative movement directions on each side of the fault, called the **sense** of displacement, are often represented by paired half-arrows (figure 9.1). If the displacement direction is steep, it is usually possible to distinguish the **upthrow** from the **downthrow** side of the fault. With a dipping fault, the side above the fault is called the **hangingwall** and the side below is the **footwall**. These two terms supposedly derive from where miners used to hang their lamps and put their feet! Most of the rich vocabulary that miners used for fault features has now disappeared, but these two terms have recently surged back into technical usage.

The size of a fault is most simply stated by its length and maximum displacement. The horizontal length at surface is readily measured from its map trace; it is the amount of displacement which causes trouble and needs to be considered very carefully.

9.3   Fault Displacement

Displacement is absolutely central to the concept of faulting, but working out by how much the rocks have

actually moved can be very awkward. The difficulty arises because the displacement seen in one view, say on a map or cross-section, is just the component of movement in that plane. It may or may not reflect the amount of movement in other directions. The problem does not arise in certain special situations involving horizontal or vertical units, but in the general case all kinds of deceptive appearances can be produced.

It is useful to consider displacement from two different aspects: slip and separation. **Slip** is the displacement of formerly adjacent *points* along a fault, in a particular specified direction and *measured in the fault plane* (figure 9.2b). Used in this way, slip is a precise term. **Net-slip** is an unambiguous measure of total fault movement. The more loose terms **strike-slip** and **dip-slip** are useful for indicating the dominant movement directions. Note that the 'strike' and 'dip' parts of these terms refer to the *fault*, not the strike and dip of the units. The disadvantage of working with slip is that it is not common to be able to establish *points* which have been relatively displaced. In practice, the points usually arise from the intersection of lines with the fault surface, but even this circumstance is not very common.

It is much more usual to see the result of *planes* having been displaced, that is, the outcrop traces on a map or section. Here, separation is the relevant term. **Separation** is the distance between two formerly adjacent *lines*, and measured in any specified direction, not necessarily in the fault plane (figure 9.2.c). Separation is therefore a much more accessible measure than slip; the snag is that it may not be particularly meaningful. Take the vertical displacement of faulted very steeply dipping beds: the separation will be small in map view, even though the vertical throw could be huge. With a fault that dips obliquely to dipping beds, the amount of separation, even when consistently measured in a particular direction, will vary according to *where* the measurement is made. It will vary, for example, from cross-section to cross-section.

Separation is therefore an important practical measurement in mapwork, but it is merely a guide to the actual amount of fault movement. It is imperative when dealing with faults on maps to keep clear in your mind the concepts of displacement, slip, and separation.

9.4   Classification of Faults

Much has been written about the best way to classify faults. Several schemes involve the stresses and mechanics of fault generation, but for our purposes a descriptive system, based on what can be observed on

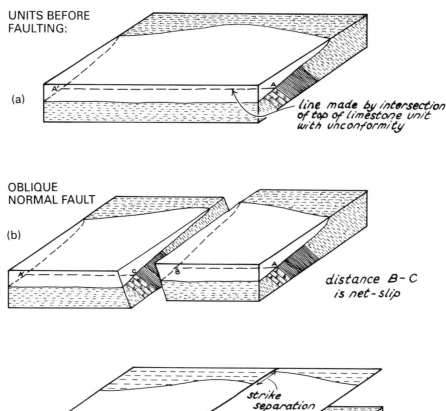

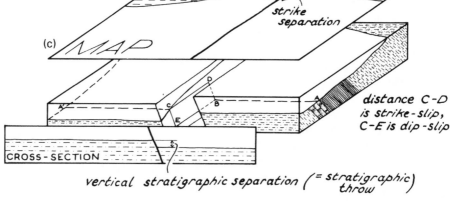

Figure 9.2 Block diagrams to show some aspects of fault displacement. (a) This shows the rocks before faulting. The unornamented unit unconformably overlies dipping units. The intersection of the top of the limestone unit (brickwork ornament) with the surface of unconformity makes a *line* (dashed). This line makes a *point* on intersecting other planes such as the edges of the block, at A and A'. (b) This shows the same rocks after faulting. The A–A' line has been displaced; the two points it makes on intersecting the fault plane, B and C, give the net-slip. (c) The same rocks as (b), showing the horizontal component of slip along the fault, C–D, the strike-slip; and the component of slip in the dip direction of the fault, C–E, the dip-slip. These three-dimensional diagrams show the displacement of *points*, and so reveal the oblique-slip nature of the fault, the amount of net-slip in the oblique direction, and the strike-slip and dip-slip components. In map and section view, however, only the *separation* of planes is seen, such as the strike separation and the vertical stratigraphic separation.

maps, is appropriate. Most practical is a scheme which divides faults according to what is judged to be the dominant slip direction, with a few descriptive terms added to enable some subdivision (figure 9.3). Thus if the main displacement seems to be along the strike of the fault we have a **strike-slip** fault, whereas a **dip-slip** fault shows displacement chiefly in the dip direction of the fault. Where the displacement is neither strike-

parallel nor dip-parallel but somewhere in between, it is an **oblique-slip fault**. Do notice that the term slip is not being employed in the rigorous sense defined in the previous section, for rarely is this fault classification applied with reference to displaced points.

For dip-slip faults a further division is useful, based on the sense of displacement with respect to the fault dip. Where the fault dips towards the downthrow side

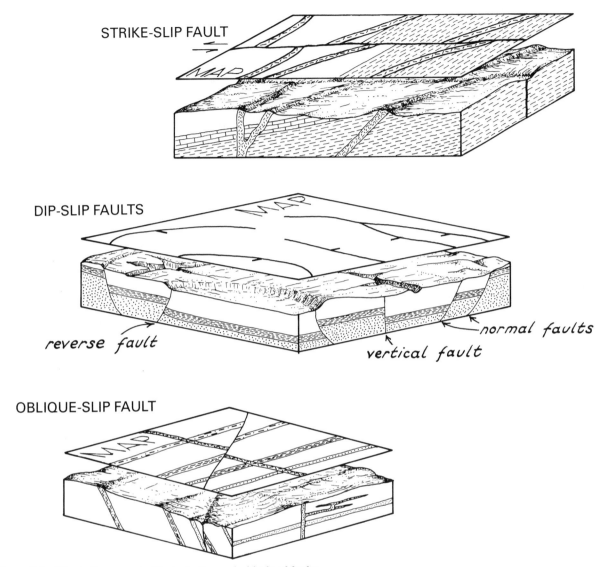

Figure 9.3 Block diagrams to illustrate the main kinds of faults.

it is a **normal fault** and where the fault dips towards the upthrow side it is a **reverse** fault. Alternatively, we could think of a normal fault as having its hangingwall downthrown and a reverse fault as having its hangingwall upthrown. The end result is the same in both definitions. The terms normal and reverse are said to come from British coal mining practice. If a coal seam was faulted out, the miners judged the kind of fault and adopted either the normal or the reverse procedures of tunnelling in order to relocate the seam. The terms have become extremely widely used and will be employed here, especially for somewhat isolated faults. Note, however, that it is a somewhat arbitrary classification. Because the name given to the fault depends on the attitude of the fault, which can vary along its length, unclear situations can arise.

Hence, reverse and normal faults are being increasingly referred to as contraction and extension faults, respectively, as discussed in sections 9.7 and 9.8.

9.5 Visual Assessment on Maps

It can be tricky deciding in the field whether or not a particular junction between two different rock units is a fault. Usually on a finalised geological map the surveyor will have indicated his decision. If the contact is thought to be faulted, it will be shown by a line of different weight or colour. There may be some symbolism to distinguish the downthrow side (figure 4.2).

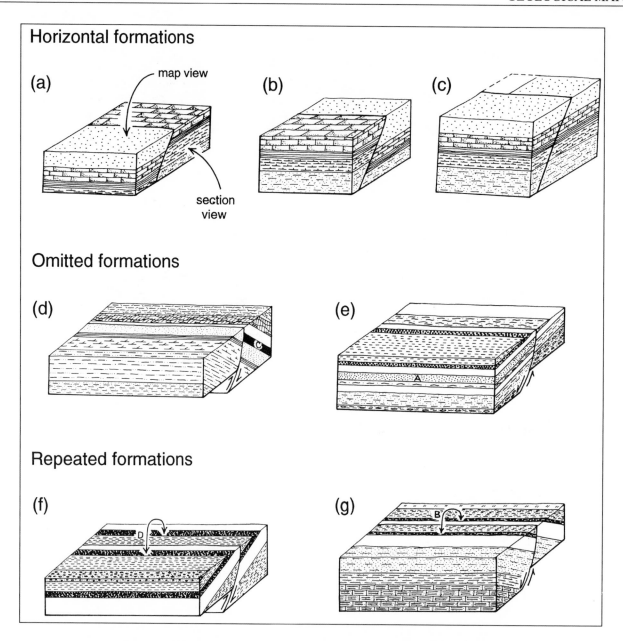

Figure 9.4 Block diagrams to illustrate some outcrop patterns arising from faulting. (a) Horizontal formations and normal fault. In map view, the younger rocks are seen on the downthrow side. (b) Horizontal formations and reverse fault. In map view, older rocks are seen on the upthrow side. Note: it is not possible to distinguish between a normal and reverse fault without knowledge of the dip of the fault. (c) Horizontal formations and strike-slip faulting. Some marker required in map view to detect the displacement; no effect in cross-section. (d) Omitted formations: unit C does not outcrop. *Reverse* fault, dipping in *opposite* direction to formations. (e) Omitted formations: unit A does not outcrop. *Normal* fault, dipping in *same* direction as formations. (f) Repeated formations: unit D outcrops twice. *Reverse* fault, dipping in *same* direction as formations. (g) Repeated formations. Unit B outcrops twice. *Normal* fault, dipping in *opposite* direction to formations.

On a less complete map, it may fall to the reader to infer the presence of faults from the outcrop patterns. Indeed, because faults in nature are rarely exposed, it is often from maps that they are detected. Formations that should be present but do not appear at surface, or that are repeated, may imply the presence of a strike-parallel fault (figure 9.4d–g). Note, however, that the absence of formations could also indicate an unconformity (the dis-

tinction between faults and unconformities was discussed in section 7.3), and that although repetition of beds is also caused by folding (section 8.3) the pattern is of a different kind (figure 9.5). If horizontal beds are involved, small-scale strike-slip faulting may not be apparent in either map or section view (figure 9.4c).

Abrupt displacements of outcrop traces on the map indicate faults which are not strike-parallel (figure 9.6).

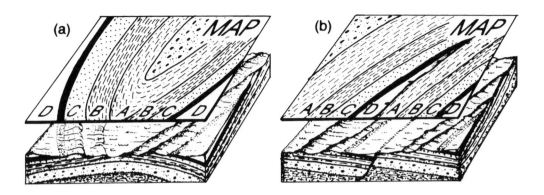

Figure 9.5   Illustration of the differing repetition patterns generated by faulting as opposed to folding.

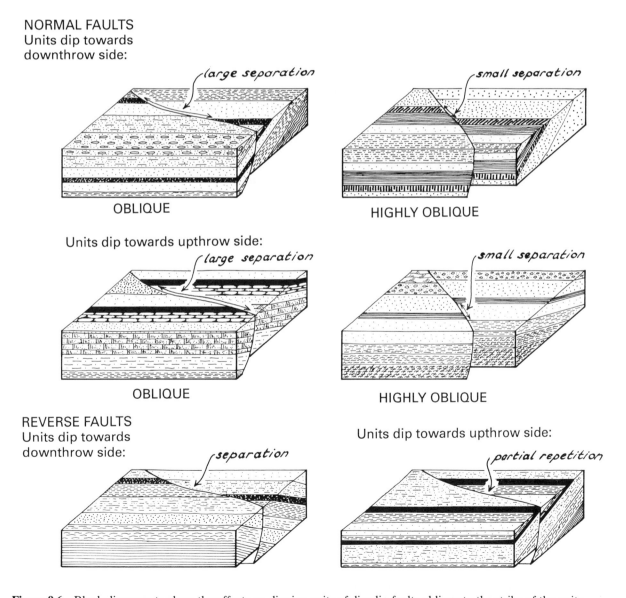

Figure 9.6   Block diagrams to show the effects on dipping units of dip-slip faults oblique to the strike of the units.

Similar effects are produced by both strike-slip and dip-slip faults if the beds are dipping. For a particular amount of net-slip, the amount of horizontal displacement decreases as the fault becomes more oblique to the strike of the beds (figure 9.6). A characteristic double displacement effect results from a dip-slip fault displacing both limbs of a fold (figure 9.7).

In general, then, the presence of a fault on a geological map is not too difficult to spot. Recognition of the kind of fault is a different matter. Horizontal displacement of beds may be conspicuous, but, as mentioned above, if the beds are dipping it could be due to either strike-slip or dip-slip faulting. It may not be possible to tell which. Similarly on a cross-section vertical displacement of beds can result from either strike-slip or dip-slip movement, *if the beds are dipping*. In short, discrimination between minor strike-slip and dip-slip faults is commonly difficult. Section 9.9.2 discusses the recognition of major strike-slip faults.

If the fault is thought to be dip-slip, then it is usually possible to identify its upthrow and downthrow sides from the relative ages of the strata on either side. In general, older rocks will be brought up on the upthrow side; younger rocks will appear on the downthrow side. Keep an eye on the bed dips though, as this generalisation can easily be undermined by any unconformities and folds which are present. Having

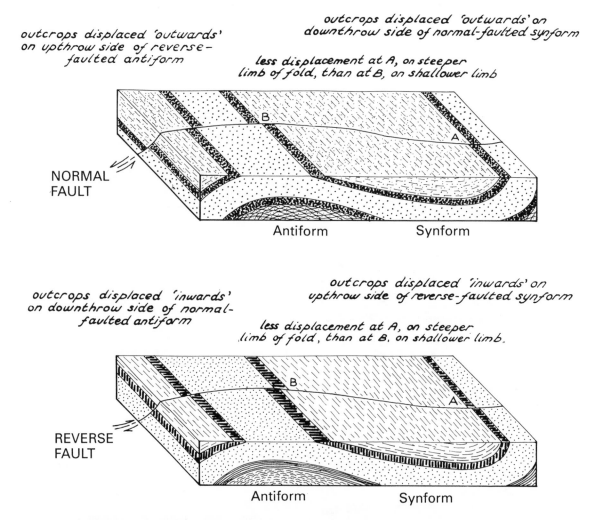

Figure 9.7 Block diagrams to show the effects of dip-slip faults on folded units. Note the opposite displacement effects on opposing limbs of the folds. The effect decreases as faults decrease in obliquity to the strike of the units.

Figure 9.8 (opposite) Some examples of interpreting faults on maps. (a) Near Dorking, Surrey, England; based on Hayward (1932). (b) Near Newtown, east Wales; based on Earp (1938). (c) Near Swinden, Lancashire, England; based on Hudson and Dunnington (1944). (d) Part of the Rangeley oilfield, Colorado, USA.

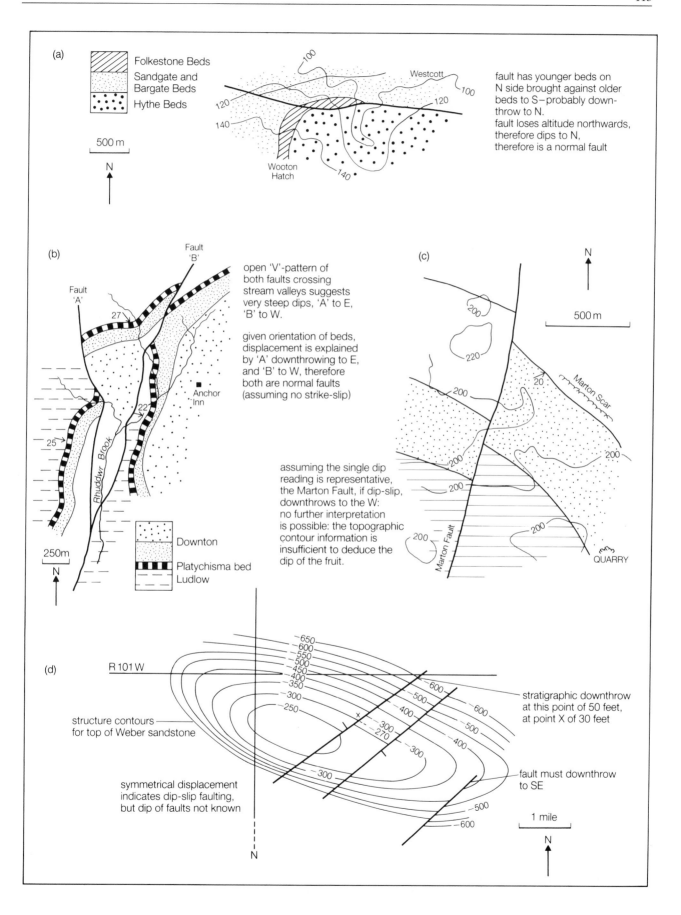

(a)

Folkestone Beds
Sandgate and
Bargate Beds
Hythe Beds

Westcott

Wooton
Hatch

fault has younger beds on
N side brought against older
beds to S—probably down-
throw to N.
fault loses altitude northwards,
therefore dips to N,
therefore is a normal fault

500 m

N

(b)

Fault
'B'

Fault
'A'

27

Anchor
Inn

22

Rhuddwr Brook

25

open 'V'-pattern of
both faults crossing
stream valleys suggests
very steep dips, 'A' to E,
'B' to W.

given orientation of beds,
displacement is explained
by 'A' downthrowing to E,
and 'B' to W, therefore
both are normal faults
(assuming no strike-slip)

Downton

Platychisma bed
Ludlow

250m

N

(c)

N

500 m

200

220

200

20

Marton Scar

200

200

200

Marton Fault

QUARRY

assuming the single dip
reading is representative,
the Marton Fault, if dip-slip,
downthrows to the W:
no further interpretation
is possible: the topographic
contour information is
insufficient to deduce the
dip of the fruit.

(d)

R 101 W

650
600
550
500
450
400
350
300
250

600

500

600

500

400

x
300
270
300

300

400

300

stratigraphic downthrow
at this point of 50 feet,
at point X of 30 feet

structure contours
for top of Weber sandstone

symmetrical displacement
indicates dip-slip faulting,
but dip of faults not known

N

500

600

fault must downthrow
to SE

1 mile

N

deduced that a fault is dip-slip and recognised the upthrow and downthrow sides, the next step is to seek information on the dip of the fault in order to find whether it is normal or reverse. The map may provide a symbol which records the fault dip. The interaction between the fault trace and the topography may provide sufficient information. For example, the principles developed for map units crossing hillslopes and valleys (chapter 6) apply equally to fault surfaces. If no information on fault dip is available, the fault should be regarded as vertical, and no further classification will be possible. Examples of the assessment of faults from maps are given in figure 9.8.

9.6   Measurements on Maps

The length of the fault trace is readily measurable on a map. A minimum value can be given if the trace continues off the map. If the trace is curved it may be necessary to follow it with a piece of thread, which is then measured tautly against a ruler.

Displacements in the horizontal are easily measured. The **strike separation**, measured parallel to the fault strike, and **offset**, measured normal to the strike of the units (figure 9.9a) are two readily cited values. The snag is, to emphasise the point again, that they may not mean very much! If the beds either side of the fault are known to dip steeply, then the horizontal separations are likely to be greatly under-representative. Conversely, faulting of shallow-dipping beds produces considerable horizontal displacement without much steep movement. Relief of the land surface can also contribute to misleading appearances. If values of the horizontal separations are all that can be obtained from the map, a comment should be made on how meaningful they are likely to be.

Clearly some measure of the steep movement is desirable, and on many maps it is possible to derive at least the vertical separation of the beds, referred to as the **stratigraphic throw**. There are two ways of approaching this. One involves the trigonometric estimation of formation depth (section 4.5, figure 4.10). Bed dips have to be known, and assumed to be constant. From the offset of a displaced bed and the tangent of the dip angle, the bed depth can be calculated and subtracted from its elevation at surface on the other side of the fault (figure 9.9b). The difference between the two altitudes is the stratigraphic throw. With faults oblique to bedding it will be necessary to project horizontally the value of bed depth to the fault trace, to bring it next to its displaced counterpart. The projection is done *along the strike* of the bed.

The second method requires the construction of some structure contours (figure 9.9c). On one side of the fault at least two contours will have to be known, so that their spacing can be established. From this, assuming constant dip, further contours can be added if necessary. Only one contour will have to be known on the other side of the fault – provided the dip of the beds is taken to be the same – from which additional contours can be sketched. Where any two contours on either side of the fault abut, the elevation difference between them is the stratigraphic throw. The most closely constrained contours are likely to give the most reliable value. If no contours meet directly then interpolation will be necessary on one side of the fault. The more structure contours that can be drawn from the information given on the map, then the greater the control and the less critical will be the assumption of constant dip.

Where two units which are not adjacent stratigraphically are brought together at the fault plane, it may be possible to establish, either by a series of measurements from the map (section 4.4) or by reference to the map key, the thickness of stratigraphy that has been excised. This value is called the **stratigraphic separation**. This thickness, being measured at right angles to the boundaries of the unit, is independent of the orientation of the fault.

Any further useful information on fault displacement requires knowledge of the dip of the fault. This may be cited on the map, or it may be possible to draw some structure contours for the fault surface itself (at least two, if constant dip can be assumed). Several further components of displacement then become measurable (e.g. Powell 1992). It also becomes possible to construct an accurate cross-section through the fault, to show the fault dip and the displacement magnitude in this vertical plane. The most useful measure is the fault **throw**. This value, in its precise sense (contrast with the usage in section 9.2), is the vertical separation measured in the true dip direction of the fault. Note that in most cases it will not have the same value as the stratigraphic throw, which involves projection along the strike of the units and does not refer to the dip of the fault.

Ordinarily in mapwork this is all the useful fault measurement that can be carried out. Unless the fault displacement is somehow known to be purely dip-slip or strike-slip, the value of net-slip still eludes us. We will not be in a position to predict accurately the displacement that will appear in other directions, say across some oblique mine adit or tunnel. Methods for deriving net-slip, all of which require further information about the displaced beds, are discussed by Ramsay and Huber (1987) and Powell (1992). Even these procedures involve assumptions, such as the amount

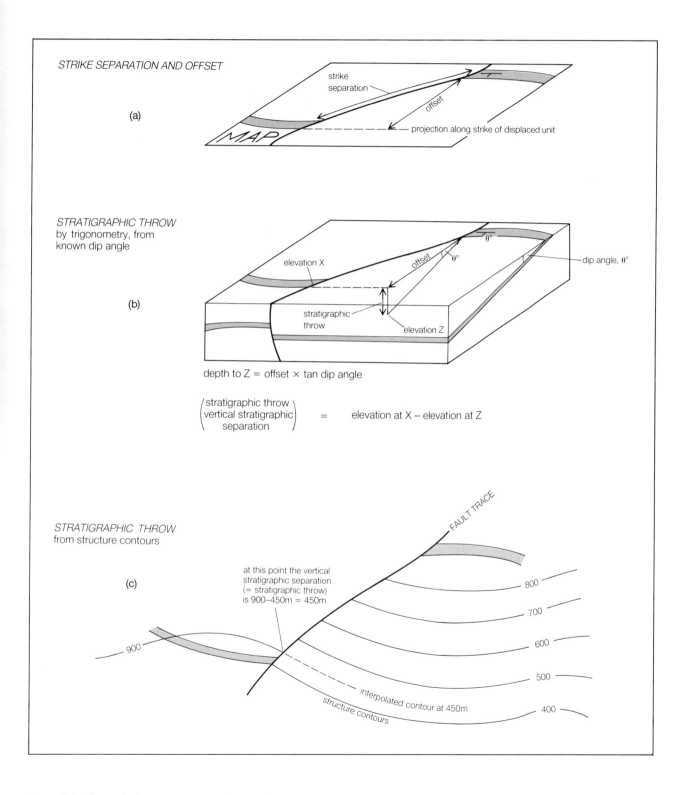

STRIKE SEPARATION AND OFFSET

(a)

strike
separation

offset

projection along strike of displaced unit

MAP

STRATIGRAPHIC THROW
by trigonometry, from
known dip angle

(b)

elevation X

offset

θ°

θ°

dip angle, θ°

stratigraphic
throw

elevation Z

depth to Z = offset × tan dip angle

$$\begin{pmatrix} \text{stratigraphic throw} \\ \text{vertical stratigraphic} \\ \text{separation} \end{pmatrix} = \text{elevation at X} - \text{elevation at Z}$$

STRATIGRAPHIC THROW
from structure contours

FAULT TRACE

(c)

at this point the vertical
stratigraphic separation
(= stratigraphic throw)
is 900–450m = 450m

800

700

600

900

500

interpolated contour at 450m

structure contours

400

Figure 9.9 Some fault measurements that can be made from maps. (a) Measurement of strike separation and offset directly from the map. (b) Derivation of stratigraphic throw by trigonometry, and (c) from structure contours. Note that the trigonometric method assumes constant dip at depth; the extent of this assumption in method (c) depends on the control on the structure contours.

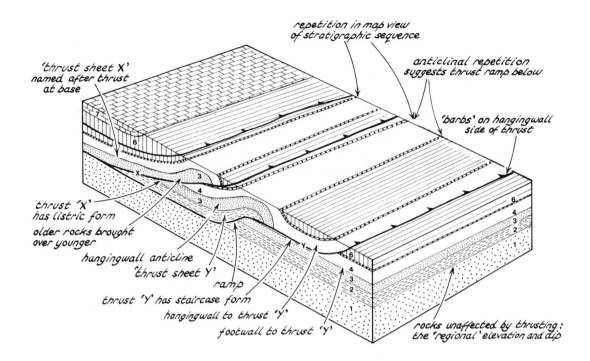

Figure 9.10 Block diagram to show the main features of thrust faults.

of slip being constant along a fault plane. And this cannot be. After all, faults start and stop somewhere, so the displacement must vary along their length.

9.7 Contraction (Thrust) Faults

Having developed the basic methods of dealing with faults on maps, we now go on to look at some special kinds of fault and how in nature they tend to occur in families. First, there are the low-angle reverse faults that are widely referred to as thrusts. They tend to occur in long, linear zones, called thrust belts. An example was mentioned in section 2.3.4. Their overall effect, along with any associated reverse faults, is to telescope together the stratigraphic sequence, producing a thick stack of beds of decreased horizontal extent. Hence all these faults are also referred to as **contraction faults**. They have attracted much attention, partly because of the intriguing problem of understanding how the rocks can actually accomplish large amounts of contraction, and partly because in recent years some thrust belts have been the scene of intensive oil exploration.

9.7.1   Characteristics

Although these structures are essentially low-angle reverse faults (section 9.4), it would be misleading to

think of this as the only difference. They involve a number of characteristics, as follows (figure 9.10).

- Thrusts rarely occur as isolated faults. Typically they are in groups, which together with splaying and interlocking smaller thrusts and folds, form the long, linear zones known as **fold-and-thrust belts** or simply **thrust belts**. Families of closely spaced splaying reverse faults called **imbricate** zones may also be involved (figure 9.11).

- The displacements on the major thrusts of the belt may be substantial – typically measured in kilometres of movement.

- The low dip of thrust faults towards the upthrow side results in more deeply buried rocks being brought up and over shallower rocks. It is a general property of thrusts that they bring older rocks over younger. The hangingwall of a thrust is commonly called a **thrust sheet** and named after the fault at its base.

- The shape of an individual thrust is not normally planar over any great distance. There are two general tendencies. One is for the thrust plane to curve smoothly, from a steeper dip at its upper end to a shallow dip at depth. This concave-upwards form is called **listric**, from the Greek word for a spoon. The other tendency is towards developing a kind of 'staircase' shape. The **ramps** commonly dip at about 30–35° and occupy a shorter distance than the

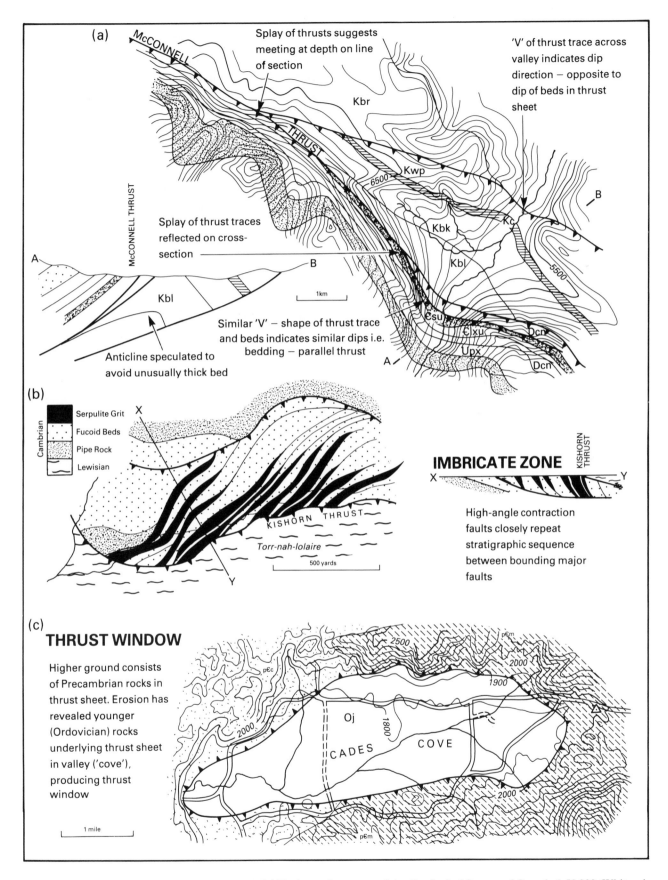

Figure 9.11 Some features of thrusts on maps. (a) Redrawn from part of the Geological Survey of Canada 1:50 000 'Whiterabbit Creek' (E) sheet 1388A, by permission of the Geological Survey of Canada. Cross-section inferred from map features. (b) Imbricate zone in map and section. Near Lochcarron, Scotland. (c) Thrust window. Cades Cove, Tennessee, USA. Redrawn from part of the 1:62 500 map in *USGS Professional Paper* 394-D, 'Geology of the Great Smoky Mountains, Tennessee', by permission of the USGS.

bedding-parallel **flats**. Some thrusts have bedding-parallel parts several kilometres long.

- A consequence of a ramp having developed is that beds moving up and over the ramp take on an anti-formal shape. The structure is termed a **hangingwall anticline**.

9.7.2 Thrust faults on maps

An isolated thrust fault will be recognisable on a map as a low-angle plane on which older rocks overlie younger ones. No other structure can look like this, except for very special situations such as an inverted unconformity! And if the various characteristics listed above are also apparent, the diagnosis of thrusting should be straightforward. In any case, it is conventional for thrusts to be indicated on maps by a special symbolism (figures 4.2 and 9.11). If you are drawing this symbol on a map, do make sure you have the barbs on the hangingwall/thrust sheet side. It is a common beginner's mistake to reverse the position.

A group of thrusts on a map may well have the effect of repeating the stratigraphy several times, and if imbricate zones are present too, the outcrop pattern can be extremely intricate. Stratigraphic repetition is also produced by extensional faulting (section 9.8, figure 9.12), but there the overall effect of bringing younger rocks over older will be lacking.

Having located and recognised thrust faults on a map, several things can be attempted by way of three-dimensional interpretation. Because of their overall low dip, thrusts are amenable to the methods developed earlier for horizontal and dipping beds (sections 6.2 and 6.3). The map trace of a thrust will tend to follow topographic contours, and will parallel them if the thrust is horizontal. The principle that formations become younger in the dip direction will have to be used with extreme caution, though, as by their very nature thrusts will be upsetting the stratigraphic succession.

If there are topographic contours on the map, or if borehole/well information is available, it may be possible to construct structure contours and sub-crop patterns for the thrust surface (cf. sections 3.4, 3.6, and 7.5). If the structure contours are closely constrained, they may bring out the listric or staircase shape of the plane.

Note that the dip of the fault at surface does not necessarily reflect the overall thrust inclination. Projection to depth on a cross-section is precarious without supporting information. It is perhaps safest to draw the thrust trace as a straight line. Some listric form could be included where the beds in the thrust sheet are not greatly different in stratigraphic age from those in the footwall. However, if basement rocks are involved in the hangingwall, the thrust must continue to a depth sufficient to have enabled the incorporation of this material. If the map shows the thrust trace to follow certain stratigraphic horizons and cross-cut others, it may be reasonable to incorporate these patterns as flats and ramps, respectively, in the cross-section. The presence of a localised anticline next to the map trace of a thrust may be interpreted as a hangingwall anticline, and the fold and ramp included on the section.

9.8 Extension Faults

In contrast to thrust belts, there are areas of the earth which have been 'pulled apart' such that they are now dominated by **extension faults**. The dominant structures are essentially normal faults, but tend to have a low dip-angle overall. The inner parts of these areas tend to be dropped down along the faults more than the outer parts, leading to subsidence and the production of **extensional basins**. These areas, too, are of great importance in oil geology – the North Sea is an example of an extensional basin.

Curiously, although on regional maps thrust belts contrast clearly with extensional basins, on large-scale maps contraction and extension faults can be difficult to distinguish. The two kinds of fault have many similarities in appearance. In a restricted area it might not be possible to match the rocks in the hangingwall with those in the footwall, or the low-angle fault may be exactly parallel to bedding, so that the critical difference of the hangingwall moving up or down will not be recognisable.

9.8.1 Characteristics

Extension faults are the exact converse of contraction faults, and yet, paradoxically, the two have a great deal in common. The first three of the following characteristics define the differences (figure 9.12) but the remainder apply equally to contraction faults.

- Extension faults are dip-slip faults which dip at a low angle towards their *downthrow* side. They are therefore normal faults of shallow inclination.
- This form requires that higher rocks are brought down next to lower – younger over older – which is the exact opposite of thrusts.
- Their overall effect is to attenuate or pull-out the stratigraphic sequence laterally, so that it becomes thinner but more areally extensive. Extensional faults therefore tend to form basins rather than linear belts.
- The faults may inter-link and have minor splay faults. The resulting basin is by no means neces-

sarily symmetrical. The variable distribution of displacements can produce complex basin shapes. Arrays of high-angle extension faults can form imbricate zones.

- The faults root into a basal, low-angle *detachment* fault. This is normally at depth, but may reach the surface at its tips. There are areas of the earth where such large, low-angle faults occur more or less singly. A famous and intriguing example forms the subject of Plate 6.
- The shapes of extension faults are highly analogous to thrusts. They can have listric or staircase form, or some combination of both.
- Like thrusts, production of these fault shapes causes the beds to fold. An anticline adjacent to an extension fault is called a 'roll-over' structure.

9.8.2 Extension faults on maps

If faults have not been marked as such on a map, the recognition of a single detachment fault could be difficult. With most extension faults, however, some of the features mentioned above will be apparent. The biggest difficulty can be the distinction of extension from contraction faults, if this is not indicated on the map by symbolism. On larger scale maps the distinction might not be possible.

The basic property of extension faults of bringing younger rocks over older may be discernible. Whereas contraction faults can *repeat* the stratigraphic

sequence – thrusts can cause the same unit to appear on a map in several different places – extension faults tend to cause *omission* of parts of the stratigraphy. Much more commonly than with thrust faults, some extension faults show a greater thickness of sediment on their downthrow side. The implication of this is that the fault did not arise after the entire sequence of sediments had formed but that it was growing at the same time as the sediments were accumulating – a process called **growth faulting**. Sediment was banked preferentially in the depression formed at the downthrow side of the fault, against the growing fault scarp.

9.9 Strike-Slip Faults

Yet other parts of the earth are dominated by zones of strike-slip faulting. Typically, these faults have very large displacements and their traces are conspicuous on maps, but correlation across the faults is notoriously difficult, even on small-scale maps, because of the large amount of movement. Even though some of the most widely known geological structures are strike-slip faults, for example, the San Andreas Fault in California, their strike-slip character can be difficult to prove.

Most of what follows refers to major strike-slip faults developed on the regional scale, but strike-slip faults can be small features, occupying only a part of a large-scale map. In this case, they may represent local

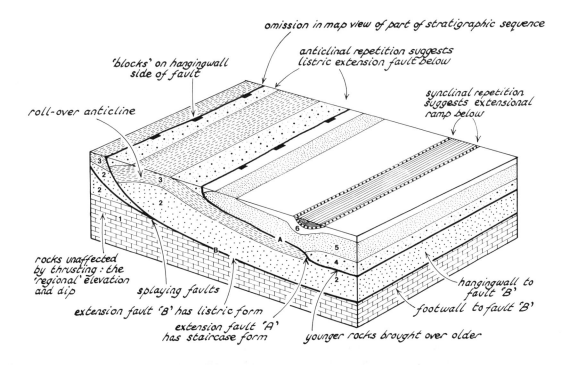

Figure 9.12 Block diagram to show the main features of extension faults.

adjustments to a nearby larger structure. Such local faults have a lot in common with the high-angle faults dealt with earlier in this chapter. The main difference is simply that the displacement was in the strike direction of the fault rather than the dip direction. Remember, though, that seeing displaced formations on a map does not mean that the structure has to be a strike-slip fault: *apparent* horizontal displacement is produced by steep fault movements acting on dipping beds (section 9.5).

9.9.1   Characteristics

Strike-slip faults show the following characteristics:

- The fault traces of strike-slip faults are typically long, measured in tens or hundreds of kilometres.
- The amounts of displacement are commonly of similar magnitudes.
- The fault surfaces tend to be very steep to vertical. Strike-slip faults tend to have straight courses for long distances. They can curve: the famous 'big bend' of the San Andreas Fault north of Los Angeles is an instance.
- There may be sub-parallel splay faults. Some faults splay and anastomose on a large scale, making broad zones of complex faulting. In fact, the geometrical details of these zones have much in common with contractional and extensional faults.
- Strike-slip faults are typically long-lived structures. Once such a major, deep-penetrating fracture is initiated, it responds relatively easily in subsequent times of stress. Consequently, many ancient strike-slip faults are still active today. However, this **reactivation** of the fault does not necessarily occur with the same kind of displacement. Strike-slip faults typically have periods of dip-slip movement in their history.
- It is usual to label a strike-slip fault with its dominant displacement sense (section 9.2), if this is known. Imagine standing on one side of the fault and looking across it; if the opposite side has moved to your left, it is called a **left-lateral** or **sinistral** fault. If the opposite side has moved to the right, it is a **right-lateral** or **dextral** fault.

9.9.2   Strike-slip faults on maps

It is usually easy enough to recognise major strike-slip faults; it is the displacement history which can be very elusive. The faults tend to make straight traces on maps. This is not only because the faults tend to be straight anyway, but because of their steepness, which allows them to cut across topographic relief. Moreover, erosion often works preferentially on the weak rocks of the fault zone, producing conspicuous straight valleys. The Great Glen of Scotland, containing the deep, elongate Loch Ness, is formed along a major strike-slip fault – the Great Glen Fault. Such features are distinctive on air photographs and satellite images, and hence tend to be shown boldly on regional reconnaissance maps.

Curiously enough, the huge displacements which can accumulate along these faults are not necessarily a distinctive feature on a map. It may be obvious that the rocks either side of the fault bear little relation to each other, but in the absence of features which match up, the displacement will not be clear. Another complicating factor is that with the common reactivation of these faults, features of different geological ages will show different amounts of displacement.

Strike-slip movement on smaller faults can be detected by the displacement of vertical features (figures 9.3 and 9.13a), and by variably oriented features being offset by the same amounts (figure 9.13c). Conversely, on a cross-section of a strike-slip fault, variably dipping formations will appear to have been displaced different amounts whereas with a dip-slip fault everything will show the same distance of vertical displacement.

Ideally, it is the *oldest* rocks in the map area that we aim to correlate across the fault, in order to detect the maximum displacement. However, the old rocks may now be covered by younger deposits. In fact, it is commonly the *younger* features that give the initial clues to the fault displacements, even if they are just the more recent movements. A good example of this is the San Andreas Fault, which, although now thought to have accomplished something like 1000 km displacement, was first detected by the offset of orange groves and present-day streams. The Hope strand of the Alpine Fault of New Zealand is thought to have displaced the bedrock by about 20 km, post-Quaternary river terraces by up to 160 km (figure 9.13b), and, in living memory, farm fences by 3 m (Freund 1971).

In the absence of suitable features either side of the fault, assessment of the displacement will be difficult. The displacement may be so large that even if such features do exist, they do not fall on a single map. Some clue, at least to the sense of displacement, may be provided by consistent offsets across any minor splays of the major fault. Otherwise, it becomes a matter of attempting to correlate regional features: granite plutons, metamorphic zones, special sedimentary environments, etc. Strike-slip faults may be dramatic features, but deducing their displacement histories is a challenge to the geologist.

Summary of Chapter

1. Faults are fractures along which the rocks have moved.
2. The displacement of rocks that is apparent from one viewpoint is not necessarily representative of

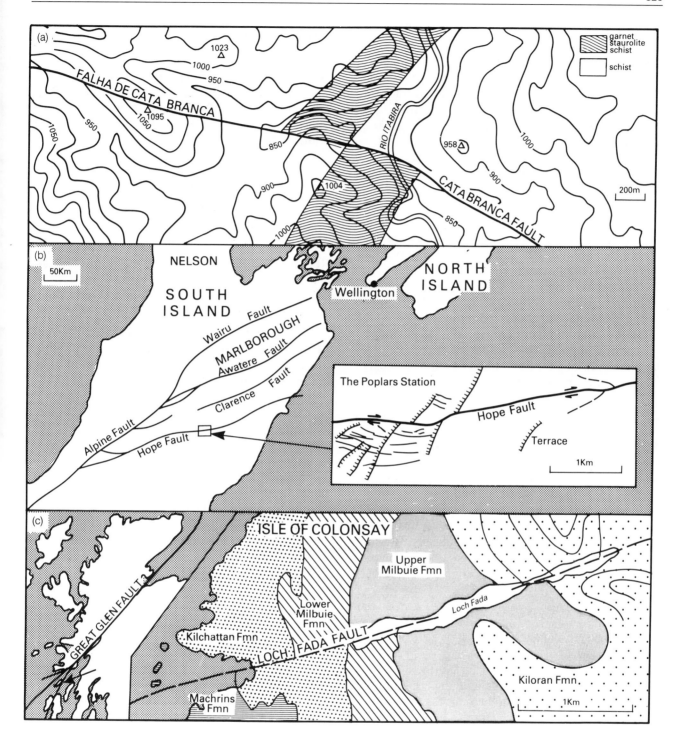

Figure 9.13   The recognition of strike-slip faults on maps. (a) The Cata Branca Fault, western Brazil. The straight outcrops, in an area of rugged topography, indicate that the schists are oriented vertically. The fault displacement must therefore be strike-slip. Redrawn from Wallace (1965), by permission of the USGS. (b) The Alpine Fault system, New Zealand. The inset shows how displaced river terraces have been used to infer as much as 160 metres of post-Quaternary movement on the Hope splay. Redrawn from Freund (1971), by permission of the New Zealand Geological Survey. (c) The Loch Fada, Colonsay, splay of the Great Glen Fault, Scotland. Variably oriented formations are displaced by similar amounts, indicating a strike-slip fault.

the actual movement. Terms have to be used carefully.

3. Slip is the displacement of formerly adjacent points, measured in a specified direction in the fault plane. Separation is the displacement of formerly adjacent planes, measured in some specified direction.

4. Faults are traditionally classified as strike-, oblique-, or dip-slip faults, the last being further divisible into normal and reverse structures.

5. Faults are usually easily detected on maps, by outcrops being displaced, omitted or repeated, but their interpretation can be elusive.

6. Contraction or thrust faults are reverse faults of low dip-angle. They bring older rocks over younger to a significant degree and decrease the horizontal extent of a rock sequence.

7. Extensional faults are low-angle normal faults, bringing younger rocks over older, and increasing the horizontal extent of a rock sequence.

8. Major strike-slip faults are typically steep, straight features, normally conspicuous on maps. They are commonly long-lived and with large displacements, though the amounts can be elusive.

Selected Further Reading

Powell, D., 1992. *Interpretation of Geological Structures through Maps*. Longman Group, London, 176 p.
    Pages 45–52 give further illustrations of deriving displacements from maps; pages 59 and 60 illustrate growth faulting.

Ramsay, J.G. and Huber, M.I., 1987. *The Techniques of Modern Structural Geology. Volume 2: Folds and Fractures*. Academic Press, London, 700 p.
    Session 23 is a thorough account of the geometry of faults, much of which is highly relevant to more advanced mapwork.

Roberts, J.L., 1982. *Introduction to Geological Maps and Structures*. Pergamon Press, Oxford, 332 p.
    Pages 131–163 discuss normal and reverse faults, and the problems of terminology.

Suppe, J., 1985. *Principles of Structural Geology*. Prentice-Hall, Englewood Cliffs, New Jersey, 537 p.
    Pages 277–289 summarise the morphology of strike-slip, thrust, and detachment faults.

Map 16   Wenlock Edge, Shropshire, England

Wenlock Edge is one of a series of escarpments (see section 13.5) in the pleasant countryside of south Shropshire, where Sir Roderick Murchison carried out much of his early work on what he was to christen the Silurian System of rocks. The simplified geological map opposite shows the rocks that form Wenlock Edge – including the richly fossiliferous Wenlock Limestone – and their displacement by a fault. Determine the stratigraphic throw this fault has produced on the Aymestry Limestone and on the Wenlock Limestone, using structure contours and using trigonometry. What do structure contours for the upper and the lower surfaces of the Aymestry Limestone reveal about the thickness of this formation?

Further examples of faults on maps

Two faults appear on Map 7, of the 'Northcrop' of the South Wales coalfield. Comment on the displacement each must have produced to account for the present outcrop pattern.

Describe the fault which occurs in the NE of Map 15 and affects the Millstone Grit.

The numerous faults in the SW part of Plate 2, of the western USA, *tend* (there are exceptions) to curve with their concave side towards their dip direction. Judging by the relative ages of the rocks juxtaposed by the faults, distinguish the areas dominated by extension faulting from those with much contraction faulting.

Note that faults appear on all the coloured maps reproduced here, and that some of the questions accompanying Plates 5 and 6, of Sanquar, Scotland, and Heart Mountain, Wyoming, USA, concern the interpretation of faults.

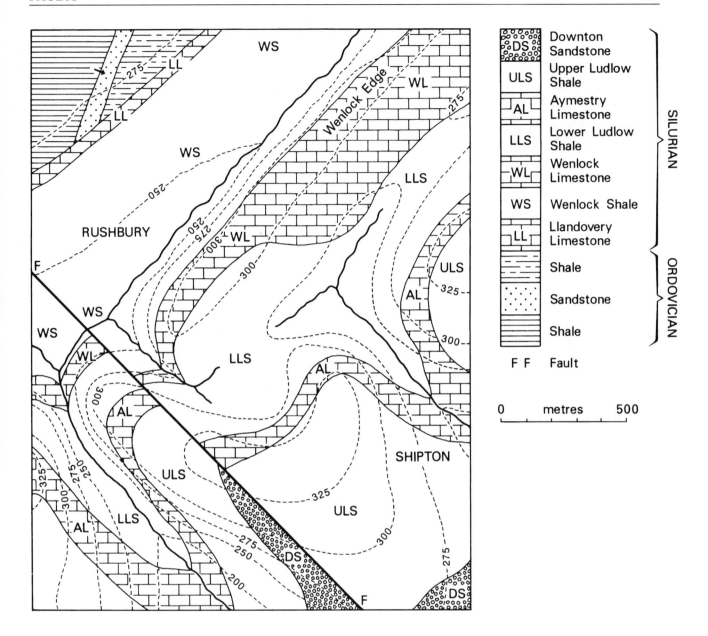

Map 17   Aspen, Colorado, USA

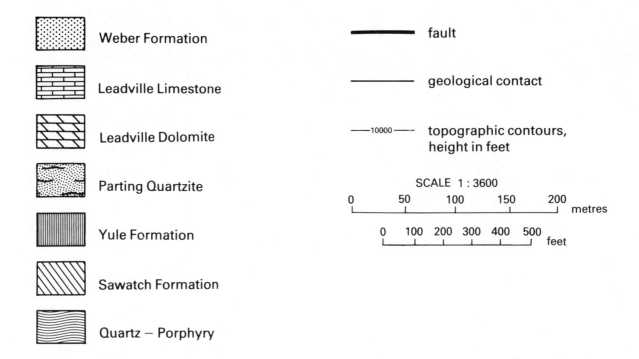

The map opposite is based on one of a folio of superb very large-scale maps published by the USGS in 1898 of the Aspen district of Colorado. The detailed study was prompted by the intensive mining for various precious metals in the district, an activity which has now ceased. The numerous faults which were once the miners' bane criss-cross the slopes now used in Aspen's new industry – downhill skiing! The area presented opposite is a part of the 1:3600 Tourtelotte Park Mining District sheet, reproduced here, with some modification, by permission of the USGS.

Is there any overall pattern to the trends of the fault traces?

Judging by the displacements seen on the map, is the Good Thunder Fault likely to be a strike-slip or a dip-slip fault?

Given that the Burro Fault is a dip-slip fault, does it downthrow to the north or south?

In which direction does the Silver Bell Fault dip? (It should be possible to draw a few structure contours for this fault.)

What is the strike separation along the Dixon Fault? What is the stratigraphic throw of the Dixon Fault shown by the top of the Leadville Dolomite? (This can be answered either by using trigonometry, or by constructing structure contours south of the fault and comparing them with their estimated counterparts to the north.)

Construct structure contours for the Dixon Fault. Draw an accurate cross-section in the dip direction of the fault, through the point where it is met by the Leadville Dolomite to the north. Project, in its dip direction, the top of the Leadville Dolomite south of the fault to the cross-section. Measure the throw of the fault.

What kind of fault is the Justice Fault?

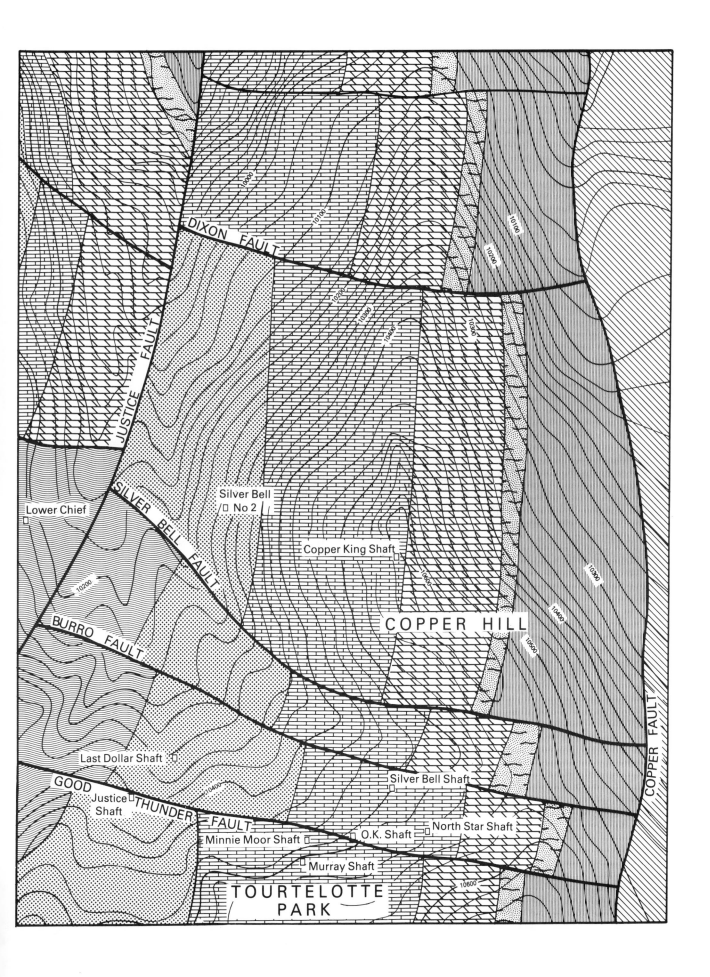

DIXON FAULT

JUSTICE FAULT

SILVER BELL FAULT

Lower Chief

Silver Bell
No 2

Copper King Shaft

COPPER HILL

BURRO FAULT

Last Dollar Shaft

GOOD THUNDER FAULT

Justice
Shaft

Minnie Moor Shaft

Silver Bell Shaft

O.K. Shaft

North Star Shaft

Murray Shaft

TOURTELOTTE
PARK

COPPER FAULT

10000

10100

10200

10300

10400

10500

10600

Map 18   Glen Creek, Montana, USA

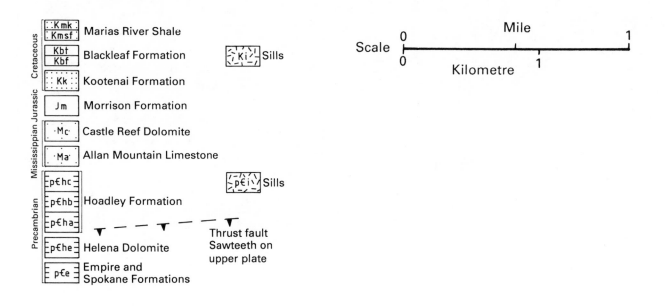

The map opposite is redrawn from the USGS map GQ-499 of Glen Creek, Montana, by permission of the USGS. It is of an area about 100 km WNW of Great Falls, in the easternmost ranges of the Rocky Mountains. The area therefore falls within the fold-and-thrust belt which dominates the eastern flanks of the Western Cordillera in much of North America, and is a northwards extension of the area of thrusting introduced briefly in section 2.3.4 and portrayed on Plate 2.

Colouring some of the stratigraphic divisions on the map, particularly units Kbf and Kbt, will help make their outcrop patterns more distinctive. What might the dashed line running through the central part of unit Kbf represent?

Looking at the relationship between outcrop and topography, and any map symbols, visualise the overall three-dimensional structure of the area.

Draw a cross-section at right angles to the dominant strike direction, to represent the structure of the area. In projecting the thrusts to depth bear in mind that they are not necessarily planes. Incorporate, for example, the symmetrical repetition of unit Kbt possibly indicating a steepening of the thrust subsurface, and any coalescing of thrusts off the section line which may reflect changes in thrust dips.

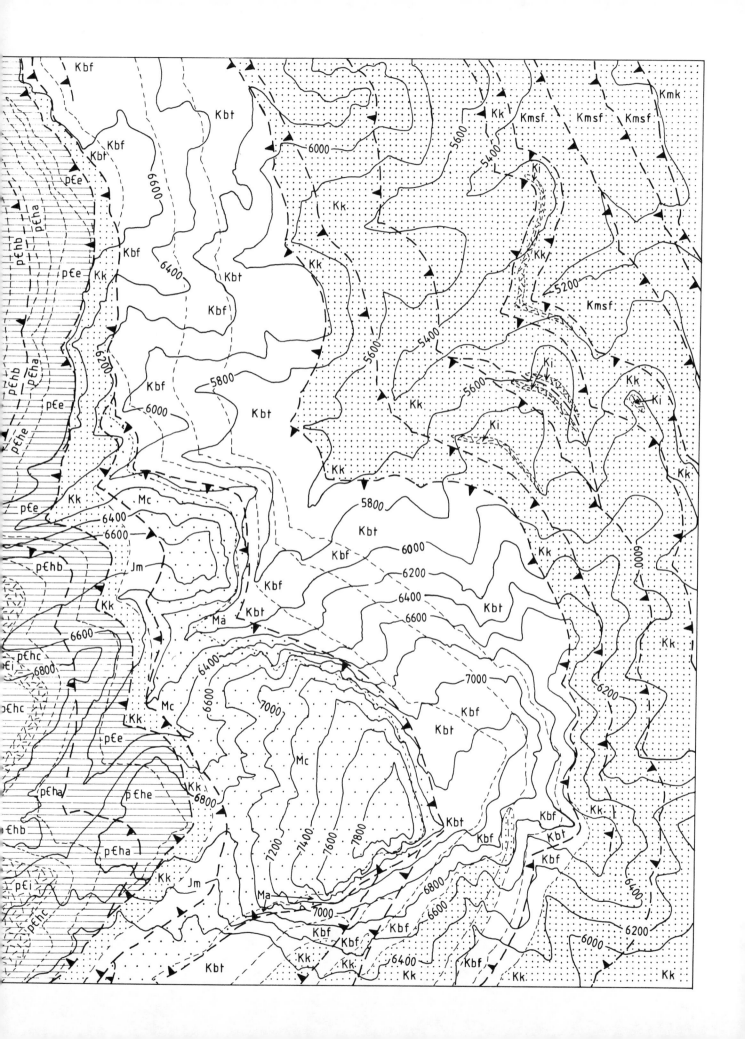

Map 19 Hamblin Bay Fault, Nevada, USA

The facing map is redrawn and simplified from a map by Anderson (1973), by permission of the USGS. It covers an area about 60 km east of Las Vegas, and is on the north shores of Lake Mead, created by the construction of the Hoover Dam, just SW of the map area. Although the area falls into a region of the western USA dominated by large-scale extension faulting (the 'Basin and Range' structural province), some parts of the region show important strike-slip faults. In the present area, an andesitic volcano and associated dykes, known as the Hamblin–Cleopatra volcano, have been dismembered by faults, particularly the major Hamblin Bay Fault.

Discuss the evidence for the Hamblin Bay Fault having dominantly strike-slip displacement. Identify the sense of movement. Estimate the amount of strike-slip displacement.

Which other faults must also have a substantial component of strike-slip displacement?

Discuss the likelihood of the faults at Saddle Mountain being strike-slip faults.

What is the overall age of faulting in the area?

Is there any relationship between the orientation of bedding and proximity to a fault? What might this imply for the actual slip direction on some of the faults?

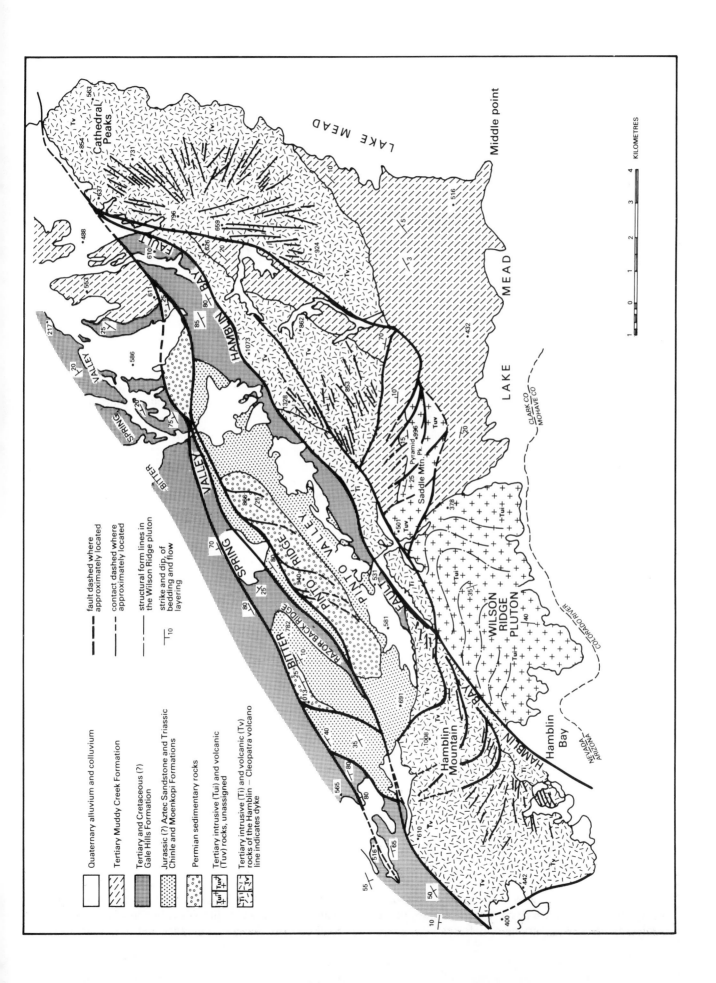

LAKE MEAD

Middle point

Cathedral Peaks

BITTER SPRING VALLEY

HAMBLIN BAY FAULT

PINTO VALLEY

RAZOR BACK RIDGE

PINTO RIDGE

LAKE MEAD

WILSON RIDGE PLUTON

Hamblin Mountain

Hamblin Bay

Saddle Mtn. Pk.

Pyramid

NEVADA
ARIZONA

CLARK CO
MOHAVE CO

COLORADO RIVER

Quaternary alluvium and colluvium

Tertiary Muddy Creek Formation

Tertiary and Cretaceous (?) Gale Hills Formation

Jurassic (?) Aztec Sandstone and Triassic Chinle and Moenkopi Formations

Permian sedimentary rocks

Tertiary intrusive (Tui) and volcanic (Tuv) rocks, unassigned

Tertiary intrusive (Ti) and volcanic (Tv) rocks of the Hamblin – Cleopatra volcano line indicates dyke

fault dashed where approximately located

contact dashed where approximately located

structural form lines in the Wilson Ridge pluton

strike and dip, of bedding and flow layering

KILOMETRES
1   0   1   2   3   4

CHAPTER 10

Igneous and Metamorphic Rocks, Mineral Deposits, and Superficial Deposits

10.1 Introduction

The preceding sections have concentrated on sedimentary rocks, which are layered and readily organised into a sequence of map units. Each unit tends to have an overall tabular shape, being bounded at its top and bottom by roughly parallel surfaces, and this allows us to make the three-dimensional interpretations discussed in the previous chapters. We have, however, to be more cautious with non-sedimentary rocks and with the sedimentary deposits that formed at the earth's surface only recently, for their shape can be much less regular. It is with these materials that this chapter is concerned.

In this brief survey of non-sedimentary rocks, the same themes constantly recur. How regular is the bulk form of the material depicted on the map? How likely is it to be tabular? These three-dimensional aspects can be awkward, but on the other hand non-sedimentary materials do lend themselves to another kind of interpretation – the chronological sequence in which they formed. This idea is introduced during this chapter, before expanding it in chapter 11.

10.2 Igneous Rocks

10.2.1 Volcaniclastic rocks

The tendency for volcaniclastic material to adopt a tabular shape is variable. At one extreme, volcanic ash which travels large distances before slowly settling will produce a continuous, thin bed-like form, whereas debris which jostles around in the throat of a volcano will form an irregular mass quite unlike a bed. In between these two extremes are many kinds of volcaniclastic rock, with varying bulk forms.

In general, fine-grained volcanic ash falls into beds, mantling the land surface on which it settles. If the topography is smooth, the top and bottom surfaces of the layer will be virtually parallel, and the normal three-dimensional extrapolations can be applied. Where the material settles on a dissected land surface, there may be rapid thickness changes, though these may not be discernible except on large-scale maps. If there are repeated volcanic explosions, a sequence of volcaniclastic layers will be built up. If there is little time for renewed erosion between the volcanic events, the dissected relief will become subdued as the successive deposits preferentially fill in the topographic lows. Figure 10.1b shows river erosion revealing a sequence of volcaniclastic deposits.

Coarse-grained material tends to travel less far, and there is usually more of it. The deposits may have a roughly tabular shape, but they will not be extensive, and if they lie on the volcano's flanks, they may have primary dips of up to 40°. Some volcaniclastic material is very localised. Patches of agglomerate on a map may be indicating choked volcanic necks; fan-shaped masses may be avalanche deposits. All these deposits can be dramatically changed by renewed volcanism or the swift erosion that can accompany eruptions. The outcrop patterns of these near-source volcaniclastic materials can be complex.

Despite all these difficulties, the field surveyor will usually have been able to divide the volcaniclastic materials into map units, and to establish their sequence of production. Hence most geological maps treat any volcaniclastic materials in the same way as sedimentary rocks, and they appear on the key in their appropriate stratigraphic position. They may be given a special ornament or colour, and perhaps a letter symbol to indicate their particular lithology, but in other ways they will be treated as a sedimentary unit of the same age.

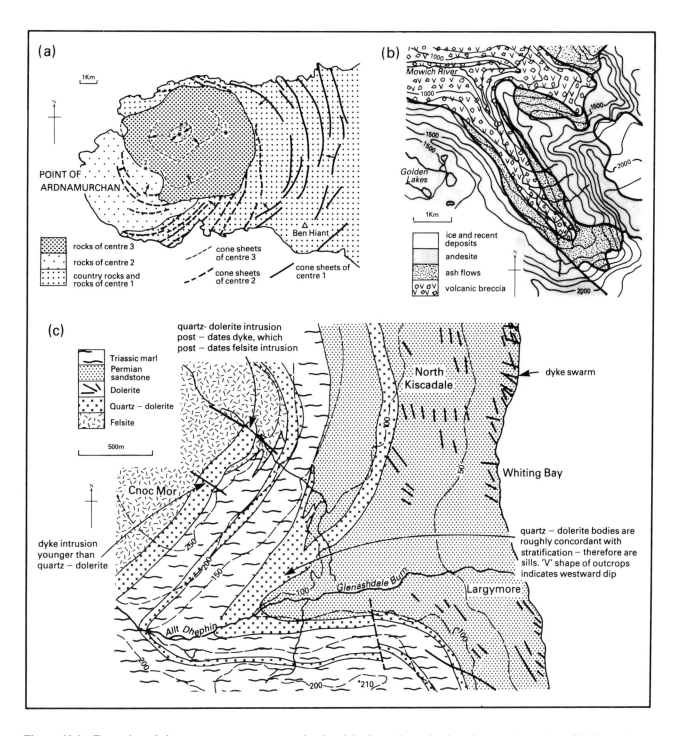

Figure 10.1 Examples of the appearance on maps of volcaniclastic rocks and minor igneous intrusions. (a) Cone sheets (arrows show dip directions), around three different volcanic centres. Ardnamurchan, western Scotland. Based on Richey (1961). (b) Rapid stream downcutting through a volcaniclastic sequence. Note that the valley gradients on the volcano slopes exceed the rock dips and hence contravene the normal 'V-rule' (section 6.3.2). Redrawn from USGS Map I-432, Mount Rainier National Park, Washington, by permission of the USGS. (c) Minor igneous intrusions, Isle of Arran, southwest Scotland. Simplified from the BGS one-inch Arran Special sheet. (a) and (c) reproduced by permission of the Director, British Geological Survey. Copyright NERC. All rights reserved.

10.2.2 Magmatic rocks

The representation of lava on maps, and its interpretation, has much in common with volcaniclastic deposits. If the lava is highly fluid, it can flow large distances, filling valleys and eventually blanketing the land surface. Thus it forms a mass of overall tabular form. The first lava may form an unconformity with the bedrock over which it flows (cf. section 2.3.2) and then a sequence of lava flows can be built up. Erosion of a series of lava flows can produce a characteristic stepped or 'trap' topography, which may be apparent on larger scale maps.

Viscous lavas are analogous to near-source volcaniclastic material. Though they will have travelled little, it may be possible from their shape on a map to deduce the direction of flow. Note that they can have abrupt terminations which can resemble faults on maps. Individual lava flows of the same type are not normally distinguished but collected together to form a single map unit. This, like the volcaniclastics, may be given a particular ornament; it may or may not appear in the key at its appropriate stratigraphic position.

Erosion of volcanic areas may expose bodies of magmatic material which failed to rise to the land surface before solidification. If the masses are irregular in form, they will appear on maps as patches, distinguishable from volcaniclastic rocks such as agglomerates only by consulting the map key to establish their igneous nature. In three dimensions the boundaries are likely to be steep, reflecting the ascent of the magma from depth, but detailed interpretation will be precarious.

Bodies of solidified magma can be fairly regular in form and indeed are commonly quite tabular, in which case they are conveniently referred to as igneous sheets. Sills, being concordant with the layers of sedimentary or other material which envelope them, appear on maps just like a regular sedimentary rock unit, and are subject to all the same geometrical principles (figure 10.1c and section 2.3.5). Note that sills need not *everywhere* be concordant to the adjacent beds. Small stretches where the top and bottom of the sill cross-cut the adjacent layering can be a helpful distinction from a lava flow.

Dykes are commonly conspicuous on maps. Being discordant sheets, they cut across the adjacent outcrop pattern, and typically being steep to vertical features, they produce on maps a characteristic narrow strip-like pattern (section 2.3.5). Patterns of dyke orientation and distribution frequency show up particularly well on maps. They may occur profusely, in **swarms** (figure 10.1c), the dykes may be **radial** to the magmatic source (section 2.3.5), or they may circle it. These may dip towards the volcanic centre – **cone sheets**

(figure 10.1a), or they may be vertical – **ring dykes**. If topographic contours are shown on the map it should be possible to interpret which of these types is present from their interaction with the dyke traces. A famous example of a beautifully executed geological map which shows all these features is discussed in section 14.3.6.

Igneous **plutons** are typically very conspicuous on maps. They form large, commonly irregular to roundish patches which bear no obvious relation to the surrounding rocks. Small-scale maps can usefully show the regional distribution of the igneous bodies, and their relative shapes and sizes (figure 10.2a). Large-scale maps may show details within the pluton, for example, lithological variations, xenolith distributions, and joint patterns (figure 10.2b). These may prompt inferences from the map on the origin and emplacement of the igneous mass.

Interpretations of the three-dimensional shape of a pluton will be precarious, but it may be possible to make some speculations. A reasonably regular boundary trace suggests some regularity in three dimensions, and you may be able to draw some tentative structure contours (figure 10.2c). The most typical configuration will be a steeply dipping, gently curved surface. Any deviations from this may be apparent from the map trace. A wriggling boundary trace would suggest a three-dimensional irregularity that would make interpretation futile. It is best, if there is doubt, to treat the pluton boundaries as vertical. Intrusive bodies intermediate in size between pluton and dykes and sills can have all kinds of shapes, with various combinations of steep and flat boundaries.

10.3 Metamorphic Rocks

The extent of any thermal metamorphism associated with an igneous body can be effectively shown on a map (figure 10.2c), but it is not always done. You should not assume, especially on small-scale maps, that because no metamorphic rocks are shown adjacent to an igneous body that there was no thermal effect. Some larger scale geological maps do show the different kinds of thermally produced metamorphic rocks, which normally appear as a series of zones, roughly concentric around the heat source. The rock type of each zone will be specified in the key, in the usual way.

An alternative approach is for the map to show not the actual kinds of metamorphic rocks, but the areas where there has been thermal change. There may, for example, be a shaded zone to represent the extent of thermal metamorphism around an igneous body (Plate 5), or different ornaments can depict greater or

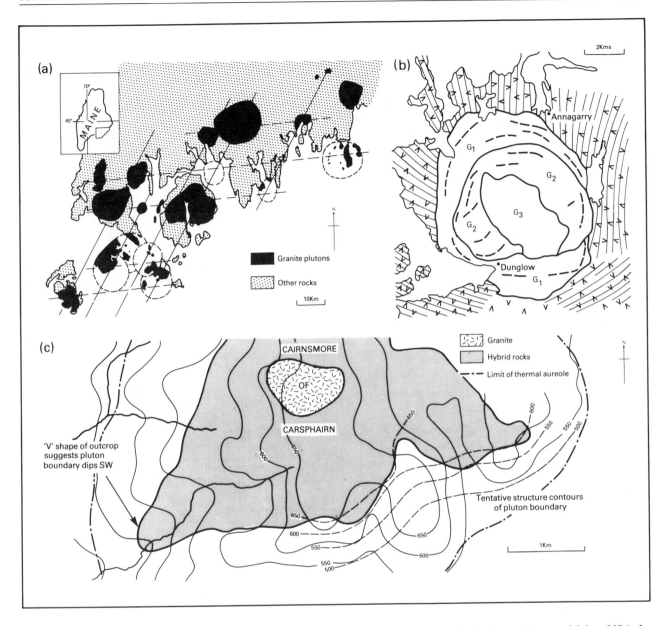

Figure 10.2 Some aspects of plutonic igneous rocks on maps. (a) Map showing areal distribution of plutons, Maine, USA. In this example, the straight dashed lines were used to suggest a reticulate distribution pattern. Based on Chapman (1968). (b) Some pluton details. The pluton comprises three granites, G_1–G_3, with xenolith alignments shown by dashed lines. The Rosses complex, Donegal, Eire. Based on Pitcher and Berger (1972). (c) The three-dimensional arrangement of a pluton boundary, Cairnsmore, southern Scotland. Note also the representation of the extent of the thermal aureole. Redrawn from Deer (1935), by permission of the Geological Society, London.

lesser degrees of change. A common device is for the map to show a line which is labelled in the key as something like 'outer limit of thermal metamorphism' away from the heat source. However the thermal metamorphism is depicted, most of the interpretive steps that can be made require further knowledge of metamorphic processes.

The representation of regional metamorphism on general geological maps presents more difficult problems. Some maps, particularly small-scale and older maps, group together diverse metamorphic rocks into one map unit, in which case little can be done in the way of interpretation. Other maps show the distribution of the different rock types.

To what extent their three-dimensional arrangement can be interpreted depends mainly on two factors. First, it depends on how well the original form of the rock masses is preserved. If they are still recognisably a succession of tabular bodies, albeit now meta-sedimentary in nature, then the same geometric

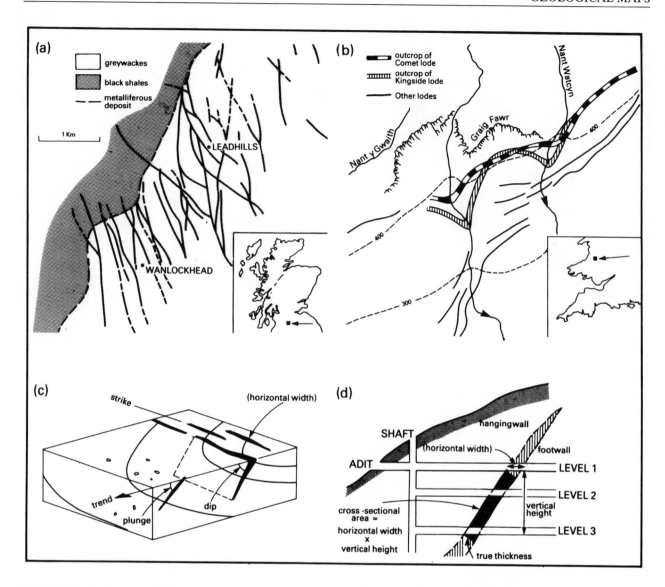

Figure 10.3 Some aspects of ore deposits on maps. (a) Narrow metalliferous deposits oblique to map units, therefore likely to be veins. Straight outcrops suggest steep dips. Note the localisation near a lithological junction, and a NNW–SSE pattern of orientation. Redrawn from Dunham et al. (1978), by permission of the Institute of Mining and Metallurgy. (b) Veins of differing orientations (lode is an old term for vein). Kingside lode makes tighter 'V'-shaped outcrop across valley than Comet lode, suggesting a shallower dip. Based on Jones (1922). (c) Similarity between orientation terminology for veins and other geological planes and lines. (d) Some mining terms commonly used on maps, and a simple method for assessing the cross-sectional area of a dipping vein.

principles as for non-metamorphosed sedimentary rocks apply. Thicknesses can be measured, outcrop patterns assessed, cross-sections drawn, etc. That the map units are of quartzite and schist rather than sandstone and shale makes no difference to the principles of mapwork. Similarly, if the original bulk form of meta-igneous rocks is still discernible, then the remarks of section 10.2 apply. Second, there is the complicating factor of deformation. Regional metamorphism is typically accompanied by ductile deformation (section 8.1), producing structures such as folds. The deformation can be intense, and it can hap-

pen repeatedly. Exceedingly complex outcrop patterns can result. Such structures on maps can be disentangled, but by methods which are beyond the scope of this book.

10.4 Mineral Deposits

Many kinds of mineral deposits are dealt with on maps by the methods already discussed. For example, gypsum beds, limestone for cement, and coal seams are effectively sedimentary rocks; slate and marble are metamorphic rocks; and many construction and road stones are igneous rocks. Metalliferous deposits, however, occur in a wide variety of forms, and it becomes particularly important in mapwork to establish any three-dimensional regularity of the body.

Disseminated and replacement ores tend to be highly irregular in shape and their manipulation on maps requires advanced techniques (e.g. see Badgley 1959, chapter 6). However, it may be useful to establish whether or not the ores have a particular areal or stratigraphic distribution, which maps and sections should make clear.

Few kinds of ore deposits will be very regular, but there may be an approximation to a shape that can be dealt with. A map should readily show if the masses are concordant with the host rocks, as in the **stratiform** deposits that are the source of much copper, lead, zinc, and iron. On the other hand, any discordance of the ore bodies can also provide useful information. Discordant tabular bodies, often called **veins**, usually have a more or less systematic distribution and orientation (figures 10.3a and 10.3b), and are commonly associated with faults. Some ores are concentrated at the meeting of different sets of faults. The commonness of the association with faults is illustrated by inclined veins having a hangingwall and a footwall (cf. section 9.2 on faults). A basic terminology for mapwork with veins is given in figures 10.3c and 10.3d. If the discordant mineral deposit is reasonably regular in form, it should be possible to make underground projections of its location, orientation, and shape, and perhaps even to make some rough volumetric estimates on ore reserves.

The kinds of maps with sufficient accuracy for useful commercial predictions are outside the scope of this book, but it is worthwhile mentioning several general points regarding maps of mineral deposits. Some ore-body maps are extremely large scale. In general, metrication is still the exception rather than the rule, and somewhat strange units such as furlongs and fathoms may be used. The topographic datum is unlikely to be sea level, but instead some prominent local feature, perhaps a mine entrance. Man-made features connected with mining operations, such as dumps and ponds, may obliterate much of the surface geology and dominate the map. This is usually not a difficulty in practice, because subsurface information may well be supplementing the surface data. In fact, many maps of mining areas omit the overall surface geology altogether and show the actual ore bodies as they appear underground, at one or more specified depths (e.g. figure 10.4).

Be careful when working with thicknesses of ore bodies on maps. The true thickness of an inclined vein is measured at right angles to its walls (figure 10.3d), analogous to a map unit (section 4.4.), and so a correction will have to be made to the outcrop width. However, mining geologists sometimes refer to the horizontal width of a vein as its thickness, as it is this value that is encountered during mining. Figure 10.3d also shows a simple graphical method of utilising the horizontal thickness of an inclined body to obtain its cross-sectional area. With knowledge of the extent of the body in the horizontal direction at right angles to the section, which will be measurable from the map, it becomes a simple matter to calculate the volume of the deposit. This would be a very rough value in terms of actual ore reserve, because other factors, particularly the percentage of usable material, would have to be taken into account to obtain a more meaningful figure.

Underground mineral deposits are usually worked by sinking a vertical or steeply inclined **shaft** which connects a series of tunnels. These are driven horizontally at a particular altitude, and are known as **levels** (figure 10.3d). Large-scale maps of areas with mineral deposits commonly show the locations of shafts and **adits**, where levels intersect with the land surface. Mine maps commonly show the geology of the ore body at each mine level. The tunnel walls may provide information in the vertical plane to help control the drawing of cross-sections. Mining geologists normally work with a whole series of cross-sections and maps for each of the underground levels, and so should have a closely controlled picture of the mineral deposit in three dimensions. Mine maps therefore differ somewhat from normal geological maps, but all the same geometric principles are employed in their use.

10.5 Superficial Deposits

Much of the bedrock at the surface of the earth is covered by recently formed loose debris, referred to as superficial deposits. Where the geological surveyor has judged these materials to be of geological significance, their distribution will have been plotted on the map. (Agricultural soil, normally formed in place from weathering of the directly underlying bedrock (section 13.3) is not normally shown.) Their nature will be explained in the map key, along with the bedrock units. Superficial deposits are conventionally placed at the top of the key, and variously labelled as superficial deposits, surface or surficial deposits, 'drift' (see below) or 'Quaternary'. They may or may not be shown as an integral part of the bedrock stratigraphic succession.

Superficial deposits are awkward to classify into map units. On some maps they are subdivided

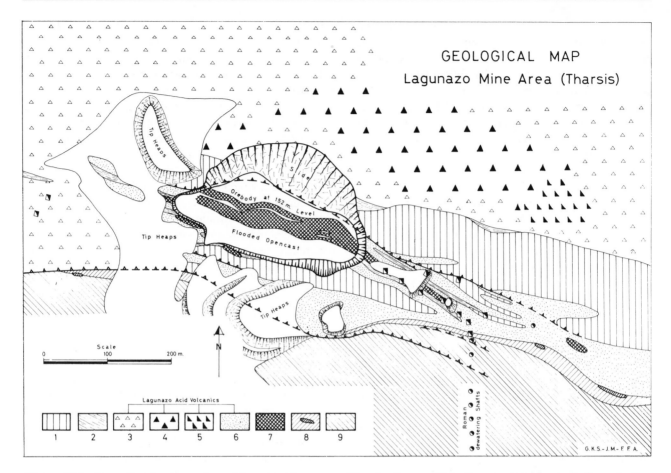

Figure 10.4 Example of a map of a mining area, Lagunazo, Tharsis, Greece. Note the amount of man-made and other superficial deposits obscuring the bedrock geology, and the ore body being shown as it appears subsurface. Reproduced from Strauss et al. (1977), by permission of Springer-Verlag, Berlin.

according to age, such as into the two series of the Quaternary systems – the Pleistocene (1.6 million to 10 000 years ago) and the Holocene (10 000 years ago until the present). Some others group the deposits according to what they are made of: their lithology. Most commonly, however, and this is where the main difficulties arise, the primary classification is based on how the deposits are thought to have formed, a sub-jective judgement on the part of the surveyor. Not only are some deposits open to various interpreta-tions, the genetic terms that are used can mean dif-ferent things to different surveyors. Some of the commonly used terms are discussed further below.

As indicated in section 1.3.2, some BGS maps, and especially the series at the 1:50 000 scale, are pub-lished in separate 'solid' and 'drift' editions. Some of the BGS 1:250 000 series are available in special Quaternary and sea-bed sediments editions. The lat-ter maps show the loose material that forms the sea-floor, classified according to grain size, that is, the proportions of gravel, sand, and mud. In some parts of the world where superficial deposits are widespread

and important, such as the Scandinavian countries and the northern states of the USA (e.g. Clayton and Attig 1993), purely Quaternary maps are common. The Geological Survey of Canada has long highlighted surficial geology mapping (Fulton 1993) and its maps include descriptive annotations with the key and sometimes detailed descriptive tables.

A particular idiosyncrasy arises with the term drift. In the early days of British geology the material at the earth's surface was thought to have 'drifted' into place during some immense flood, even – some believed – the biblical Flood of Noah. The deposits were there-fore referred to as 'drift'. (Other terms from this period of thinking such as 'diluvium' are still seen on some geological maps, especially ones issued in Ger-many.) With the growth of polar exploration in the early nineteenth century, floating ice became a more fashionable agent for having carried and then dumped the material, but the sediments were still called drift. Then, with the recognition in the middle of last cen-tury of the widespread, on-land, Pleistocene glacia-tion, the deposits became known as 'glacial drift'. This usage continues today in many English-speaking countries although very often with the glacial prefix dropped: superficial deposits may be referred to in the key simply as drift. This usually means deposits of glacial origin, but not always. On BGS maps drift is more or less synonymous with on-land superficial deposits and it may or may not involve material of glacial origin.

Superficial deposits fall into 10 genetic groups, listed below. Many of the types of lithology in each group will be familiar to you – sand, gravel, scree, and peat, for example – but the less common terms, in bold type, are briefly explained. Representative types of each group, including the symbols conventionally used on BGS maps, are shown in figure 10.5. There are other superficial deposits in addition to those listed below; their nature will normally be explained on the map key.

- Mass-movement deposits. These consist of materials that are transported downslope under the influence of gravity. Scree, more commonly called talus in the USA, fits into this category, together with **colluvium**. This is a mixture of fallen coarser fragments and finer material repeatedly washed down a hillslope. It is also called **hillwash** and – particularly by the BGS – **head**. Especially important in some applications of maps (section 12.5) is the portrayal of the masses of relatively sudden bulk movement known as landslides and (especially in the UK) landslips. Where the stratigraphy can still be mapped, the slipped material is referred to on BGS maps as **foundered strata**.
- Wind deposits. **Loess**, the silty sediment blown large distances, would be included in this group. Some maps distinguish wind-blown sand and some mark sand-dunes.
- Organic material. This group includes peat and shell sands.
- River deposits. Gravel, sand, silt, and clay can be deposited by rivers. They are collectively referred to, especially where they are fine-grained sediments such as silt and clay, as **alluvium**. Some maps indicate alluvial fans and different levels of river terraces that arose in earlier river conditions. They are called alluvial cones and flood-plain alluvium if the deposits are forming now.
- Lake deposits. These commonly comprise silt and clay and for them the term **lacustrine** is often used. Maps of arid areas may show **playa** deposits, formed in short-lived lakes.
- Marine deposits. Sediments produced by the sea are of very variable lithology. Moreover, they can be modern deposits or the result of older events. Different environments are distinguished on some maps, such as tidal, mudflat, swamp, beach, etc.
- Interglacial deposits. These comprise Pleistocene materials known to have formed between periods of ice presence, irrespective of whether they formed in a river, marine, or other setting.
- Glacial deposits. Three kinds of material dominate this group: sediments from streams formed in direct association with ice sheets – **glaciofluvial deposits**; the jumble of material dropped at the margins of the ice – **moraine**; and **glacial till**. The last term, often simply called **till** and in Britain sometimes called **boulder clay** to emphasise the large fragments distributed within it, is the unstratified, unsorted material that forms directly from entrainment within ice. It may be prefixed to indicate a special mode of formation, such as lodgement till, or a material that dominates the lithology, as in chalky till.
- Spring deposits. **Tufa**, calcium carbonate precipitated from groundwater, occurs in some places in sufficient amounts to be shown on maps. Similar deposits from hot springs are marked as **travertine**.
- Artificial deposits. Into this group fall those areas where human activity has significantly altered the earth's surface. Some BGS maps show **made ground**, which includes spoil heaps associated with

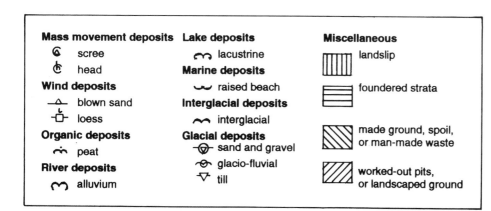

Figure 10.5
Representative types of superficial deposits, showing the symbols conventionally employed on maps of the BGS.

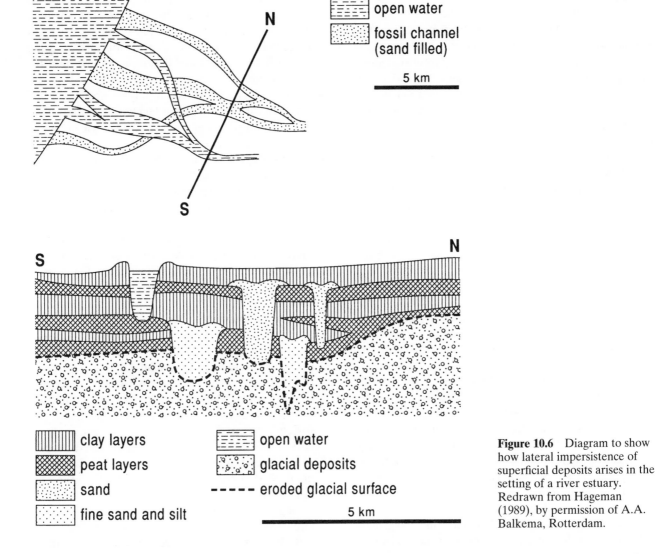

Figure 10.6 Diagram to show how lateral impersistence of superficial deposits arises in the setting of a river estuary. Redrawn from Hageman (1989), by permission of A.A. Balkema, Rotterdam.

mining and other man-made accumulations of waste, and **worked ground**, where excavations such as opencast pits have been infilled. **Landscaped ground** indicates extensive remodelling of the land surface for one reason or another.

A further group of features is included on some maps:

- Morphological features. Some maps indicate local physical features associated with the superficial deposits, and this is particularly true of glaciated upland areas. Such things as meltwater channels, glacial striations, and drumlins may be noted by various symbols: they will be explained in the key.

A number of maps in this book show examples of superficial deposits. One instance is Plate 3, which shows various mass-movement, wind-blown, and river deposits, and distinguishes between deposits of river terraces at three different levels. Figure 10.4 shows

superficial deposits resulting from mining. It has been emphasised from the outset that geological maps are themselves interpretive – formation boundaries, for example, cannot normally be seen on the ground. This is particularly true of the bedrock shown on maps, because the boundaries are commonly covered by superficial deposits, but it is also true of the superficial deposits themselves. In many parts of the world they are covered by agricultural soil or by concrete and other man-made devices, and the surveyor has to interpret their extent also.

When it comes to interpreting from maps the three-dimensional arrangement of superficial deposits we use the same principles as for sedimentary bedrock formations, but there is a problem. *A characteristic of most superficial deposits is their marked lateral variability and inconsistency.* A particular unit may well be very localised in its distribution and it may vary greatly in thickness, which means that extrapolating

Figure 10.7 Simplified geological map and cross-section of Schouwen-Duiveland, an island in the Rhine estuary, Netherlands, to illustrate the lateral variability of superficial deposits. Redrawn from Hageman (1989), by permission of A.A. Balkema, Rotterdam.

units to depth, contouring surfaces, etc., will be greatly hindered. This difference from sedimentary formations in the bedrock comes about for three main reasons.

First, most of the superficial deposits shown on maps were formed on land, where the surface is uneven and the depositional processes tend to be localised. Many bedrock sedimentary formations, in contrast, formed on the relatively flat sea-floor where the deposits tend to blanket large areas, giving greater uniformity. Few of the sediments formed below the sea during the Quaternary have so far been uplifted.

Second, because superficial deposits formed relatively recently and are easily accessible, they tend to be subdivided with a precision not attempted for bedrock formations. Even a single advance or retreat of an ice sheet can give rise to an intricate interleaving of tills and glaciofluvial deposits that may well be portrayed separately. Even though bedrock formations may contain variations they tend to be included within map units rather than shown separately.

Third, most on-land deposits are loose and weak, and are soon exposed to weathering and erosion, much more so than submarine sediments. In fact, during the Quaternary much of the northern European and North American landscape underwent particularly vigorous erosion because of the widespread glacial and periglacial (near-ice) conditions. Moreover, where a change in conditions caused such erosion to affect newly deposited material it was made even more vulnerable by the sparsity of vegetation. Thus, a particular localised deposit was likely to be reduced in extent still further. Take loess, for example. This wind-blown dust can blanket large areas of land and be relatively sheet-like in form (though its base may be uneven if it was laid down on an uneven topography), but it is weak and very prone to erosion. So despite relatively widespread and consistent deposition for an on-land sediment, it is typically preserved only in irregular masses, and appears in sporadic patches on maps.

An illustration of the practical problems arising from the lateral variability of superficial deposits is given in figures 10.6, 10.7 and 10.8. Figure 10.6 depicts how local impersistencies arise in the setting of a river estuary and figure 10.7 gives an example from the Netherlands. Detection of ancient river channels such as those shown in figure 10.7 being present below surface is important not only for deciphering the past history of the river system but also the siting of construction. The routing of motorways and pipelines, for example, will have to take into account such irregularities in the foundation conditions. Hageman (1989) recounts how an excavation for a new pumping station in the west of the Netherlands

was planned on the basis of the consistent layering shown in figure 10.8a. The possibility of sand-filled channels being present, such as depicted in figure 10.8b, was not considered until excavation punctured one, leading to the violent release of pressurised water and expensive remedial work (figure 10.8c). Such lateral impersistencies arise not only in estuaries but in most of the on-land situations where sediment is deposited.

Clearly, therefore, such considerations will have to be borne in mind when we are working with superficial deposits in the environmental planning context. Chapter 12 explores this problem further, such as in assessing aggregates for construction (section 12.3) and near-surface foundation conditions (section 12.4). And we have to remember the special character of superficial deposits as we turn now to a new and different aspect of geological map interpretation: that of reconstructing geological histories from maps.

Summary of Chapter

1. Volcaniclastic rocks are commonly presented on maps in the same way as sedimentary units, though the extent of possible three-dimensional interpretation is much more variable.
2. The distributions of magmatic bodies show well on maps.
3. Tabular masses of magmatic rocks such as lava flows, sills, and dykes can be analysed using the same three-dimensional principles as for sedimentary rocks, but the methods are of limited application to irregular bodies.
4. Maps are useful for showing the areal extent of thermal metamorphism, but representation of regional metamorphism requires special treatment.
5. Metalliferous mineral deposits can be stratiform, and treated like sedimentary units, or irregular in form.
6. Some three-dimensional interpretation of mineral veins may be possible, and their relationships with faults discerned.
7. Large-scale mineral deposit maps may employ special techniques of representation.
8. Superficial deposits, the loose geological debris at the earth's surface, are interpreted using the same principles as bedrock formations but there are complications.
9. Local timing relationships in superficial deposits can be subtle and three-dimensional interpretations are complicated by the characteristic marked lateral variation and impersistence of the units.

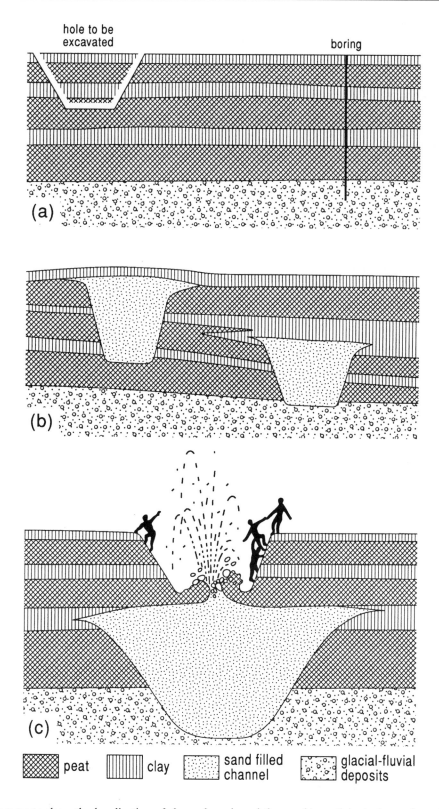

Figure 10.8   Diagrams to show the localisation of channel sands and the need to anticipate them when planning engineering work. (a) A borehole away from the proposed excavation site can provide the required depth information if the sequence is regularly stratified, unlikely in a fluvial setting. (b) Conceptual diagram of a more likely subsurface arrangement, indicating the need for careful pre-excavation mapping. (c) Inaccurate planning can lead to the excavation tapping localised, water-rich channel sands. Redrawn from Hageman (1989), by permission of A.A. Balkema, Rotterdam.

Selected Further Reading

Catt, J.A., 1988. *Quaternary Geology for Scientists and Engineers*. Ellis Horwood, Chichester, 340 p.

Chapter 6 deals with Quaternary features on maps.

Compton, R.R., 1985. *Geology in the Field*. John Wiley, New York, 398 p.

Includes chapters on volcanic, plutonic, and metamorphic rocks, parts of which are relevant to mapwork.

Roberts, J.L., 1982. *Introduction to Geological Maps and Structures*. Pergamon, Oxford, 332 p.

Chapter 6 includes a discussion on the portrayal of igneous rocks on maps.

Warren, W.P. and Horton, A., 1991. Mapping glacial deposits in Britain and Ireland. In: Ehlers, J., Gibbard, P.L. and Rose, J. (editors), *Glacial Deposits in Britain and Ireland*. Balkema, Rotterdam.

Pages 379–387 trace how the topic of the chapter title has evolved.

Map 20   Honister Slate Mine, Cumbria, England

The map opposite is of the underground workings which until very recently were used to extract the Ordovician Honister Slate for roofing material. The map, adapted and redrawn, is used by permission of Sir Alfred MacAlpine plc.

Four different levels are shown, connected by a NW–SE incline. The elevations of the floors of the levels are indicated, in metres above a local datum. The levels have worked along the base of the Honister Slate, which has a reasonably consistent orientation and a remarkably constant thickness of 10 metres. Construct structure contours for the top of the Honister Slate.

What is the overall strike and dip of the unit?

Construct an accurate cross-section to show the true dip of the unit and the location of the different levels.

Consider the following, largely hypothetical, mining programme. It is proposed to open a new level SE of level 3, to be called no. 2 level. There has to be at least 10 m horizontal distance from other levels to avoid the risk of roof collapse. Indicate on the section and map your proposed location and route for level 2.

What is the disadvantage of continuing level 4 in the SSW direction being followed when it was terminated? What might be a better direction in which to continue this level?

No. 5 level was stopped when it met fault rocks and the slate unit was absent. If the level had been continued past the zone of fault rock, would it have met rocks above or below the Honister Slate? Indicate on the map the direction in which the level should be driven in order for it to retrieve the slate unit.

Indicate on the map, on the basis of the locations of the fault rock and any displacements of the structure contours, where the fault trace is likely to run. What is the stratigraphic throw of the fault?

In the south branch of no. 6 level, the fault can be seen to dip at 70° to the east. Draw a cross-section across the fault at this point, in the true dip direction of the fault, and indicate the top of the slate unit either side of the fault. What is the throw of the fault? Why is it that the value differs slightly from the stratigraphic throw?

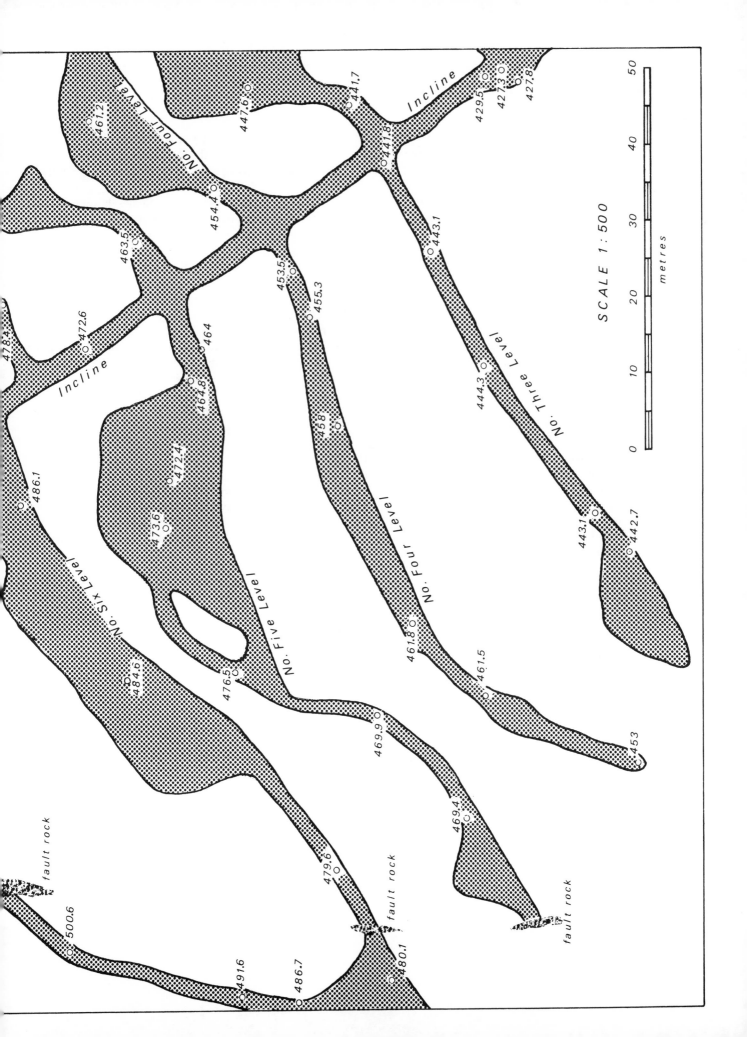

No. Four Level

Incline

461.2

447.6

441.7

429.5

427.3

427.8

454.4

441.8

463.5

443.1

472.6

478.4

443.1

442.7

Incline

464

455.3

453.5

444.3

No. Three Level

464.8

458

486.1

472.4

473.6

No. Six Level

No. Four Level

461.8

484.6

476.5

No. Five Level

461.5

469.9

453

fault rock

500.6

469.4

491.6

479.6

fault rock

486.7

fault rock

480.1

SCALE 1 : 500

0   10   20   30   40   50

metres

Map 21   Gwynfynydd gold mine, North Wales

This map, adapted by permission from the unpublished surveying of John Ashton, PhD, is of the Gwynfynydd mine, 4 km north of Dolgellau, North Wales. A number of gold-bearing quartz veins are sufficiently rich in gold to have tempted miners over many centuries and, by tradition, to have provided the material for numerous Royal wedding rings. The veins, given individual names such as Chidlaw, Collett, Main, and New veins, commonly occur along faults, which displace the host Cambrian sedimentary rocks.

The map opposite shows the workings of the Chidlaw Vein at three different levels: Nos. 1, 2, and 6. Good potential control on the three-dimensional geology is provided by the information recorded at these three different elevations.

Construct an accurate N–S cross-section across the eastern part of the map to incorporate the information at all three levels. Project the Chidlaw vein between the levels. What is the overall orientation of the vein? (Note that a N–S section is not necessarily in the true dip direction of the vein.)

Indicate on the section the location of the Trawsfynydd Fault at No. 6 level. Project the course of the fault up to No. 2 level. (Note that the cross-section

is oblique to the fault.) How well does the projected location compare with that inferred from level 2 on the map, i.e. how planar is the Trawsfynydd Fault?

If the Gamlan Flags–Clogau Shales boundary were projected onto the cross-section, what approximate orientation would its trace show?

Extrapolate on the map the Gamlan–Clogau junction northwards from No. 2 level to where it should occur at No. 6 level. (Remember that the levels are at different elevations. By analogy with the trigonometric method for finding the depth to a dipping unit knowing the horizontal distance and dip angle (section 4.5 and figure 4.10a), here, knowing the dip angle and elevation difference, the horizontal distance can be calculated.) Any discrepancy between the predicted location of the junction and that shown on the map can presumably be accounted for by a fault, seeing as the geology either side of the parts of level 6 fails to correspond.

Using the predicted and map locations of the junction and its orientation, and assuming that the known orientation of the Chidlaw vein represents that of the fault, find (a) the stratigraphic throw, and (b) the downthrow shown by the fault.

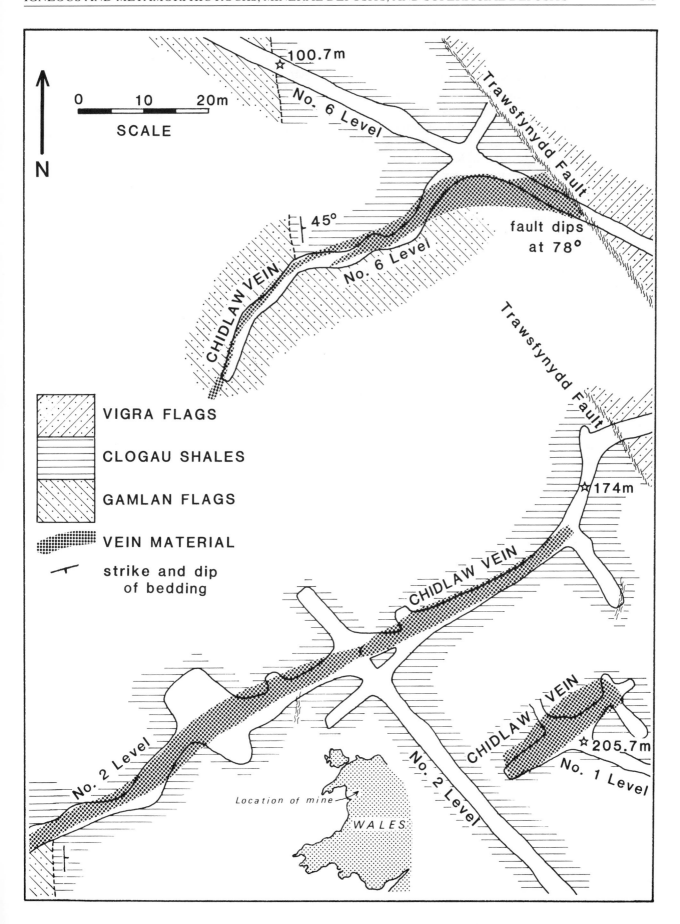

Map 22 Woore Moraine, Cheshire, England

The superficial deposits shown on Map 22 occur in the Cheshire Plain, between Stoke-on-Trent and the Welsh Borders, and include part of the Woore glacial moraine. The map and associated information is based on the work of Moseley (1979), with the author's permission. The bedrock of Keuper Marl is Triassic in age and falls within unit 90 seen in the extreme NE of Plate 1, though the area of Map 22 is some 20 km further to the north. Topographic contours are shown, in metres. Thicknesses recorded from four boreholes are listed in the box (values given in metres). Both the map and the borehole data illustrate the intricate sequences and rapid lateral variations characteristic of superficial deposits, and the difficulties of distinguishing between similar deposits with the precision expected for these young materials. The map key shows lithologies only: there could be more than one deposit of each type of material.

Study the borehole data carefully, especially the relatively complete sequence seen in borehole A. Note that because the shellfish *Corbicula* lives in fairly warm freshwater, the clay that contains it is likely to represent an interglacial period. Try to visualise the three-dimensional arrangement of these materials. Sketch a cross-section, say NW–SE or W–E, to show how the intervals seen in the boreholes might correlate and how they might be arranged.

Where is the thickest development of sand that remains to be exploited?

BOREHOLE A surface altitude: 60 m OD		**BOREHOLE B** surface altitude: 87 m OD	
peat	3.3	till	37
plastic clay	2.6	Keuper Marl	
till	3.3		
sand	6.5		
Corbicula clay	2		
sand	2.6		
till	16.7		
Keuper Marl			

BOREHOLE C surface altitude: 65 m OD		**BOREHOLE D** surface altitude: 88 m OD	
sand	7.3	till	3.2
till	23.7	sand	1.8
Keuper Marl		till	3
		sand	

Further examples of non-sedimentary materials on maps

Describe the form of: the banded rhyolite mass on Map 6, of the Boyd volcanics, Australia; the base of the quartz porphyry that appears on Map 17, of Aspen, Colorado, USA; and the various igneous bodies that are present on Map 19, of the Hamblin Bay Fault, Nevada, USA.

Note that igneous and metamorphic rocks and a metalliferous vein appear on Plate 5, Sanquar, Scotland. Plate 7, of Marraba, Australia, and the questions given there, are very largely concerned with igneous and metamorphic rocks.

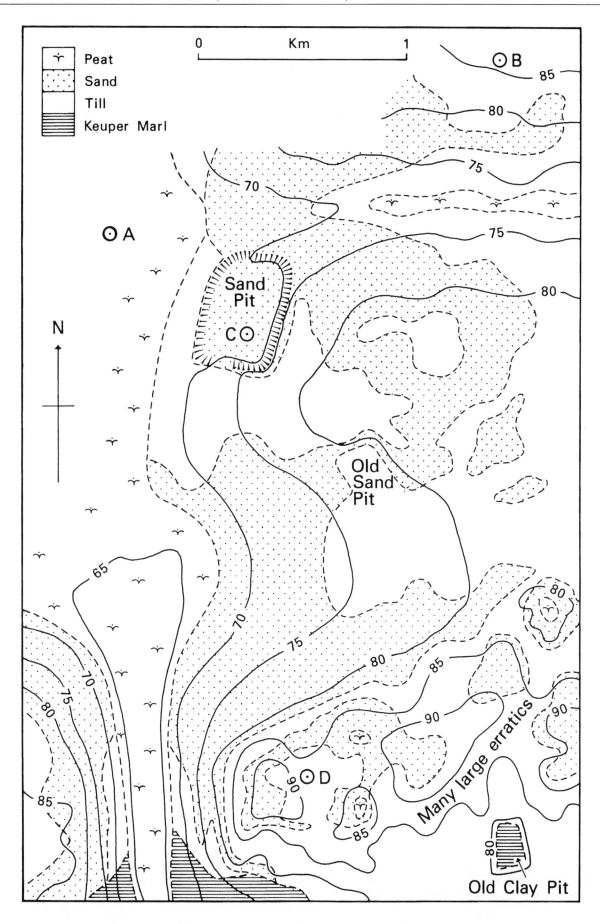

CHAPTER 11

Geological Histories from Maps

11.1 Introduction

The previous chapters have developed methods for the three-dimensional interpretation of maps. We now turn to looking at maps in a further dimension – the geological past. For the outcrop patterns on maps can reveal much about how the formations evolved through time; they enable interpretations to be made on the geological history of the map area, even the reconstruction of past geological environments (figure 11.1). The timings that are derived are usually relative, although if the map key gives time-stratigraphic information it may be possible to assign absolute dates to events.

This chapter looks in turn at the kinds of interpretations that can be made from sedimentary successions, deformed rocks, non-sedimentary materials, and superficial deposits, but interpreting the geological past typically involves dealing with the interaction of a variety of geological processes. This is one reason why geological maps are so central to geology.

Reconstruction of the geological evolution of a map area is of great academic importance, for surely the understanding of how an area became that way must be one of the fundamental objectives of geology, but it can be of applied significance too. If the geological circumstances which prompted the formation of some material of commercial interest can be deduced, it becomes easier to discover further occurrences. The petroleum industry, for example, makes much use of reconstructing past sedimentary environments, to help focus on those areas where oil might be found. Knowledge of how superficial deposits were laid down enables better predictions of their form and extent.

It will become apparent that often the history of features seen on the map can be interpreted in more than one way. This is because we are dealing with map *interpretation*. The geologist has to weigh in his or her mind the evidence seen on the map, and decide his or her preferred interpretation. There is no right or wrong solution available to us. Even experienced geologists may disagree on the most likely interpretation. However, nature tends to do things in an uncomplicated manner. You should aim for the interpretation which explains most features in the simplest way.

11.2 Sedimentary Rocks

The idea that sediments are laid down on top of each other, forming a pile with successively younger deposits higher up, seems almost self-evident to us now, although it does appear to have dawned only about two hundred years ago. Because of it, we are in a position to work out the relative ages of a series of

Figure 11.1 (opposite) Examples of the interpretation of geological history from maps. (a) Annotated map of Republic, Michigan, USA. (b) Sedimentation and folding perhaps related to faulting. Localisation of coal measures (d^{c1}–d^{c3}) adjacent to fault, restriction of folding to Carboniferous rocks, and increased dips towards fault all suggest folds are local responses to fault movement, perhaps synchronous with sedimentation. Near Alloa, Scotland. Redrawn from the BGS 1:50 000 Stirling sheet, by permission of the Director, British Geological Survey. Copyright NERC. All rights reserved. (c) Lenses of limestone pebble conglomerate within (now metamorphosed) mudrocks may represent distributary channels on a submarine fan. Lake Bomoseen, Vermont, USA. (d) Suite of granite plutons post-dates folding, a common sequence in orogenic belts. Hill End, Australia. Based on Powell et al. (1976). (e) Variations in outcrop width of 'Smiddy Ganister' sandstone are not paralleled by other units, therefore probably represent primary thickness variation, perhaps reflecting deposition in channels. Moor House, North Pennines, England. Based on Johnson and Dunham (1963).

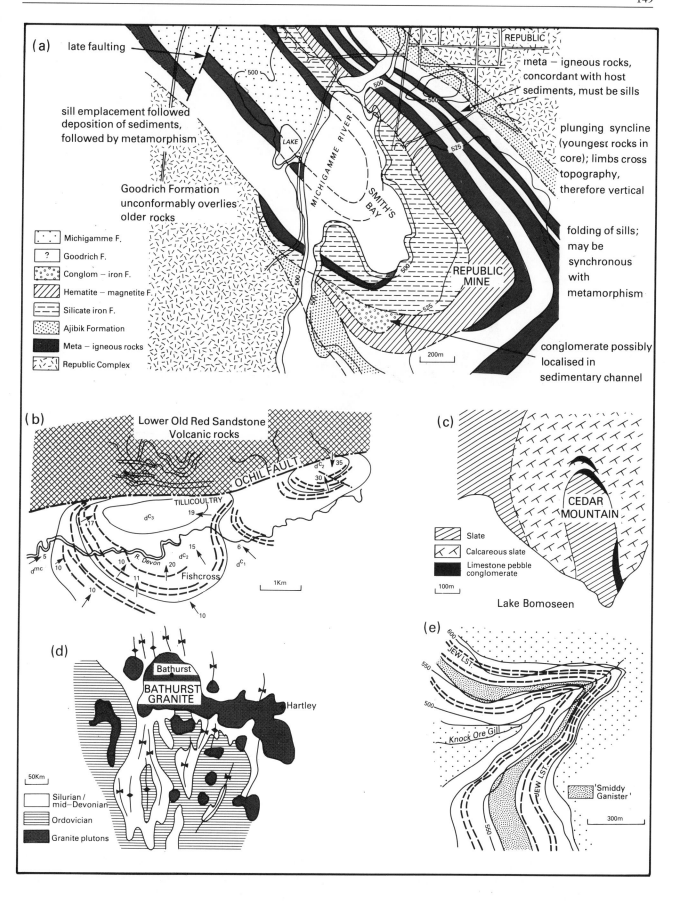

(a)

late faulting

sill emplacement followed
deposition of sediments,
followed by metamorphism

Goodrich Formation
unconformably overlies
older rocks

REPUBLIC

meta – igneous rocks,
concordant with host
sediments, must be sills

plunging syncline
(youngest rocks in
core); limbs cross
topography,
therefore vertical

folding of sills;
may be
synchronous
with
metamorphism

conglomerate possibly
localised in
sedimentary channel

MICHIGAMME RIVER

LAKE

SMITH'S BAY

REPUBLIC MINE

200m

Legend:
- Michigamme F.
- ? Goodrich F.
- Conglom – iron F.
- Hematite – magnetite F.
- Silicate iron F.
- Ajibik Formation
- Meta – igneous rocks
- Republic Complex

(b)

Lower Old Red Sandstone
Volcanic rocks

OCHIL FAULT

TILLICOULTRY

d^{c_2} 35
30

d^{c_3} 19

17

d^{mc} 5

10

10   R. Devon   20 d^{c_2} 15

11   Fishcross   6 d^{c_1}

10

10

1Km

(c)

CEDAR
MOUNTAIN

Legend:
- Slate
- Calcareous slate
- Limestone pebble conglomerate

100m

Lake Bomoseen

(d)

Bathurst
BATHURST
GRANITE

Hartley

50Km

Legend:
- Silurian / mid–Devonian
- Ordovician
- Granite plutons

(e)

600
JEW LST
550
500

Knock Ore Gill

JEW LST

550

'Smiddy
Ganister'

300m

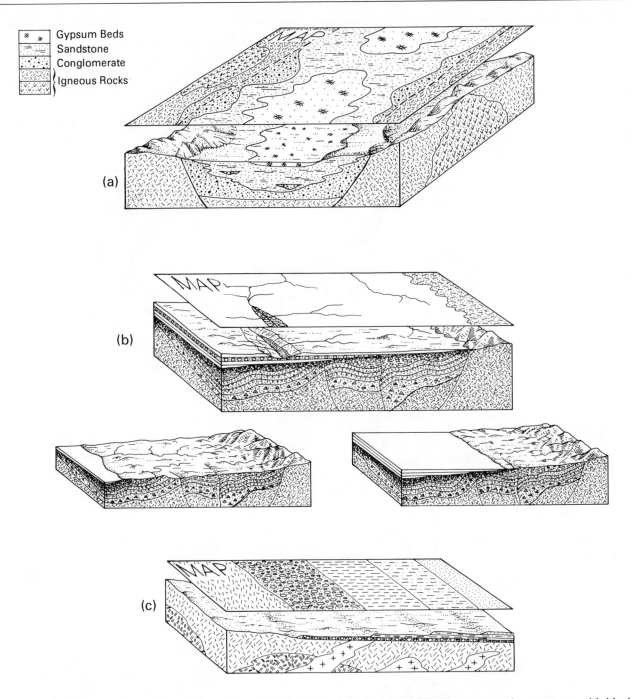

Figure 11.2 Aspects of geological history that may be interpreted from maps. (a) Sedimentary rocks on a map, with block diagram showing an interpretation of the depositional environment as a saline subaerial basin with alluvial fans. (b) Map showing unconformity with overlap (see section 7.4). The white unit extends further than the unit with pebbly ornament (circles), which is only revealed in present-day river valleys. Block diagrams show interpretation as a marine transgression. As the sea encroaches further, successively more extensive sediments are deposited. (c) Unconformity, with possible offlap. The map distribution of the sedimentary rocks could be due to present-day erosion fortuitously acting on once equally extensive units, but probably represents a marine regression. As the sea recedes from the land, successive deposits become less extensive.

sedimentary units shown on a map. If the beds are flat, those units that are topographically higher will be younger, and dipping beds become younger in the dip direction. The sediments above an unconformity must be younger than the rocks below.

However, the above statements assume that the succession is 'right way up'. Rock successions can be tilted past the vertical, leading to inversion of the sequence. This happens particularly in areas of much folding, where entire fold limbs can become overturned. If the map shows signs of intense folding, or if the area is known to be within an orogenic belt, watch out for inverted successions. The map may indicate the **younging direction** of the formations by a symbol (figure 4.2) though examples are outside the scope of this book.

Where units grade laterally or intertongue with each other, they must have formed at the same time. Such relationships may be indicated on the map key. Sedimentary units may vary in thickness across the map area. Make sure that this is a real variation, and not an effect of topography or bed dip (section 6.3), or of faulting. Thickness characteristics can give information on the environment in which the sediments were deposited. For example, conspicuous fluctuations may indicate deposition on a highly uneven floor or in channels (figures 11.1a and 11.1c), whereas consistent thicknesses reflect stable sedimentation.

Knowledge of how different associations of sedimentary rocks come about can lead to reconstructions of the sedimentary environments (e.g. figure 11.2a). If an unconformity is present, analysis of the relations between the formations may prompt interpretations of the geological history of the area (figures 11.2b and 11.2c; section 7.4). The relationships between sediments and faults can be informative, especially where a unit is not simply displaced by the fault but changes character across it. For example, sediment can be thicker on the downthrow side of the fault, which implies that the fault was forming at the same time as sedimentation. Incoming sediment accumulates preferentially in the downthrown depression, against the fresh fault scarp (figure 11.1b). Some sedimentary units, and especially sands forming on the front of a river delta, are present on both sides of the fault but with differing thicknesses and amounts of displacement. The interpretation here is that the fault was moving at the same time that the sediments were accumulating. The feature is termed a growth fault (section 9.8).

The patterns of the lines on isopach maps (section 4.4) are much used in the oil industry to derive interpretations on sedimentary environments and possible accumulations of hydrocarbons. Areas of thinning may indicate features of positive relief during sedimentation, and thickening reflects negative areas and increased amounts of deposition (figure 11.3). A concentric bull's-eye pattern with increasing thicknesses towards the centre shows a basin, with the thickest isopachs depicting the deepest parts. Elongate, closely spaced isopachs indicate deep, linear troughs; those that wander somewhat randomly, broadly spaced, may denote a stable shelf environment. Abrupt changes in isopach spacing can reveal hidden faults, especially growth faults. Indeed, one great value of

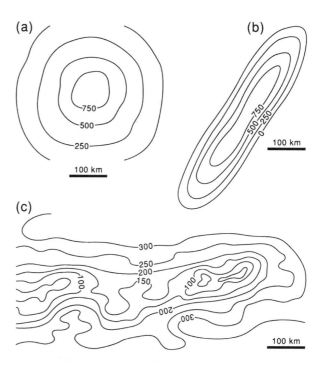

Figure 11.3 Examples of patterns of isopachs (units in metres) indicating different depositional settings. (a) Concentric thickening towards a centre, implying a circular basin of deposition. (b) Elongate, closely spaced isopachs, indicating a deep, linear, depositional trough. (c) Spaced, irregular isopachs, indicative of a broad shelf environment.

isopach maps is that they can reveal and help locate features that are not apparent from other kinds of map. If isopachs are available for more than one formation, and the stratigraphic ages are known, it becomes possible to bring in the time dimension, and deduce from the changes of thickness with time the rates of processes such as subsidence and uplift.

11.3 Deformed Rocks

The detection and description of folds and faults on maps were discussed in sections 8.3, and 9.5. However, it may also be possible to make interpretations on the *timing* of the deformation. The first thing that can be said from a map is that the material must have been there before the deformation. That is, the folding or faulting must post-date the age of the rocks. Quite how much younger is a difficult matter. Deformation can occur soon after sediments were deposited, or long after they have been turned into rocks; the distinction would be virtually impossible to detect from a map. The duration of the stratigraphic divisions of most maps allows the formation of the rocks and their deformation to be accomplished within the same stratigraphic interval. Thus it is certainly possible to have, say, rocks shown on the map as middle Permian in age deformed by a middle Permian fault. Of course, if rocks younger than Permian are affected by the same structure, then the deformation, too, has to be younger.

This key principle for interpreting timings from maps – cross-cutting relationships – was introduced in section 2.3.5. If, continuing with the above example, beds of Jurassic age pass across the fault without displacement, then the fault movement had ended before Jurassic times. Although the principle is applicable to sedimentary rocks – an angular unconformity, for example, is a cross-cutting relationship – it is especially useful in interpreting igneous and metamorphic rocks and deformation sequences.

If one fault displaces another fault, then the former is the younger structure. A folded fault means that the fold is the younger structure. If, however, a fault which displaces a series of beds is folded, but to a lesser degree than the beds, it means that the faulting took place *during* the folding. Sometimes the relative timing has to be deduced through indirect logic. Suppose two separate igneous intrusions are known from the key to be of the same age. If one of them intrudes beds which overlie a fault, and the other is faulted, there must be two ages of faulting. Cross-cutting relationships are applied to map interpretation in endless permutations.

Regarding cross-cutting faults, there can be ambiguities. Make sure that the relationship is a sharp truncation. If the faults simply merge, they could have formed synchronously. If a unit is faulted but the adjacent one is not, this *may* indicate a depositional break (disconformity), in that there was enough time for faulting of the lowermost unit before formation of the upper one. On the other hand faults do have to end somewhere. Movement along a fault is greatest along its central part, and decreases outwards such that the fault tip may coincide with the boundary of the unit. Support for this interpretation may be provided by variable displacement along the fault. Many of the same considerations apply to folds. Also, an area, especially one within an orogenic belt, can be subjected to more than one period of folding, although analysis of this situation is beyond this book.

If we are interpreting geological histories from a map that shows folds and faults, we naturally want to assign stratigraphic ages to their production. Cross-cutting relationships and reference to stratigraphic ages in the key may allow this, but only to the extent that the relations are present on the map. Too often the fault of interest does not reach the rocks of critical age, or disappears off the edge of the map! Another separate structure that looks similar may provide the evidence. The question then arises as to what extent different structures can be grouped together as being of the same age. In general, groups of faults are likely to be different in age from groups of folds because the conditions during rock deformation tend to favour either ductile or brittle structures (section 8.1). There are, however, many exceptions to this, for example, the hangingwall anticlines related to thrusts (section 9.7.1) and the roll-over anticlines resulting from extension faults (section 9.8.1).

But how are groups of synchronous structures recognised? Folds of the same age tend to have similar characteristics. The axial surfaces are likely to be roughly parallel, and the fold attitudes and styles alike (section 8.2). Fold scale depends primarily on bed thickness, so is a less direct guide. If, for example, the map reveals a number of NE–SW-trending, reclined to recumbent, moderately plunging non-cylindrical folds, they may well be of similar age. If folds are also present which are, say, E–W trending, upright cylindrical folds, it is likely that these structures formed in a different period of deformation. The relative ages of the two periods of folding may be discernible from the stratigraphic ages of the rocks affected. Similar considerations apply to faulting, in that one *kind* of fault tends to dominate at any one time, but there are many factors which can complicate these generalisations.

11.4 Non-Sedimentary Rocks

Detailed analysis of the genesis of non-sedimentary rocks is not usually possible on standard geological maps, but some interpretation might be possible. As discussed in section 10.2.1, investigation of volcaniclastic rocks has to take into account both the sedimentary and igneous processes that are involved. A map of a volcanic centre may prompt inferences in the way the ashes and tuffs accumulated, for example, a large *volume* of volcaniclastic material reflects major or protracted eruption whereas a large *area* covered by the material indicates energetic, probably violent explosion. But it may be difficult to establish these two amounts from a map, not only because the bulk form of the ejected material could be highly irregular, but also because such deposits are vulnerable to erosion. The type of volcanic rock may give clues to the environment of formation, e.g. pillow lavas indicate sub-aqueous, usually marine, eruption. The rocks may be cross-cut or inter-leaved with magmatic bodies, prompting interpretations of how the volcanic activity progressed through time. Figure 11.4 provides an example.

The thing that igneous rocks lend themselves to on maps is the relative dating of emplacement. An intrusive igneous body has to be younger than the rocks it has intruded, and any deformation structures

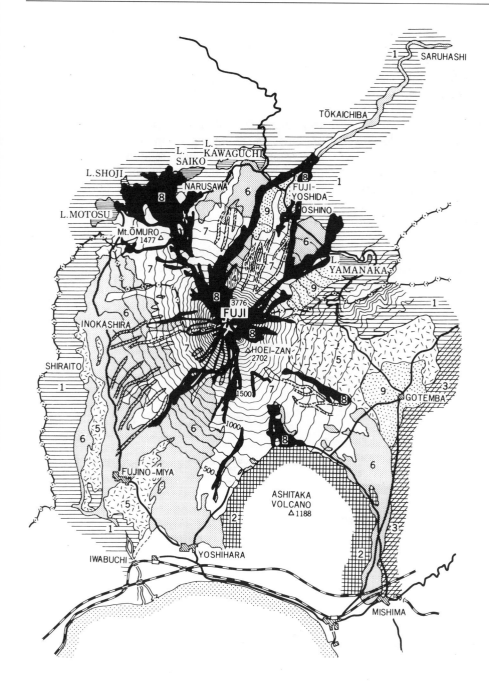

Figure 11.4 Geological map of the Fuji volcano, Japan, based on Tsuya et al. (1988). The graceful conical form of Mt Fuji belies a complex volcanic history, some of which is evident even from this simplified map. The youngest units (9, ash and alluvial fan deposits and 8, lava flows) represent the most recent eruptions and cross-cut the ejecta of unit 7, of which much of the modern cone is built. This cross-cuts the older ejecta of unit 6. Unit 5 consists of still older ejecta, though the map relations are ambiguous. All the above units are centred on the present Fuji. Units 4, 3, and 2, however, are centred on earlier volcanic centres; the Ashitaka deposits (unit 2) show this clearly. Unit 1 consists of Tertiary formations on which the long-lived volcanic complex was constructed. Reproduced by permission of the Geological Survey of Japan, Agency of Industrial Science and Technology.

it cross-cuts (figure 11.1d). Conversely, it must be older than any features by which it is cross-cut. Any deposits which overlap onto the igneous body will also give a minimum age for emplacement. If the map covers an igneous complex, it may be possible to erect a detailed scenario of intrusive events. If something about the composition of the rocks is known from the key, it may be reasonable to speculate on the broad features of magmatic evolution.

Nevertheless, these methods may give a large stratigraphic time interval, and it may not be possible to determine exactly when the intrusion took place. For the same reason the field surveyor may have been

unable to pinpoint the stratigraphic age of the intrusive masses. Therefore, it is normal for igneous rocks to be shown on the map key with little or no indication of age, and to be listed separately from sedimentary, volcaniclastic, and extrusive igneous rocks. It then falls to the reader to deduce what he or she can from the map about the age of the bodies.

Metamorphism of sedimentary rocks will usually have taken place long after deposition of the sediments, to allow time for their burial. Knowledge of the metamorphic conditions may allow some speculation on the time interval, but rarely can this be specified from maps alone. The map key may provide some information, but usually cross-cutting relationships are the only direct guide, i.e. metamorphic rocks being overlain by non-metamorphosed rocks. Igneous intrusions, assuming they formed at some depth, can be subjected to metamorphism at any time subsequent to their emplacement. It is common during regional metamorphism for the rocks also to be subjected to ductile deformation (e.g. figure 11.1e). Therefore, if folds are present on the map, they may well be synchronous with any regional metamorphism. If thermal metamorphism is indicated on the map note that the igneous mass which caused it may be partly or even completely buried. The present outcrop does not necessarily reflect the size and shape of the heat source at depth.

Regarding mineral deposits, the rocks in which they are found should disclose the overall nature of the mineralisation, whether they are sedimentary deposits, for example, or related to magmatism. But a general geological map is unlikely to reveal much about the conditions of mineralisation.

11.5   Superficial Deposits

Deriving geological histories from superficial deposits on maps is analogous to interpreting their three-dimensional arrangement (section 10.5). While the principles remain the same as for sedimentary bedrock formations, their unusual nature presents special difficulties. As discussed in section 10.5, the problems arise because most superficial deposits that appear on geological maps were laid down locally on land and tend to be weak materials, and hence are vulnerable to a range of weathering and erosion processes. And we attempt to interpret all these interacting processes with much greater precision than for ancient eras.

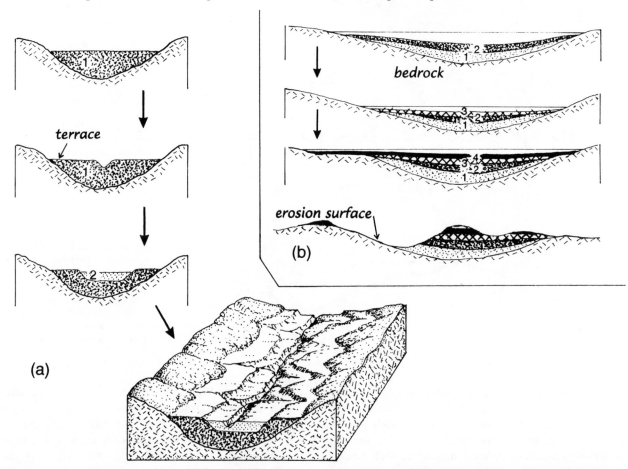

Figure 11.5 Diagrammatic sketches of some complexities in interpreting geological histories from superficial deposits. (a) The formation of two pairs of river terraces, the *older* pair being physically *higher* than the younger ones. Deposition of an alluvial flood-plain (1) is followed by a period of downcutting. Formation of a new, lower, flood-plain (2) is again followed by downcutting, leaving the river terraces shown in the block diagram. (b) Interrelationships between lake deposits in response to fluctuating water levels (with vertical exaggeration). Subsequent disappearance of the lake followed by erosion (bottom sketch) reveals different sequences in different places. To the left of the sketch, deposit 4 rests directly on bedrock, in the centre the sequence is complete, to the right deposit 3 rests on bedrock.

Sorting the deposits into their relative ages is a case in point. In principle, just as with sedimentary rocks, younger materials are deposited on top of older ones. But downslope mass-movements, for example, can transfer young sediments downhill to anomalously low positions. River terraces have to be interpreted carefully. Terraces commonly form by the upbuilding and spreading out of alluvium, followed by downcutting. The river may later begin to form a new terrace at a lower elevation. The cycle can repeat, to form a series of terraces, the oldest of which will be physically the highest (figure 11.5a) and the youngest will be at the lowest elevation. The same kind of thinking applies to marine terraces forming in response to lowering of sea level with respect to the land – the oldest terraces are normally the highest. Further difficulties arise in tectonically active parts of the earth's surface, and figure 11.6 illustrates river terraces that formed at the same time but are now at different elevations, because of differential uplift of the land.

The same applies with the principle of cross-cutting relations: older features are cross-cut by younger ones. However, it is not uncommon on a large-scale geological map to see, for example, the boundary of a patch of boulder clay cross-cutting peat in one area and, not far away, peat cross-cutting the boulder clay.

The interpretation may be that the two patches of glacial deposit formed synchronously and the two areas of peat at different times, or vice versa, or there may be no time equivalence between the two areas. Deducing which superficial deposits formed at the same time, where only the lithology is known, can be tricky. Even deposits formed in a single setting may show intricate interrelationships (figure 11.5b). However, the surveyor may have noted features during his or her fieldwork that have enabled the deposits to be listed in overall chronological sequence in the key, which may also give additional information on age relationships.

It may be possible from the relationships on the map to deduce something about the processes that gave rise to the deposits. The superficial deposits most widely encountered on maps of Western Europe and North America involve river and glacier processes. These kinds of depositional environments and the typical resulting arrangements of the deposits are shown in figure 11.7. Figure 11.8a shows how some different kinds of glacial till can be related and figure 11.8b indicates some possible relationships between head deposits.

The deposits portrayed on a map may allow inferences about overall environmental conditions. Glacial and periglacial materials clearly imply a colder climate than interglacial deposits. Marine deposits

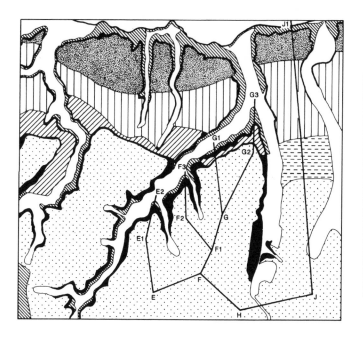

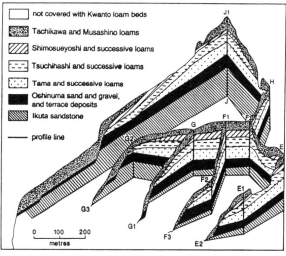

Figure 11.6 Redrawn part of the Geological Survey of Japan 1:5000 geological map and fence diagrams of the Gotanda-gawa basin, in Kawasaki. The area is to the NW side of Tokyo Bay, between Tokyo and Yokohama. The sequence of 'loam beds' consists of sands and gravels deposited on successive terraces each with a particular volcanic ash blown in during an eruption of the Fuji volcanic complex to the west. Even in this small area the terraces vary in altitude and show differing relationships, because of ongoing tectonic activity. Matching of terraces relies on key marker bands of volcanic tephra and pumice. The fence diagrams (called profiles here) show reconstructions of the deposits before erosion. Used by permission of the Geological Survey of Japan, Agency of Industrial Science and Technology.

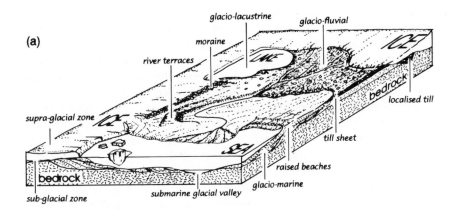

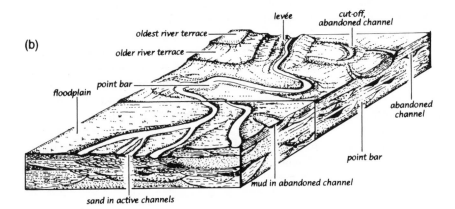

Figure 11.7 Illustrations of two settings in which superficial deposits are commonly formed. (a) The glacial and near-glacial environment. (b) The river environment. Features shown in the rear half of the diagram are telescoped closer to those in the front half than would occur in a real, more extended, river.

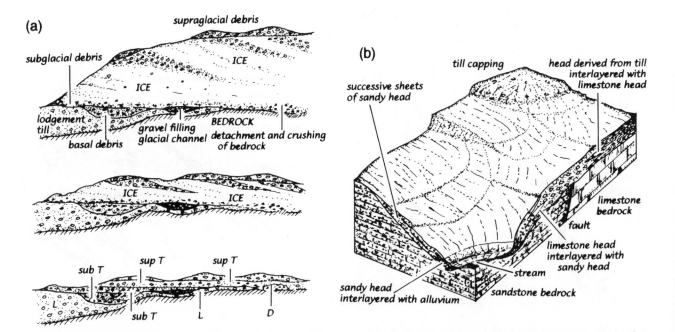

Figure 11.8 Illustrations of interpretations of superficial deposits. (a) Till sequences. Bottom diagram shows deposits which could be interpreted as shown (sub T = subglacial till; sup T = supraglacial till; L = lodgement till, the 'localised till' of figure 11.7a; D = deformation till). Upper diagram shows the original distribution of the deposits, with respect to the ice sheet; the middle drawing shows an intermediate stage, during ice melting. (b) Head distributions. The composition of the head reflects the upslope bedrock and, in the case of the deposit at the right of the diagram, a till deposit capping the hill. Redrawn from Catt (1986), by permission of Clarendon Press.

that are now on-land, unless in a tectonically active part of the world such as Japan or California where uplift may have occurred recently, may reflect the influence of climate on sea level. A cooling climate increases the amount of seawater locked up in ice and hence reduces sea level, exposing previously submerged marine deposits. However, you will have to consider the overall setting and relationships in order to distinguish this from the converse situation: melting in a warming climate decreases the weight of ice bearing down on land, which responds by rebounding upwards, also exposing marine material.

Wind-blown deposits suggest that little vegetation was around at the time of their formation, to help bind the particles together. While this may be due to a lake or a sea-shore setting, which should be apparent from nearby deposits, most commonly it indicates arid conditions. Peat deposits imply formation in water-rich conditions. Suitable circumstances come about in the final stages of infilling of a lake, in a marine or alluvial backswamp, or at high altitudes where the rainfall is greater than the soil can drain away. It may well be possible to distinguish the circumstances from the map. Figure 11.9 illustrates some interpretations of superficial deposits made from a standard geological map.

11.6 Reading a Geological Map

With the possibilities of interpreting geological histories outlined above, in addition to all the three-dimensional aspects of earlier chapters, it will by now be clear that there is a great deal to be achieved by looking at a map (e.g. figure 11.10). The analogy with reading a book is not a distant one. But unlike a book, when looking at a map there is no obvious place to start and finish!

It is therefore wise to adopt a systematic procedure. A series of steps is listed below, as a guide. The exact approach will depend on the purpose for which the map is being consulted, and the kind of geology involved. When making a full inspection of a map, it is usually worth while to jot down notes and make sketches as you work through the steps.

- Note the map scale. Establish the regional location of the area. Orient the map (i.e. determine the north direction).
- Note the main topographic features. Trace the main drainage and watersheds. Note the topographic contour interval. Visualise the relief.
- Examine the key very carefully. Note the variety of rock types and their stratigraphic distribution, the extent of information provided in the key, and any

symbolism used. If the map is black and white it may be useful to add colour to some of the units.
- Assess the overall stratigraphic distribution, and the main outcrop patterns. (For example: are there any major unconformities? Are the map units horizontal, uniformly dipping, folded, faulted, or some combination of these?) Make a sketch map which simplifies the geology but brings out the major features of the area.
- Establish the main dip and strike directions. (You may need to sketch some structure contours to gauge the strike.)
- Confirm the stratigraphic succession if it is provided in the key. Note any thickness variations or peculiarities. Watch for minor unconformities.
- Deduce the main structures. Note the main points of shape and orientation (sketching structure contours may help), and any cross-cutting relationships. Can the structures be grouped, for example according to style or stratigraphic age?
- Note any igneous rocks (shapes of bodies, relative ages) and any features concerning metamorphic rocks.
- Sketch or construct cross-sections as appropriate. Make any necessary corrections. Make any required measurements.
- Sketch any particularly significant relationships apparent on the map in order to emphasise them. Illustrate any significant structure contour patterns.
- List the main stages in the geological evolution of the area.

11.7 Writing a Map Report

Sometimes a map is consulted out of general interest, sometimes for a specific reason. It is occasionally necessary to prepare a report of the geology of an area as apparent from the map for another person. For example, a company may be considering investigating the area, and, before committing any money, requires some preliminary geological information. You, as a geologist, may be asked to summarise the geology from the available maps.

Such a report is mainly a matter of formalising in writing the steps listed above. A conventional order for dealing with the various topics has grown up, as follows:

ABSTRACT – summarises the main factual points made in the text of the report.

INTRODUCTION – provides a brief introductory statement: gives the map location, scale, any characteristics, and the main geological features.

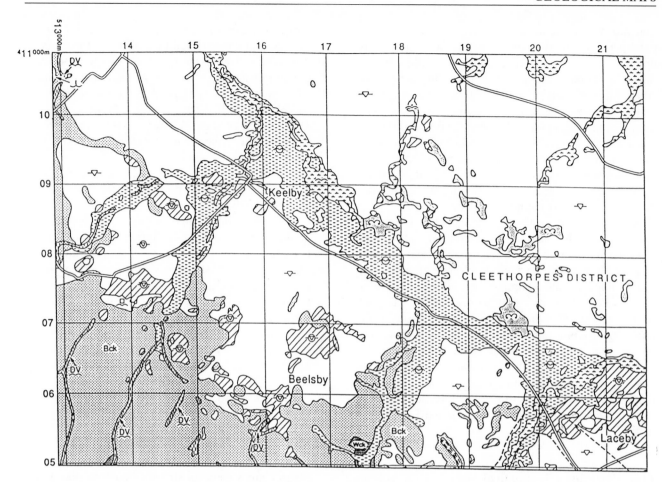

Figure 11.9 Some interpretations made from a published geological map, here redrawn in simplified form. The bedrock (Bck) is Chalk; V-rules suggest it to be dipping very gently towards the NE. The tiny patch of 'Interglacial Gravel Beach Deposits' is overlain by glacial deposits and so must be the legacy of an early interglacial period. The glacial till thins inland, its westward boundary roughly paralleling the modern-day coast. The till is presumably the result of the ice sheet that moved down the North Sea, which encroached only a little way inland. The area therefore marks the westward limit of the direct influence of North Sea ice. The rounded patches of glacial sand and gravel are likely to be the result of ponding beneath the melting ice sheet, whereas the distinctly elongate shape of the fluvioglacial deposits indicates channels eroded at the ice-sheet base. The dry valley deposits may have formed while the chalk bedrock was frozen and able to support surface streams. Although the alluvium appears to be forming in association with modern-day streams, its distribution coincides closely with the fluvioglacial deposits. Presumably the valleys eroded by the sub-glacial streams are still being utilised by rivers today, in a NW–SE direction near Keelby and a SW–NE direction at Laceby. Part of the BGS 1:50 000, Solid and Drift, Grimsby Sheet 190, redrawn by permission of the Director, British Geological Survey. Copyright NERC. All rights reserved.

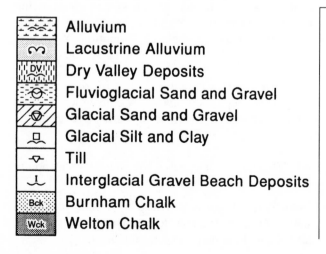

Alluvium
Lacustrine Alluvium
Dry Valley Deposits
Fluvioglacial Sand and Gravel
Glacial Sand and Gravel
Glacial Silt and Clay
Till
Interglacial Gravel Beach Deposits
Burnham Chalk
Welton Chalk

STRATIGRAPHIC SUCCESSION – briefly mentions each significant unit *in chronological order* (oldest first), grouping into systems if appropriate. Mentions lithologies and their distributions, thicknesses, unconformities, and any special features.

STRUCTURE – mentions any dip variations; describes folds, faults. May well use cross-sections and sketches.

IGNEOUS AND METAMORPHIC ROCKS

ECONOMIC GEOLOGY

GEOMORPHOLOGY AND SUPERFICIAL DEPOSITS

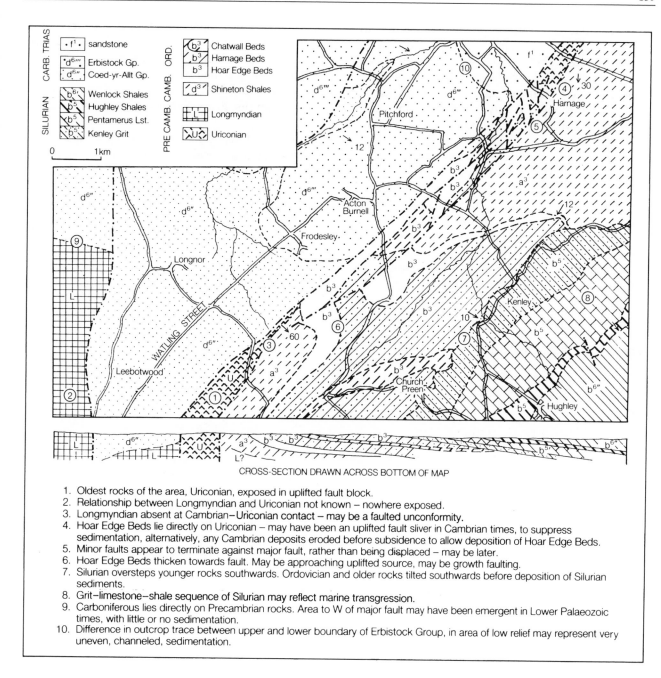

Figure 11.10 Part of the BGS one-inch Shrewsbury sheet, greatly simplified, with examples of observations relevant to interpreting the geological history of the area. Redrawn by permission of the Director, British Geological Survey. Copyright NERC. All rights reserved.

GEOLOGICAL HISTORY – lists the main events in the geological evolution of the area, in order, to serve as a summary.

This general order is, of course, modified as appropriate to the task in hand. Subheadings can be used also. It is vital that the report is rigorously organised. Equally important is that the writing is concise and to the point. Distinguish 'fact' from inference. For

example, if you are discussing palaeoenvironments, timings of events, etc., distinguish clearly between these *interpretive* aspects and your *observations* from the map. Unless you have been instructed otherwise, you should assume the reader is a trained geologist, but completely unfamiliar with the map area.

For some purposes certain aspects of the geology may require particular attention. If the report is meant to address some specific question, it is important that the answer is provided straight away, at the beginning of the report. The sheer quantity of information derivable from a map can be impressive. Nevertheless, company managers are not usually interested in wading through pages of geological reasoning, brilliant though it may be. They want direct geological answers to their question!

Summary of Chapter

1. Sedimentary successions lend themselves to the interpretation of relative ages and, to varying extents, the environments in which they were deposited.
2. Cross-cutting relationships between features are fundamental in deducing relative ages.

3. Structures of similar appearance and orientation may have formed at the same time. Folds and faults may fall into groups of differing ages.
4. The relative timing of igneous intrusion can commonly be established, but details of magmatic and metamorphic histories may be elusive from a map alone.
5. Superficial deposits present special difficulties because of their localised nature and the precision expected but may contain information on depositional environments and climate change.
6. It is wise to adopt a systematic procedure for studying geological maps.
7. Reports on the geology of a map area should follow the conventional sequence of topics; they should be rigorously organised and concisely written.

Map 23   Baraboo, Wisconsin, USA

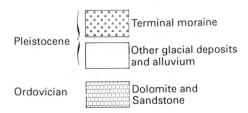

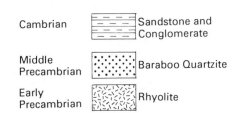

The map opposite is of a geologically famous area in central Wisconsin, USA. Early observations on the rock structures of the district were influential on the development of structural geology, and the list of teachers and students who have been drawn to the area has been referred to as a 'Who's Who' of American geology.

Little of the early Precambrian basement is exposed. The middle Precambrian rocks are represented by the upstanding ridges of the tough Baraboo Quartzite, and associated units are known from mine and well data. They are depicted on the cross-section by dashed lines. Note that the courses of the boundaries

between the various bedrock units have been continued on the map below the terminal moraine.

Summarise the geological history of the area, paying particular attention to the structure of the Precambrian rocks and its relation with that of the Lower Palaeozoic rocks. Consider the glacial history. Why might the drainage pattern, with several gorges cut *through* the ridges of Baraboo Quartzite, be so little related to the Precambrian units? Suggest an explanation for the location of Devil's Lake. The Wisconsin River is thought to have once flowed approximately N–S through the Lower Narrows and the gorge containing Devil's Lake. Account for its change in course.

Further examples concerning geological history

Many of the black and white maps reproduced here lend themselves to the interpretation of geological histories, particularly Maps 6, 10, 12, 19, and 22. All the

maps reproduced in colour incorporate interesting geological histories, most particularly Plates 5, 6, 7, and 8.

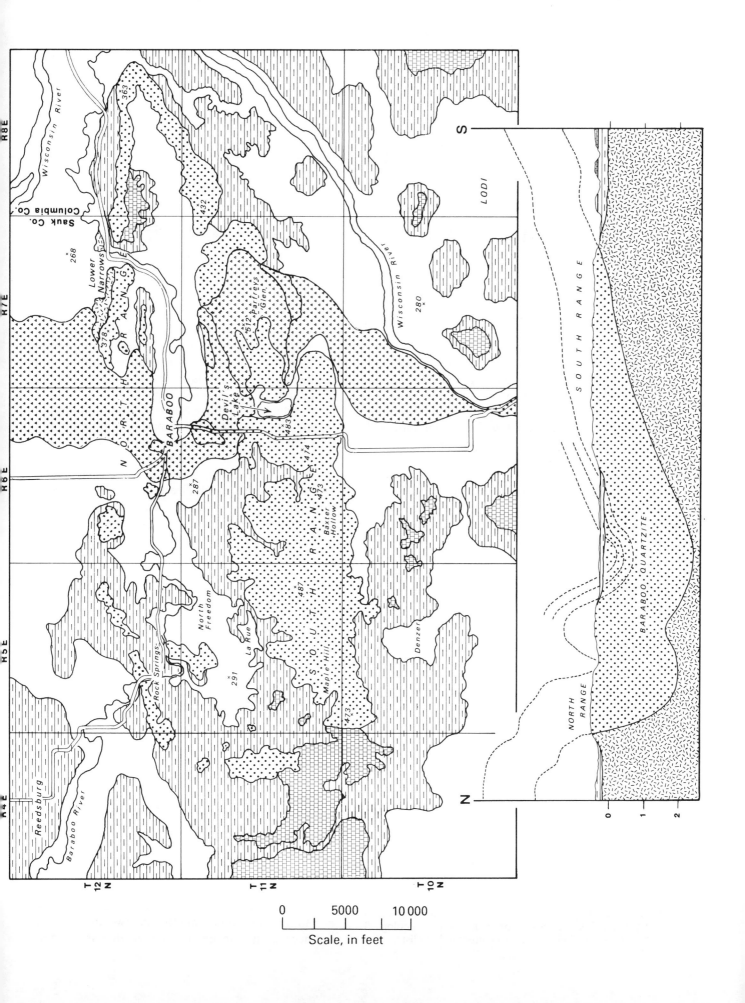

Scale, in feet

0   5000   10 000

N

BARABOO QUARTZITE

NORTH RANGE

SOUTH RANGE

S

LODI

0

1

2

Wisconsin River

Sauk Co.
Columbia Co.

363

432

Lower
Narrows

× 268

378

512
Partreys
Glen

Devil's
Lake

× 483

287

414

473
×

Baxter
Hollow

487

North
Freedom

La Rue

291
×

473

Maple Hill

Denzer

280
×

Wisconsin River

Rock Springs

Reedsburg

Baraboo River

NORTH RANGE

SOUTH RANGE

T
12
N

T
11
N

T
10
N

R 4 E   R 5 E   R 6 E   R 7 E   R 8 E

CHAPTER 12

Maps in Environmental Geology

12.1 Introduction

The recent burst of concern for environmental matters has led to new applications of geological maps, of three main kinds. First, maps remain the best way to portray the distribution of geological data of environmental relevance, such as groundwater pollution and sites of land subsidence. Second, maps allow us to explore possible connections between the geology of an area and some of its environmental issues. An example is the possible correlation between geology and the concentration of radon gas in houses. Third, an important new application is to highlight on maps those aspects that should be considered in the sensible planning of future land use. Such documents have variously been termed 'engineering geology maps' (even where they have been produced for environmental rather than purely engineering reasons), 'geoscience maps', and 'applied geological maps'. In the USA they have been called 'geological maps for

planning', and in Australia, 'land system survey' or 'integrated land resource survey' maps. They are most widely referred to as 'environmental geology maps'.

The first edition of this book contained a couple of paragraphs on the nature of environmental geology maps; now for many government surveys as well as private consultancies the preparation of maps that highlight the geological factors relevant to planning is a high priority (McCall and Marker 1989). It is possible that geological maps of future years will find their widest application in this direction. The following sections outline some of the geological aspects that contribute to environmental awareness and that are typically involved in environmental planning, emphasising the way they are communicated on maps.

12.2 Hydrogeological Aspects

Subsurface water is known as groundwater and its study forms part of the subject of hydrogeology. Groundwater is important directly, such as in its amount and suitability for a particular purpose, and indirectly through its influence on concerns ranging from the stability of slopes to the health of the population. Consequently, although hydrogeological maps are well established they are adapting to meet new environmental needs (e.g. Coe 1981; Erdelyi and Galfi 1988).

Conventional hydrogeological maps may include information on the character of the water itself, such as spatial variations in chemical composition, but mainly they summarise the geology of an area and emphasise how the distribution of water holding and transmitting properties varies. Figure 12.1 summarises the relative abilities of some common geological materials to store and to transmit water: note that the two attributes do not necessarily coincide. Formations that are capable

WATER HOLDING	WATER TRANSMITTING
well-sorted gravel	fissured clay
porous lava	silt
cavernous limestone	tuff
well-sorted sand	well-sorted sand
poorly sorted sand	poorly sorted sand
poorly sorted gravel	poorly sorted gravel
sandstone	well-sorted gravel
fractured crystalline rock	faulted sandstone
silt	porous lava
tuff	sandstone
faulted sandstone	cavernous limestone
fissured clay	fractured crystalline rock
dense crystalline rock	dense crystalline rock

Figure 12.1 Table to show the relative water-holding and water transmitting capacities of some common geological materials, decreasing from top to bottom in both lists.

of both are known as aquifers. Hydrogeological maps make much use of contour lines, essentially using the principles developed in chapter 3. Figure 12.2 gives some hypothetical examples and provides a commentary. Figure 12.3 is a real example of how the maps are used to show structure contours for important aquifers (figure 12.3a) and isopach maps for the aquifer thicknesses (figure 12.3b), enabling calculation of the potential for water storage. Contours can be interpolated from known water levels in wells to derive maps of the water table (figure 12.3c). Among other things, combining the information from a geological map with knowledge of the distribution of water at depth enables predictions to be made on where springs are likely to develop (figure 12.4), a critical factor where construction is proposed.

The data contained in conventional hydrogeological maps can be abstracted and highlighted on maps directed at environmental planners. The information may be chemical or physical, or some combination of both. Water chemistry can be crucial for sensible planning; the maps may indicate polluted areas, for example, for which remediation is needed. The groundwater of some areas may constrain where construction can be allowed, as some building materials react with certain chemicals in water. Correll and Dillon (1993) reviewed how the risks are assessed in Australia. Properly planned development will ensure

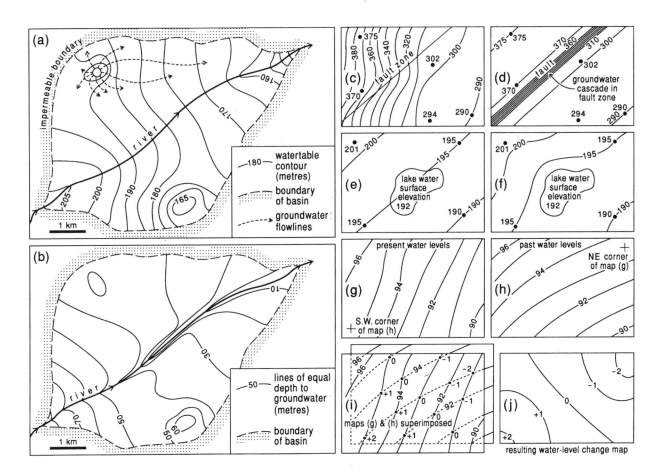

Figure 12.2 Examples of contouring methods and their significance on maps dealing with groundwater, based on Davis and DeWiest (1966). (a) Contour map of the water table in a hypothetical groundwater basin. The spacing of the lines reflects the 'head', or tendency for the water to flow, and the ability of the aquifer to transmit it. The form of the lines gives a preliminary indication of the direction of flow, which is normally at right angles to the contours, down-gradient, as indicated by the dashed lines in the NW of the area. (b) A depth-to-water map of the same groundwater basin as in (a), obtained by subtracting the elevations in (a) from topographic contours of the land surface. (c) – (f) Examples of contouring a water table from spot, well-level data, showing some differences in concept from a geological surface. From the five spot heights shown in (c), contours for a geological surface might be drawn as shown in the figure, with the fault accounting for some apparent swing of the lines but nothing more. However, for a water table, because faults commonly act as water conduits (e.g. see figure 12.4), the pattern shown in (d) is more likely. In another hypothetical example, the four spot heights shown in (e) could give, for a geological surface, the contour pattern shown in (e), with the surface having been eroded away at the site of the lake. However, this is not possible if the heights are of a water table; the interpretation shown in (f) is more likely. (g) – (j) Derivation of a water-level change map, by taking present (g) and past levels (h) and subtracting the difference (i), to allow the contours shown in (j) to be derived.

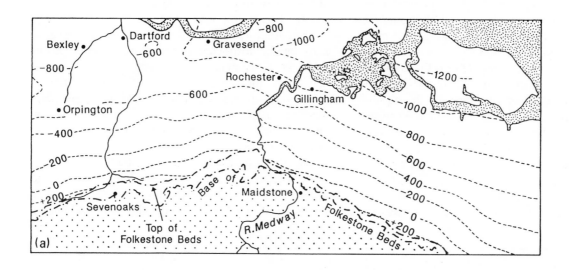

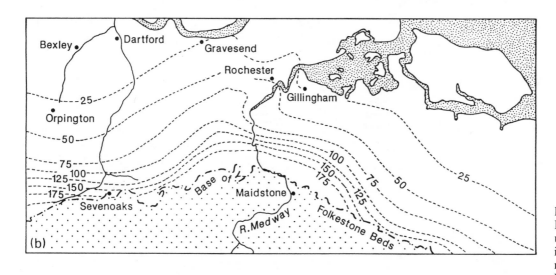

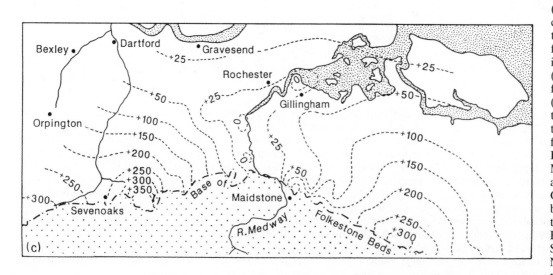

Figure 12.3
Examples of the use of contour lines in hydrogeological maps: the Folkestone Beds, southeast England. (a) Structure contours for the top of the formation, an important aquifer. (b) Isopach map for the formation. (c) Contour map of the local water table. Redrawn from the Hydrogeological Map of the Chalk and Lower Greensand of Kent, by permission of the Director, British Geological Survey. Copyright NERC. All rights reserved.

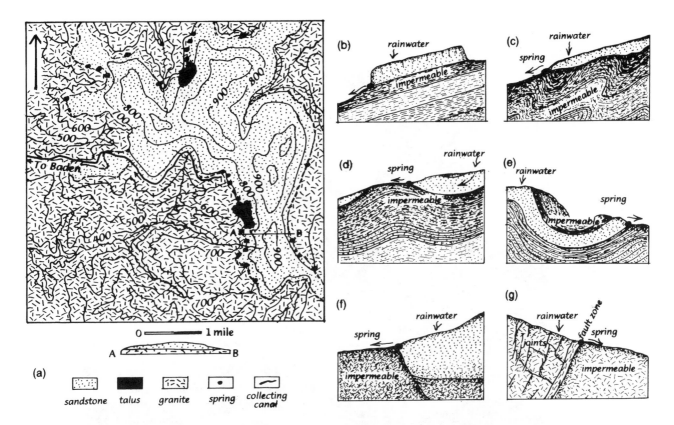

Figure 12.4 Common situations giving rise to groundwater springs, interpretable from geological maps. (a) Geological map and cross-section of an area at Baden-Baden, Germany. Rainwater percolates down through the sandstone until reaching the underlying granite, which is impermeable. (b) Pervious formation overlies impermeable rocks, as in (a), but dip of beds causes groundwater to concentrate at one side of outcrop. (c) As in (b), but with springs arising at a surface of unconformity. (d) and (e) Springs arising in folded formations. (f) and (g) Springs arising in faulted formations. (g) comprises igneous rock, itself impermeable, but rainwater percolates down through interconnected joints before exiting along a relatively permeable fault zone.

that sources of desirable water will not be jeopardised. Pure water is generally sought for drinking water whereas certain impurities are beneficial for some commercial purposes. Scotch whisky, for example, is supposed to benefit from water that has picked up certain trace elements by passing through granite.

Another new kind of hydrogeological map is one showing the vulnerability of groundwater to pollution (e.g. de Smedt et al. 1987). This can be shown both in general terms and as a response to a particular process, say the leaching of nitrates from fertilisers. In both cases the pollution susceptibility tends to relate very closely – even on the regional scale – to geology: compare the map shown in figure 12.5 with a standard geological map of southeast England. Vulnerability assessment becomes especially important where locations are sought for the disposal of wastes. The water-transmitting properties of formations, long shown on hydrogeological maps in order to assess groundwater availability, are now used in a converse way: in planning waste disposal sites. Although sites

are engineered to contain the waste in isolation, they are best located where the groundwater is not susceptible to pollution and in geological materials that will help confine any accidental leakage. Section 12.6 discusses waste disposal further.

12.3 Economic Deposits

Right from the start, geological maps have shown the distribution and arrangement of materials of potential economic use (sections 10.4 and 14.2). A first important point in environmental planning is to protect future access to such resources, even if there is no intention to utilise them in the foreseeable future. There are numerous tales of useful geological materials having been rendered inaccessible by unplanned development. However, the main point of this section is to mention maps that not only show potentially economic deposits but highlight the environmental matters that would arise from their exploitation.

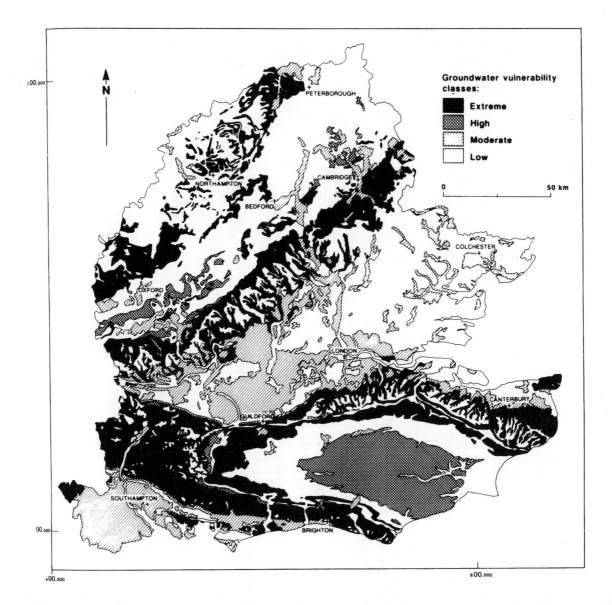

Figure 12.5 Example of a groundwater vulnerability map, for southeast England. Areas of extreme and high vulnerability tend to be associated with outcrops of important limestone and sandstone bedrock aquifers, whereas moderate and low vulnerability areas occur on certain superficial, especially clay-rich, deposits. In the study from which this map is taken, the groundwater vulnerability was linked to routes of transporting hazardous waste in an attempt to quantify pollution potential in tanker kilometres. Reproduced from Hiscock et al. (1995), by permission of the Geological Society, London.

Some physical constraints	Some planning constraints
Presence of:	Area already utilised
Thick superficial deposits	Need for buffer zones
Overlying rock	
Solution features	*Presence of:*
Unusable clay layers	High-grade agricultural land
Rapid variation in rock quality	Special landscape areas
Faults, joints, and deep	Sites of Special Scientific
weathering	Interest
High groundwater levels	Recreational areas
Remoteness from markets	Aquifer production zones

The extraction of metalliferous, industrial, and energy-generating minerals can have severe environmental implications, but the issues tend to be localised because of the very patchy distribution of these materials. Much more widespread are problems arising from the exploitation of the geological materials used

Figure 12.6 Some constraints on further extraction in northeast England of Magnesian Limestone, a material widely used as a dimension stone and an aggregate, based on an analysis by Clarke et al. (1987), by permission of the Geological Society, London.

in construction, and especially roadstone, sand, and gravel – collectively known as **aggregates**. These materials occur widely and production sites are numerous, in order to mimimise transport costs. They are consumed in enormous quantities. The US, for example, uses road and building stone at a rate equivalent to almost 5 tonnes a year for each individual citizen; each German who lives for 70 years will indirectly have used 460 tonnes of sand and gravel (Archer et al. 1987). As with all mining, each stage of the extraction, processing and transport of aggregates can harm the environment. Planning of development therefore seeks some balance between locating sources of suitable aggregates while ensuring the impact of their extraction on the environment is acceptable.

Although rocks are quarried to give large blocks and slabs for building – so-called **dimension stone** – the big demand is for aggregates. In the UK over 99% of geological material extracted is used for aggregate. About half of this rock is quarried and then crushed; the remainder is sand and gravel, mainly dug from alluvial and glacial deposits. Constructing a road, for example, requires material varying from coarse-grained, strong, well-drained and compactable ballast for its base to finer material that is resistant to wear and tear for the top surface (e.g. McLean and Gribble 1992). The latter is less widespread, as it must in addition bond well with bitumen while having adequate resistance to impact, abrasion, and – in order to reduce skidding – polishing. Much used for this purpose are basalt, gabbro, and other igneous rocks, non-foliated metamorphic rocks, and some tough limestones and sandstones. A geological map will show the distribution of such rocks; if the map shows superficial deposits then possible sand and gravel resources may be discernible. Some maps are published specifically to show aggregate resources, such as those in the Resource Assessment series of BGS and those produced under contract (e.g. Crimes et al. 1994). Geological maps produced for planning purposes will include construction materials if development is involved (e.g. de Mulder and Hillen 1989).

The river and glacio-fluvial deposits that may contain suitable aggregate can vary abruptly in thickness and in lithology, so the map will need to be of adequately large scale and detail for such resources to be assessed. Section 10.5 emphasised the characteristic lateral impersistence of these materials. Boreholes or excavations might be needed to supplement the map information to enable meaningful structure or thickness contours to be drawn. Deposits of outcrop less than about 1 km² in area may be only marginally economic unless they are unusually thick or the need is great. Environmental and resource assessment

maps may show the thicknesses of superficial deposits, using isopachs, and there may also be some indication of the variation in quality (e.g. size, sorting and hardness of the grains, and purity of the deposit), the accessibility (e.g. type and thickness of overburden), and an assessment of the ease of extraction. Removal of any covering overburden is a major cost in exploiting superficial deposits.

At the same time, maps for planning will show any constraints on extracting the construction materials. These may include safeguarding land valuable for agriculture or recreational purposes, or of outstanding aesthetic or scientific value – many sand and gravel deposits happen to be important wildlife habitats. Though depiction of such things on a map involves subjective judgements, they are highly relevant factors. There may be potentially harmful implications for surface and groundwater. Planning maps may portray an assessment of the ease of restoration of the land when extraction ceases. Figure 12.6 lists some of the factors relevant to planning further extraction of the Magnesian Limestone in northeast England, a formation quarried for hundreds of years as an excellent dimension stone and for crushing as an aggregate. The desirability of the material has to be balanced against the constraints, some of which are listed in the table. Good planning considers all the relevant issues together and strives to reach the optimum balance.

12.4 Construction Conditions

Large-scale geological maps can yield information which allows, at the planning stage, the delimiting of areas of differing suitability for construction. Such preliminary information is a cheap and sensible aid to planning. Engineering geology maps highlight construction conditions (e.g. Radbruch-Hall et al. 1987). They can be technical in content and may focus on a particular proposed construction. Geological maps to which engineering data have been added have the drawback that geologically defined formations may not have consistent engineering properties. More useful for environmental purposes are maps where the geology has been subdivided into units that reflect not stratigraphic age and the like but reasonably uniform engineering properties or behaviour. Bedrock may be dealt with separately from the superficial deposits. Such maps represent some compromise between showing technical detail and simple summaries of recommended usage.

The kinds of properties that are relevant to construction include ease of excavation, strength, load-bearing capacity and behaviour in the presence of water. In a very general way, sands and gravels can provide suitable construction sites – provided the hydrogeology is suitable – but clays commonly give difficult conditions. Some of these may be discernible from a geological map (e.g. de Mulder 1994). For example, irregular, lobate boundaries at the foot of a topographic slope may well imply landslipping (section 12.5), and clayey material deposited in a past glacial lake may well take on the very weak condition known as 'quick' if it is disturbed. Also, clay deposits are prone to undergo large amounts of shrinking and swelling as humidity changes (section 12.5). Essential for construction design is knowledge of the thickness of superficial deposits (sometimes called overburden in this context) that overlie the bedrock (often called 'rockhead' by engineers). This is variously presented as thickness, isopach, depth to bedrock or rockhead, or rockhead contour maps (section 4.4). Construction is sometimes proposed on areas of made- and worked-ground (see section 10.5) but such materials can give awkward and unpredictable conditions.

Sedimentary rocks are very variable in engineering properties. Mudrocks vary a great deal in strength and some contain sulphate, which in solution can degrade some concrete. Limestone presents special problems because of its solubility. The presence of solution features such as sink-holes and caves, which can be hidden, has a major impact on construction. Sandstones generally provide good construction sites, especially where they are strong, but they can be expensive to work. The same is true of igneous and metamorphic rocks and special attention has to be paid here to the degree of weathering.

But in many ways, an overriding consideration is the uniformity of the material. Wherever there are abrupt variations in geology, through well-developed stratification or faults in the bedrock, or in type and thickness of superficial deposits, there are likely to be awkward conditions for construction. Hence, extensive construction is preferably located on areas of consistent geology. Airport sites provide an example. Heathrow Airport, London, is located on superficial deposits of reasonably uniform properties associated with wide river terraces, and Mirabelle Airport, Montreal, is on a wide plain of glacial till. Neither has experienced particular foundation difficulties. In contrast, the new airports at Osaka, Japan, and at Chek Lap Kok, Hong Kong – both sited in the sea – have had to overcome intricate variations in engineering properties resulting from buried submarine channels, slumps, and surfaces alternately submerged and exposed to air.

Similar considerations apply to underground excavation, say for storage or transport routes. The bedrock needs to be strong and lacking planes of weakness; again abrupt changes in geology cause diffi-

culties. Figure 12.7 illustrates a contrast between construction in a relatively simple, uniform, and well-known area with a region of poorly known and complex geology. In both examples, geological maps and cross-sections were fundamental in the siting and the design of the construction. However, in the latter example the complexities of the geology required the initial design to incorporate expensive engineering precautions. Then, as the work progressed, understanding of the geology had to be revised and expensive modifications had to be made to the design of the construction.

12.5   Hazards

Geological hazards are widespread, and in some parts of the world they are dramatic. Maps are routinely used to portray the distribution of the risks and results of phenomena such as volcanic eruptions, earthquakes, and tsunamis. Elsewhere, less striking but none the less real geological hazards are increasingly being communicated through environmental geology maps. Such hazards may be natural or due to human activities.

Vulnerability to flooding is a parameter shown on some environmental geology maps. It is an important aspect of planning: it has been said that floods probably affect more individuals and their property than all other hazards put together. Floods occur when the channel cut by the river is temporarily unable to cope with the influx of water gathered from the catchment area. They are most common in the lower reaches of rivers where a flood-plain has developed. This may be marked on a map or recognised from the spread of alluvial material around the river. Whether or not there are river levees present – either natural or man-made – a flood-plain will eventually be subject to flooding and planning has to recognise this. The problem is in estimating the likelihood of flooding and balancing this against desirable uses of the area.

From the record in the alluvial deposits, and from the regional precipitation, runoff, and river discharge, flood-frequencies can be estimated and hazard zones established. River depths are not normally shown on geological maps so be careful if you are using the width of a river to estimate its discharge. Discharge increases in proportion to the square of the width of the channel, but generally river channels become wider relative to their depth with increasing distance downstream. Any development of the channel zone must not decrease the discharge area or rate. The flood hazard zone of greatest risk, say where floods are expected every 1 to 20 years, could be used for

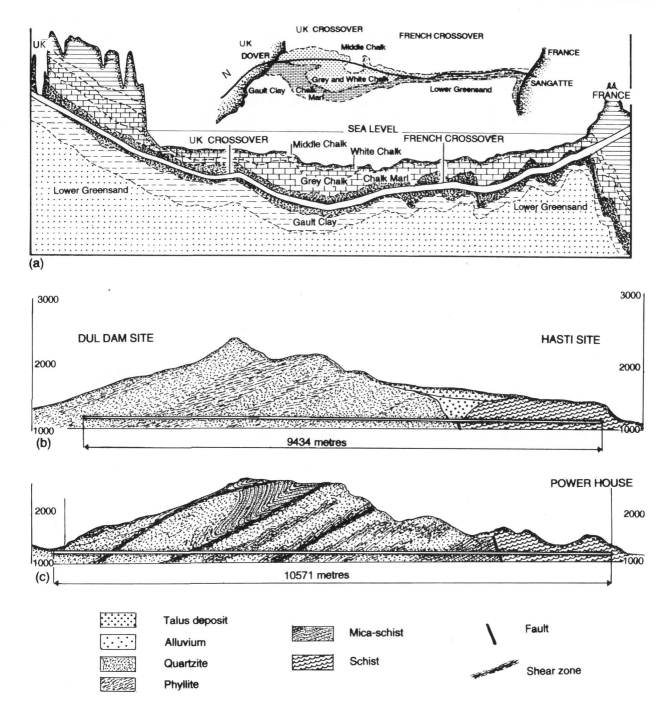

Figure 12.7 Illustrations of the advantage of a consistent, predictable geology for construction sites. (a) The Channel Tunnel, between England and France. Sketch map of the bedrock geology and geological cross-section, showing the location of the Channel Tunnel almost wholly within one formation suited to tunnelling, called the Chalk Marl. (b) The Dul Hasti hydroelectric project, Kashmir, India, simplified from Winter et al. (1994). The cross-section shows the geology as it was understood at the commencement of the project: a sequence of interbedded quartzite and phyllite faulted against schist and gneiss. A river valley, now filled and covered by superficial deposits, was thought to have been eroded along the fault. A tunnel was to connect a reservoir with the hydroelectric powerhouse in a straight line, penetrating just the lowest portion of the river deposits which were to be specially reinforced. Construction began and revealed not a buried valley but sediments accumulating across an active fault (c). There was no feasible alternative but to loop a longer tunnel around the active depression. The structures in the quartzite and phyllite turned out to be much more complex than anticipated, and further special treatments were needed. The geological maps and sections had to be continuously updated as the work proceeded, and the construction grew in complexity and cost.

agricultural or recreational purposes. Construction might be allowable in zones of next greatest risk, but only with appropriate precautions in place. People talk of places with '50 year floods, 100 year floods' and so on, meaning that *statistically* those areas will experience one significant flood in every 50 years or in every 100 years, etc. Flood vulnerability maps portray this information and form the basis of flood management strategies.

A similar kind of approach is used in coastal areas to indicate areas of greatest risk from marine action. Coasts with rock platforms and cliffs, which should be distinguishable on a geological map, will be undergoing erosion. It could be rapid if the cliffs are formed of superficial deposits. The resistance of bedrock depends not only on the strength of the rock itself but on the presence of weaknesses such as joints and, with sedimentary rocks, on the orientation of bedding.

While this in itself may not significantly accelerate the rate of erosion, as the fallen material may offer some protection, it is clearly a matter to be accounted for in planning.

Slope instability is related closely to the local geological conditions, either in the bedrock or, more commonly, in the superficial deposits (figure 12.8). Most large-scale geological maps explicitly show areas that have suffered landslide (see, for example, Plates 3 and 6) but environmental geology maps are more concerned with the future likelihood of movement. Some simply distinguish different steepnesses of topographic slope but this alone is insufficient as different geological materials have differing stabilities. Some show areas where slopes coincide with the dip of sedimentary formations, an arrangement especially vulnerable, and highlight the steep slopes. Yet other maps classify slopes according to overall stability, and

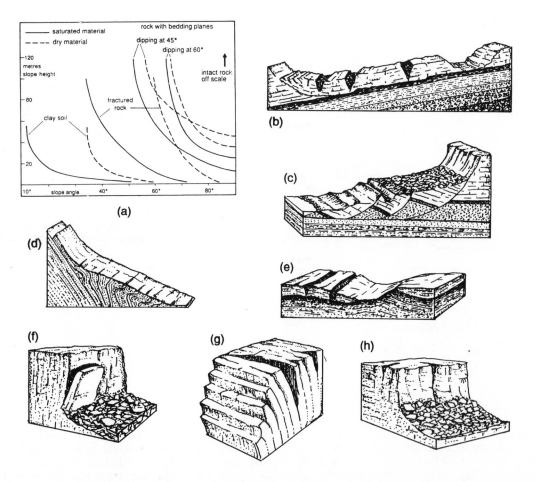

Figure 12.8 Examples of geological arrangements, potentially interpretable from maps, that tend to give slope instability. (a) Maximum slope height and slope angle for a stable slope in the materials indicated. Situations to the right of each curve, that is, greater height or slope angle, are unstable. (b) Formations dipping in the same direction and at the same angle as the topographic slope. Any weak beds will make the formation vulnerable to downslope sliding. (c) Cliff cut in horizontal strata. (d) Steep slope parallel to very steeply dipping formations. (e) Horizontal, tougher strata draped over bulging weaker beds tend to become fissured and move slightly downslope, a process known as cambering. (f, g, and h) Steep slopes cut into homogeneous and strong but well-jointed rocks can give rise to falls of slabs or blocks of the rock.

may attempt a semi-quantitative assessment of the risk of landslide, dividing the area into hazard zones. Where horizontal or shallow-dipping strata involve soft clay horizons in an area of dissected topography, slow creep may give rise to 'valley bulges' and on hills to the effect known as 'cambering' (figure 12.8e). BGS maps depict such slipped strata – around the city of Bath, for example – as 'foundered beds' and affected areas pose major construction problems. Areas at risk from slippage of these different kinds may have to be restricted to agriculture, forestry or, perhaps, recreation.

Besides movement down a slope, land may subside vertically, through natural causes such as dissolving away of parts of the underground strata, and as a result of mining and drilling activities and settlement over disused waste-disposal sites. The former is a common problem in areas of limestone (figure 12.9a) and other soluble materials. Figure 12.9b gives an example involving salt. Subsidence associated with mining and drilling is very variable in nature. It depends on the material involved, as, say, ore deposits and coal are extracted in very different ways to water, oil, and gas, and the resulting effects are correspondingly different. Moreover, a single commodity can be mined in a host of different ways, according to its quality, its arrangement in the ground, and the local custom of the time. Each method will leave a shape and distribution of underground voids that will appear different in map view from other methods (figure 12.10).

Planning in areas of past mining activity must therefore pay attention to where underground conditions may be hazardous, and maps attempt to portray the subsidence risk. Unfortunately, because details of past mining activities have not always been kept, the locations of underground voids can be unknown and difficult to detect. Figure 12.11 illustrates three different ways of dealing with this problem on maps. Note that in addition to subsidence, mining areas and sites of

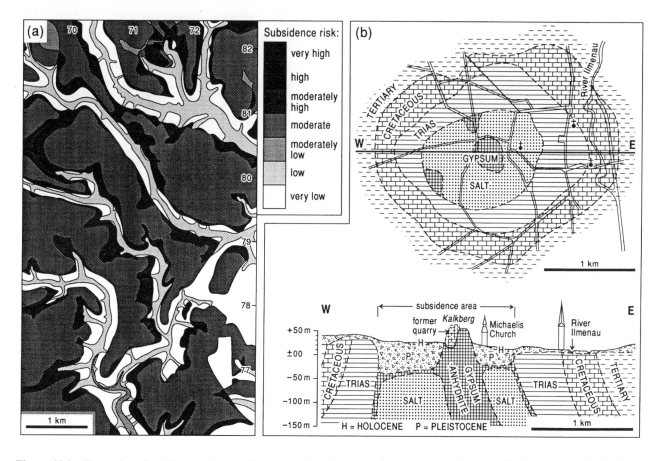

Figure 12.9 Examples of subsidence due to dissolving of underground formations as a hazard. (a) Assessment of subsidence hazard due to natural solution features in limestone, near Reading, Berkshire, England. Redrawn from Edmonds et al. (1987), by permission of the Geological Society, London. (b) Geological map and cross-section, Luneburg, Germany. The town is built over salt which has forced its way upwards, deflecting the adjacent strata. The top of the salt is at a depth of 40–70 metres and has long been mined. Resulting subsidence, as much as 7 cm a year, has damaged buildings and led to flooding. Redrawn from Aust and Sustrac (1994), by permission of Oxford University Press.

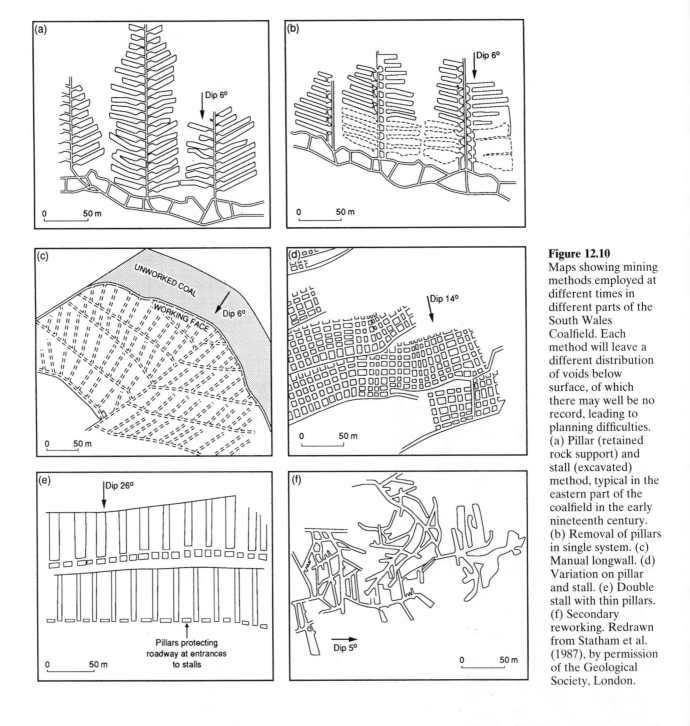

Figure 12.10
Maps showing mining methods employed at different times in different parts of the South Wales Coalfield. Each method will leave a different distribution of voids below surface, of which there may well be no record, leading to planning difficulties. (a) Pillar (retained rock support) and stall (excavated) method, typical in the eastern part of the coalfield in the early nineteenth century. (b) Removal of pillars in single system. (c) Manual longwall. (d) Variation on pillar and stall. (e) Double stall with thin pillars. (f) Secondary reworking. Redrawn from Statham et al. (1987), by permission of the Geological Society, London.

Figure 12.11 (opposite) A series of maps and block diagrams to show different ways of presenting data on subsurface voids left by past mining, where knowledge is incomplete. (a) Geological map of a hypothetical area. Note that information is available on workings of coal seams A, B, and Y, but not for X even though it is thought to have been worked. (b) Schematic map showing where voids are possible. Such a map may be suitable for reconnaissance appraisal but it does not distinguish between known and suspected workings. (c) and (d) Schematic maps showing known mining within 30 metres of the bedrock surface and total known mining, respectively. Single and multiple seams are distinguished, and specified in (c). Speculation is minimal and considerable detail is shown, but two maps were required to present the information and extra knowledge (e.g. borehole data) was needed. Such detail may not be appropriate for most planning purposes. Moreover, the information is confined to *known* workings, even though other voids may be present. (e) Schematic map emphasising known and inferred workings at different depths. While this may appear to provide the information relevant to planning, it involves assuming that where a coal seam is present it is likely to have been worked, and hence suggests that almost the entire area is underlain by voids. Reproduced from McMillan and Browne (1987), by permission of the Geological Society, London.

(a)

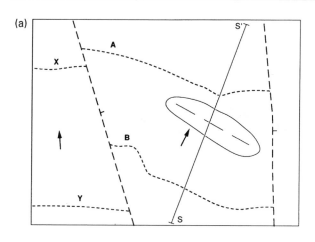

- - - - - Coal outcrops

- - ⊥ - Fault

→ Direction of dip

⬭ Drumlin: thick drift deposits in excess of 20m. Elsewhere thickness averages 5m

S ⊢——⊣ S' Line of section

Mine plans available for Seams **A**, **B** and **Y**. Seam **X** known to have been worked 'in the area' but no plans available.

Intervening strata generally less than 30m between seams **A** and **B**; generally more than 30m between seams **X** and **Y**.

(b)

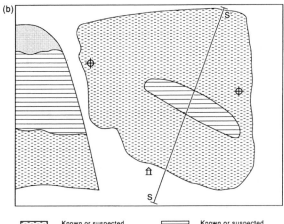

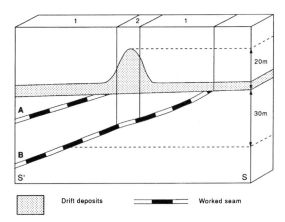

▦ Known or suspected shallow mining generally within 30m of surface

▥ Known or suspected shallow and deep mining

▤ Known or suspected mining generally more than 30m below surface

⊕ Pit shaft, abandoned

⊩ Mouth of mine abandoned with orientation

▦ Drift deposits ▬ Worked seam

1 Known or suspected shallow mining generally within 30m of surface

2 Known or suspected mining generally more than 30m below surface

(c)

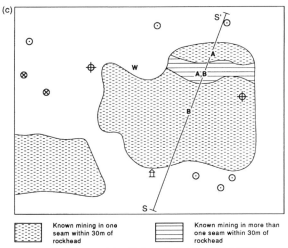

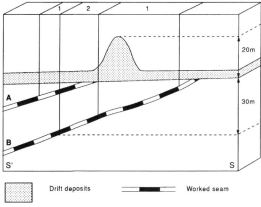

▦ Known mining in one seam within 30m of rockhead

▤ Known mining in more than one seam within 30m of rockhead

⊙ Borehole encountering old mine workings within 30m of rockhead

⊗ Borehole encountering old mine workings more than 30m below rockhead

⊕ Approximate site of abandoned pit head

⊩ Mouth of mine adit

W Waste of older uncharted workings

A Seam A

B Seam B

▦ Drift deposits ▬ Worked seam

1 Known mining in one seam within 30m of rockhead

2 Known mining in two seams within 30m of rockhead

(continued over)

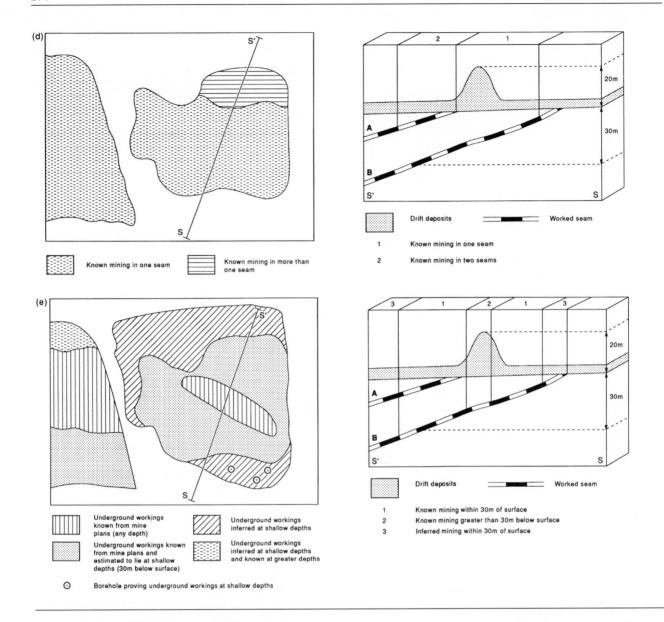

(d)

Known mining in one seam

Known mining in more than one seam

Drift deposits · Worked seam

1 Known mining in one seam

2 Known mining in two seams

(e)

Underground workings known from mine plans (any depth)

Underground workings known from mine plans and estimated to lie at shallow depths (30m below surface)

Underground workings inferred at shallow depths

Underground workings inferred at shallow depths and known at greater depths

⊙ Borehole proving underground workings at shallow depths

Drift deposits · Worked seam

1 Known mining within 30m of surface

2 Known mining greater than 30m below surface

3 Inferred mining within 30m of surface

waste disposal can also be exposed to a risk of underground explosion. Figure 12.12 portrays a tragic example of poor geological siting of a landfill. There is also the risk in such areas that groundwater has become polluted or even corrosive, the so-called aggressive groundwater. Hazards and risks of this kind are also included on environmental geology maps.

A much more subtle hazard but one that affects many property owners – and is currently giving rise to a dramatic increase in insurance claims – is that of shrinkable clays. Clayey soils alternately swell when they become wet in rainy seasons and then shrink on drying. As a result, buildings, especially those with shallow foundations, are alternately levered upwards and then lowered down, an effect referred to as **heave** and **subsidence**. If your house has cracks running through the walls this may well be the reason (Freeman et al. 1994). A geological map indicates where the problem is likely to be serious because it will show areas of clay-rich bedrock and clayey superficial deposits. The issue in environmental planning is to try to minimise construction in areas of thick clay deposits and to ensure that the building codes for any new buildings that are allowed incorporate the appropriate precautions.

The relation between certain medical conditions and geology is well established. For example, increased incidence of heart disease has been related to calcium-bearing bedrock and hard water, and concentrations of trace elements derived from bedrock such as cobalt, selenium, and molybdenum can cause nutritional diseases (e.g. Appleton et al. 1996). Geological maps help reveal such correlations. Of particular concern has been the distribution of our

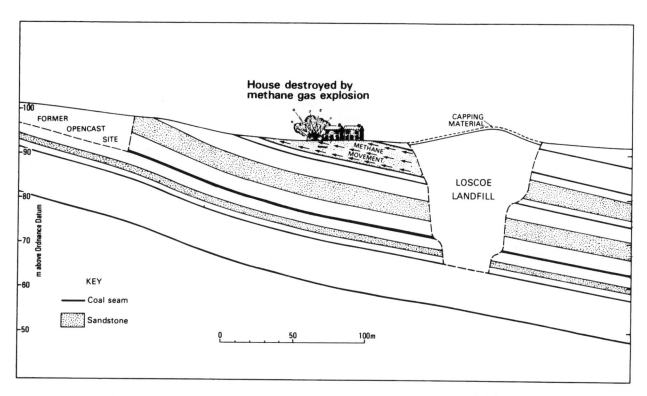

Figure 12.12 Schematic cross-section through the landfill site at Loscoe, Derbyshire, UK, from Stenestad and Sustrac (1992). In 1986 methane gas accumulating in the landfill site began to leak along a permeable formation, as shown. The gentle dip of the formation caused the gas to reach surface some distance from the site, below houses. Oxidation reactions involving the methane gradually increased the temperature until the gas spontaneously ignited. The explosion completely destroyed one house and severely injured its three occupants. Reproduced by permission of Oxford University Press.

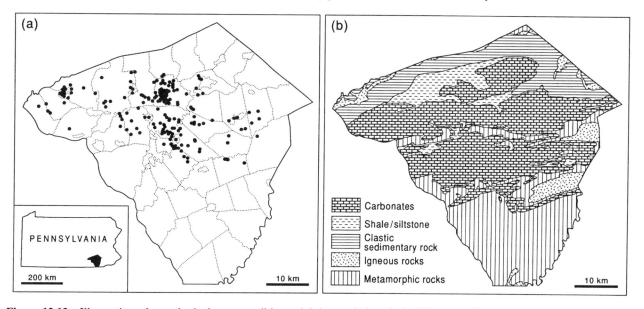

Figure 12.13 Illustration of a geological map possibly explaining variations in health: the incidence of radon gas, suspected to cause cancer, in Lancaster County, Pennsylvania. (a) This shows the distribution in sample houses of radon levels above the sample median. (b) Generalised bedrock geological map. Based on Geiger and Barnes (1994).

exposure to the natural gas radon, which some studies have linked with the incidence of cancer. Radon tends to be concentrated in certain kinds of bedrock, such as granite and mudstone and superficial deposits derived from them (e.g. Hudak 1996), and so exposure is greater in such areas. Where such materials are at depth, the ability of the overlying bedrock to transmit the gas is highly influential. Figure 12.13 shows increased radon levels where fissures have allowed the gas to leak from depth, and similar effects can occur around faults (e.g. Varley and Flowers 1993). The BGS has produced maps showing the radon risk of certain areas and the USGS has compiled what it calls 'radon potential' maps, for example, for the counties around Washington, DC.

12.6 Waste Disposal

Much routine disposal of domestic waste utilises land-fill sites, called sanitary landfills in the USA. These days the planning requirements are strict and are aimed at totally isolating the waste, apart from some controlled pumping of the fluids and gases that are generated, from the site's surroundings. Ideally, sites in geologically stable settings and in impervious materials are sought, in order that the landfill is not disturbed and that the spread of any fluids that might accidentally escape from the site is curbed. Clay-rich sedimentary, massive igneous and metamorphic rock formations are most likely to provide sufficiently impermeable materials. Sandy materials, limestone, and rocks with much jointing, in contrast, are not.

However, rarely are new sites pinpointed from ideal considerations. More often it is a matter of having to extend a present site or convert an existing excavation, such as a disused quarry or railway cutting. The geological suitability will have to be evaluated and will influence the design and specification of the installation. Existing maps might provide the necessary information on any geological instability at the site, such as faulting, landslip, or subsidence, together with some indication of any threat to the adjacent groundwater system. A geological map will also provide information on another relevant factor. The site will have to be lined with impermeable material and although plastic-based membranes can be purchased they are expensive, difficult to install, and rendered useless by a puncture. If deposits of clay are available nearby, they may be sufficiently impermeable to do the job. Clay is relatively cheap, easy to work, and is self-healing. Any candidate material should be apparent on a reasonably large-scale map showing superficial deposits, and may make all the difference to the economic viability of a proposed site.

All these criteria will have to be applied even more rigorously if the proposed waste is itself in some way hazardous. For example, it has been suggested that the foundation material should possess even lower permeability thresholds and be at least 15 metres thick, and that no groundwater well should be within 2 km of the proposed site. The possibility is sometimes mooted of disposing toxic liquid wastes by injecting them into deeply buried geological reservoirs. Geological maps and sections help provide initial estimates of the geometry and capacity of such potential reservoirs before detailed tests are carried out. The mapwork is closely comparable to assessing hydrocarbon reserves, but there the prospect is to pump *out* the relatively light and rising material that ideally has been trapped near surface, say in an antiformal structure. In waste disposal, the idea is to pump *in* relatively heavy fluids, and so synformal structures of suitable formations that outcrop are sought, so that the liquid can be injected from the surface. From there the waste descends into the reservoir, which must be tightly confined above and below by impermeable formations.

Figure 12.14 illustrates an example of this approach, where possibilities of deep, subsurface disposal were assessed in the 'Valley-and-Ridge' fold province of the Pennsylvanian Appalachians, discussed in a different context in section 13.8. Three structures were initially looked at, but one revealed some faulting which offered escape routes from the potential reservoir, and another contained significant reserves of coal. Geological maps showed that these factors were absent from the third structure, which revealed a very suitable basinal arrangement and was named the Dunning Cove syncline (figures 12.14a and 12.14b). Moreover, it contained a porous formation, the Tuscarora sandstone formation, that was confined above and below by poorly permeable shale-bearing units. Isopach maps showed that the confining formations were present in thicknesses likely to provide adequate barriers to leakage (figures 12.14c and 12.14e), and that the sandstone would provide a reasonably capacious reservoir (figure 12.14d.). From the dimensions on a larger scale map of the periclinal Dunning Cove syncline, together with the thickness and structural relief shown by the formation in cross-sections, Hardaway (1968) calculated that the Tuscarora sandstone in this structure was capable of storing about a cubic kilometre of liquid waste and suggested it was worth closer analysis.

Some countries have had to put in much effort to try to find suitable sites for storing radioactive waste underground. In general, areas have been examined that have stable, low permeability rocks such as poorly jointed granite, clay-rich sediments, or rocksalt. Geological maps show the occurrence of such materials, and will help define those sites where the necessary very intensive further examination might be carried out. Numerous facets of the geology need to be considered in a detailed examination. Figure 12.15 shows simplified versions of a few of the scores of maps produced by the body responsible in the UK for

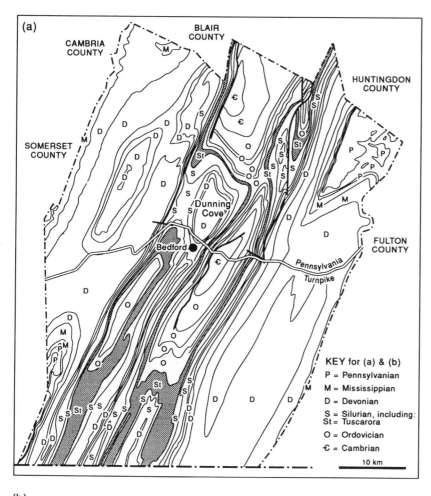

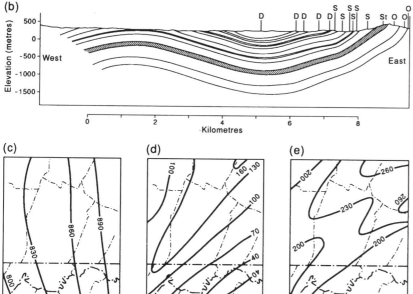

Figure 12.14
Examples of geological maps used in planning possible subsurface storage of liquid wastes, from Hardaway (1968). (a) Generalised geological map of Bedford County, Pennsylvania. Note the overall pattern of plunging and periclinal folds, with some localised faulting. Key omits subdivision details of stratigraphic systems apart from Tuscarora Formation. (b) W–E geological cross-section across Dunning Cove. (c) Isopach map of Upper Ordovician formation underlying Tuscarora Formation. (d) Isopach map of Tuscarora Formation (Lower Silurian). (e) Isopach map of Middle Silurian formation overlying Tuscarora Formation. See section 12.6 for further explanation. Redrawn by permission of the American Association of Petroleum Geologists.

Figure 12.15 (over) Illustrations of maps (redrawn here) produced by UK Nirex Ltd. as part of investigations into a potential repository site for radioactive waste at Sellafield, northwest England. (a) Isopach map of superficial deposits, thicknesses in metres. Data not available for northernmost part of area. PRZ = boundary of potential repository zone. (b) Nature and origin of superficial deposits, from a map entitled 'Surface Geology. Quaternary Domains'. ALV-T = post-glacial terrestrial deposits. GLO-G = mostly gravels, GLO-S = mostly sands. THN-C = thin till overlying Carboniferous bedrock. SOL-C = solid bedrock,

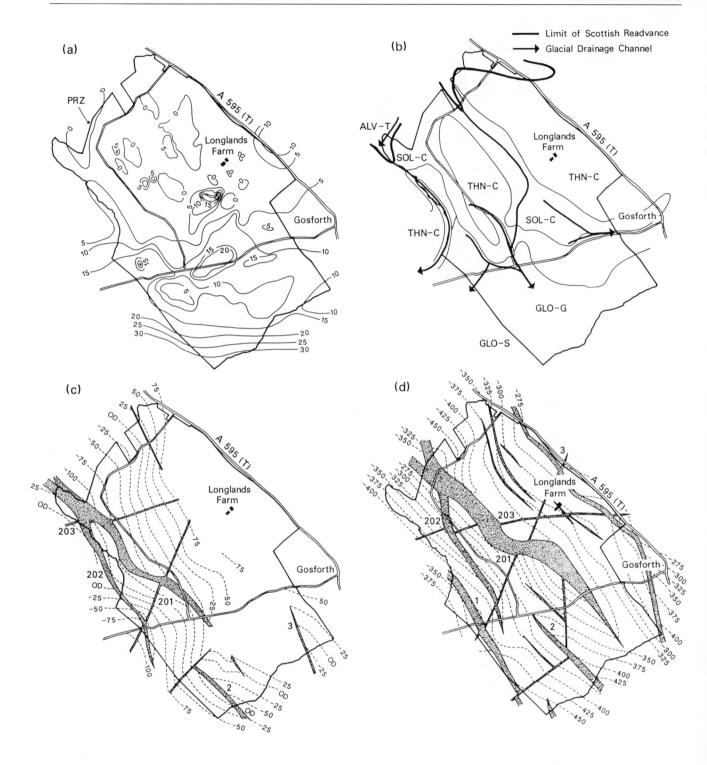

here of Carboniferous age. (c) Structure contour map for the base of the Calder Sandstone (see key in (e)). Contours in metres relative to a fixed datum. Stippled zones indicate that contoured surface is not seen in map view, because of displacement on a dipping fault. Note that for the purposes of this project each fault has been numbered. (d) Structure contour map for the base of the St Bees Sandstone. Although the overall pattern resembles the higher surface shown in (c), the altitudes are less and the differing locations of similarly numbered fault zones reflect the fault dips. (e) and (f) Geological cross-sections along the lines C–D and A–B, shown in (h). Note from the structure contour maps, (c) and (d), that the two cross-sections are roughly in the dip and strike directions, respectively, of the bedrock formations, and hence little dip angle is apparent in (f). (g) 'Rockhead contours' in metres, relative to a fixed datum, i.e. elevation of the bedrock surface. (h) Contours, in metres relative to a fixed datum, of the 'saline transition zone', i.e. the elevation of the boundary between fresh and salt groundwater. Redrawn by permission of UK Nirex Ltd.

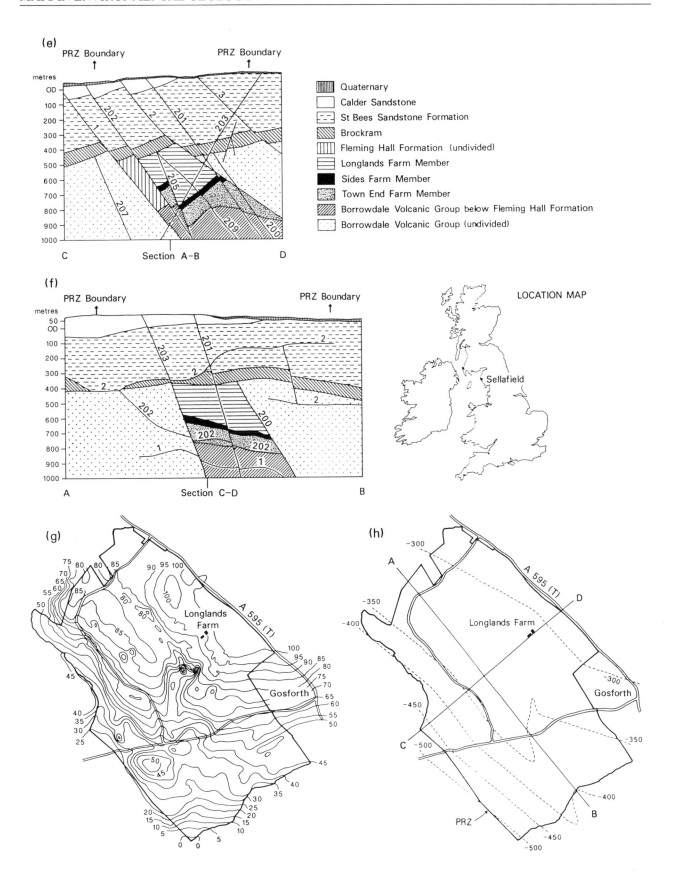

(e)

Section A–B

Quaternary
Calder Sandstone
St Bees Sandstone Formation
Brockram
Fleming Hall Formation (undivided)
Longlands Farm Member
Sides Farm Member
Town End Farm Member
Borrowdale Volcanic Group below Fleming Hall Formation
Borrowdale Volcanic Group (undivided)

(f)

Section C–D

LOCATION MAP

Sellafield

(g)

Longlands Farm

Gosforth

(h)

Longlands Farm

Gosforth

PRZ

such studies – Nirex UK Ltd – in its analysis of the Sellafield site, northwest England.

12.7 Planning Advice

Any one of the aspects mentioned in the preceding sections might be crucial for a particular area but it should not be dealt with alone. Almost certainly it will interact with other geological factors, and good planning considers all the various aspects together. The area in figure 12.16 illustrates how a number of the aspects mentioned interact together, in this case in an area no bigger than 25 km². Quarrying of pebbly beds in the Triassic sandstones is expanding, as they make good aggregate. Yet the public water supply of the area comes from these same sandstones, so clearly it

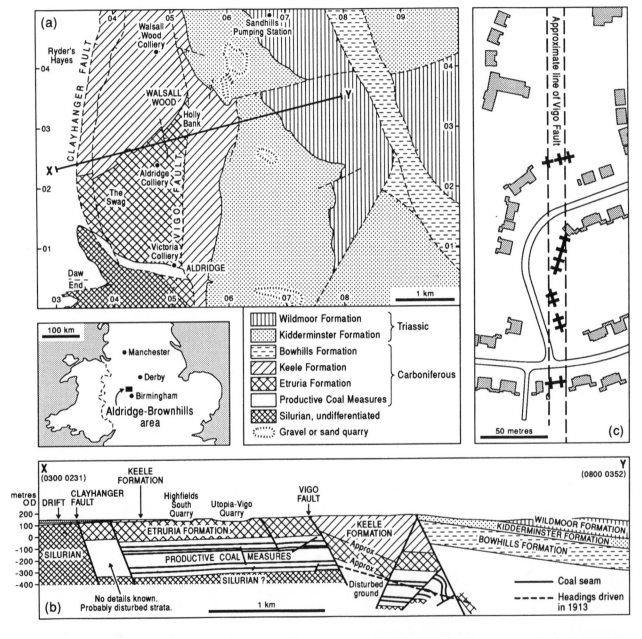

Figure 12.16 Interaction of geological factors in environmental planning: an example from the Aldridge–Brownhills area, England. See text for discussion. (a) Sketch map of the generalised bedrock geology of the area, showing line of cross-section (X–Y). Insets show location and key. (b) Geological cross-section. Only representative coal seams are shown. (c) Subsidence effects, apparently related to underground collapse along the Vigo Fault. Redrawn from Wilson (1986), by permission of the Geological Society, London.

has to be protected from any future development. There are other good economic deposits in the area, such as the important brick-making reserves of the Etruria Formation, though in places these are covered by thick superficial deposits. Even recently, some accessible reserves were covered by new housing. Construction conditions are very variable: some bedrock and areas of till, sand, and gravel provide sufficiently stable foundation conditions for most normal constructions.

However, other superficial deposits show variable foundation conditions and mask an uneven bedrock surface. There are buried glacial channels and some peat, a particularly awkward construction material. The mudstones in the Etruria Formation are prone to landslip; they currently threaten a canal and an existing landfill site. But the major hazard to be incorporated in planning is subsidence. Mining for coal and limestone used to be intensive in this area and it has left a legacy of major subsidence problems. Many of the old underground workings are unrecorded. Substantial damage to buildings has already occurred, especially where the workings appear within 15 metres of the surface. At such sites loading by modern structures would cause catastrophic collapse. There are two major faults in the area and underground failure seems to be concentrated along them (figure 12.16c). Levelled dumps of old mining waste are extensive, and though some have been landscaped and set aside for recreation, others are the sites of industrial development. Domestic wastes have been disposed of in disused brick quarries. Toxic liquid wastes have been injected into deep, disused coal workings, beneath the impervious Etruria Formation – the same formation that is being exploited for brick clay. It is no easy task to reconcile even these geological factors, let alone the often competing social and economic pressures, but the first step is to have the information available.

There is obviously a limit to how much relevant geological information can be communicated clearly on a single map, so the trend is for the geological advice to be broken into individual topics, each of which is dealt with on one map with a specific end use in mind. A suite of such maps for a particular area then provides the geological background for informed planning. Figure 12.17 lists the topics covered by the maps that comprised an early effort of this kind by the BGS. This kind of approach is now firmly established at the BGS and, although the individual themes differ from area to area, the number of these folios of maps for environmental planning is rapidly expanding. Because each map focuses on a single topic, the information is more accessible to planners. Many writers, including Robinson and Spiker (1978) in the USA,

1. Location of the area	13. Areas of shallow undermining
2. Topography	
3. Location of boreholes	14. Areas of potential landslip
4. Superficial deposits	
5. Lithology of superficial deposits	15. Opencast workings
6. Engineering properties of superficial deposits	16. Hardrock aggregate resources
	17. Limestone resources
7. Thickness of superficial deposits	18. Brick and tile clay
8. Depth to water in superficial deposits	19. Mudstone for brickmaking
	20. Hydrogeology
9. Thicknesses of sand and gravel deposits	21. Underground storage potential within 100 m of the surface
10. Bedrock geology	
11. Lithology of bedrock	22. Sand and gravel potential
12. Rockhead contours	23. Foundation conditions
	24. Groundwater resources

Figure 12.17 List of themes of the suite of maps in the Glenrothes Environmental Geology folio (Nickless 1982), which provided a model for later BGS environmental planning maps.

Gozzard (1985) in Australia, Dearman (1987) in the UK, Wolff (1987) in Norway, and de Mulder (1988) in the Netherlands, have urged that environmental geology maps be made digestible to all involved in planning and not just those with some geological training. Tyler (1995) used the example of planned land development in the San Francisco Bay area to illustrate how the environmental geological information could be presented in a very simple way. Indeed, some USA environmental geology maps are published in two forms, one that includes the hard data and the other aimed specifically at a non-technical audience. Environmental maps may be accompanied by an explanatory report, and some are richly illustrated with colour photographs, all of which is meant to encourage planners to incorporate geological matters in their deliberations.

There is some temptation to shortcut the process and simply present a map summarising the geologically based advice, but there are at least two reasons why the geological data have to be communicated properly. One is that environmental maps can be used for a variety of purposes and the advice is not necessarily the same in all circumstances. The second is that advice can carry with it liability, as discussed by McCall and Marker (1989), and the geological surveyor will not wisely step beyond his or her professional expertise.

Environmental geology maps are now being published in many parts of the world, with emphases reflecting the particular needs of the region, the nature of the geology, and the varying balances of interest in, for example, planning for development, engineering aspects, and conservation. McCall and Marker (1989) provided a review of the progress to that date and catalogued representative examples. Hagan (1994) summarised the role of state surveys in the USA. There are numerous publications that document case histories of the use of maps in environmental geological planning, including Culshaw et al. (1987), McCall and Marker (1989) and Lumsden (1994) and textbooks of environmental geology. They illustrate further the need to integrate the relevant geological information into a better understanding of the environment and into the planning process. While some of the instances they cite are models of good planning, other developments failed to incorporate geological factors, and the results were disastrous.

Summary of Chapter

1. Geological maps are finding rapidly increasing applications in portraying the distribution of geological data of environmental relevance and in exploring possible connections between the geology of an area and some of its environmental issues.
2. Hydrogeological maps emphasise how the distribution of water holding and transmitting properties varies in an area, and may contain information for environmental planning purposes.
3. Maps may highlight constraints on the extraction of mineral deposits such as aggregates; they may delimit areas of differing suitability for construction and sites of geological hazard.

4. Maps aid in the sensible planning of landfill sites, the below-ground storage of toxic waste, and assessing possible long-term repositories for radioactive waste.
5. Environmental geology maps present and integrate the geological factors relevant to planning future land use and understanding the environment better.

Selected Further Reading

Catt, J.A., 1988. *Quaternary Geology for Scientists and Engineers*. Ellis Horwood, Chichester, 340 p.

Chapter 7 deals with the economic and constructional significance of Quaternary features.

Lumsden, G.I. (editor), 1994. *Geology and the Environment in Western Europe*. Oxford University Press, Oxford, 325 p.

Not specifically about maps, but includes many examples and case histories.

McCall, J. and Marker, B. (editors), 1989. *Earth Science Mapping for Planning, Development and Conservation*. Graham & Trotman, London, 268 p.

A comprehensive review of environmental geological maps and their applications, up to 1989.

Monmonier, M., 1997. *Cartographies of Danger: Mapping Hazards in America*. University of Chicago Press, 363 p.

Examples from across the USA illustrate how maps are used to depict hazards such as those associated with earthquakes, volcanic eruptions, coastal erosion, floodplains, and groundwater pollution.

Woodcock, N., 1994. *Geology and Environment in Britain and Ireland*. UCL Press, London, 164 p.

A succinct account of environmental geology in the British Isles.

CHAPTER 13

Wider Uses of Geological Maps: Understanding the Landscape

13.1 Introduction

Viewing our physical environment in a more integrated way has, in addition to the planning approaches discussed in the previous chapter, revived interest in the interaction between geology, scenery, culture and society – landscape in its broadest sense. And because the nature and distribution of the bedrock of an area have a fundamental influence on these things, geological maps take on a further use. Here, it is mainly the geological materials at surface – maps in the two-dimensional sense – that are more relevant than the three-dimensional matters that occupy much of this book. The idea of using geological maps to trace how outcrop influences landscape is by no means a new one, and Fortey (1993) has entertainingly restated it for the general reader. His introduction touches on the significance of the geological map for explaining such varied things as why in some areas cows thrive and carrots grow well, why particular building materials have been used in some districts, and why wild-flowers flourish in certain places. He concludes: 'the basic tool then for understanding the foundations of landscape is the geological map'. The present chapter introduces how this tool can be used.

An initial general illustration is provided by the overall landscape of Great Britain. The geological map (figure 13.1) shows that the SE half of the country is dominated by younger rocks. These softer materials have produced lowlands of high agricultural quality with easy communications between sprawling, progressive cities of brick. In contrast, the older rocks to the NW have produced upland areas, veneered thinly with acid soils fit only for limited livestock, with tortuous communications between scattered villages – built from the local bedrock – and where older cultures still survive. Such influences in the UK are

Figure 13.1 Simplified geological map of the UK to show the outcrop of Carboniferous and older rocks (ornamented) and rocks younger than Carboniferous (white), a division that affects many aspects of the country's landscape.

explored further by Woodcock (1994), who presented maps of the influence of geology on a wide range of environmentally related activities.

Scotland provides a further illustration. A 'Highland line' is often visualised as dividing the Scottish Highlands from the remainder of Scotland. The

'Highlands' have been defined in various ways: physically – on the basis of land over 300 m in altitude (figure 13.2a); economically – according to land use; commercially – on the distribution of malt whisky distilling (figure 13.2b); and culturally – by the distribution of ancient clans (figure 13.2b) or the speaking of the Gaelic language (figure 13.2c). All of these aspects show a close coincidence with the bedrock geology, as seen on a geological map (figure 13.2d). Note, incidentally, that although bottled water from Scotland tends to be labelled with Highland images of tumbling streams and snow-dusted mountains, all the main bottling plants derive their water not from streams on the dominantly Precambrian rocks of the Highlands but from groundwater in younger rocks.

In the United States, New England tends to have thin soils, strewn with glacial boulders, in valleys between steep-sided mountains. Further south, say in Pennsylvania, the seaboard plains have deeply weathered bedrock with well-drained, fertile soils. Thus it is no coincidence that the New Englander has been caricatured as 'a scrawny, sallow, Scrooge-faced fellow who copes through miserly thrift and native guile' whereas a Pennsylvania farmer was pictured as 'a jolly, rotund fellow with an apple-cheeked wife of ample girth, cheerfully preparing mountains of highly calorific food' (Conzen 1994). In other words, a starting point for understanding a nation's physical and cultural divisions is an examination of its geological maps.

13.2 Physical Landscape

Section 2.2 introduced the idea that geology influences the relief of an area. Other things being equal, stronger rocks are more resistant to erosion and form higher land. Figure 13.3 summarises in a very general way the resistance of common rocks to erosion and indicates the typical kinds of landforms that result. Lithology is not the sole factor, though. Deformation and metamorphism tend to subdue the toughness contrasts between sedimentary formations. The presence of joints (section 9.1) can have a weakening effect. And because with time the particles of rocks tend to become more strongly bonded together, older rocks tend to be tougher. Figure 13.4 shows how the resistance to erosion of some rocks common in the UK varies in a general way with stratigraphic age.

An area consisting of an inclined, stratified sequence of different sedimentary rocks is likely to show marked differences in topography. A common situation is for relatively weak rocks such as mudstones and shales to underlie valleys and for tougher sandstones and limestones to form upstanding ridges. Land surface that slopes in the same direction as the dip of the beds commonly follows the dip angle whereas land crossing the strata is often steeper. The effects are seen in the right-hand part of figure 6.7 and in figure 7.2. The scarp-and-vale topography of parts of lowland Britain is due to these relationships. Landforms of the southwestern USA such as buttes,

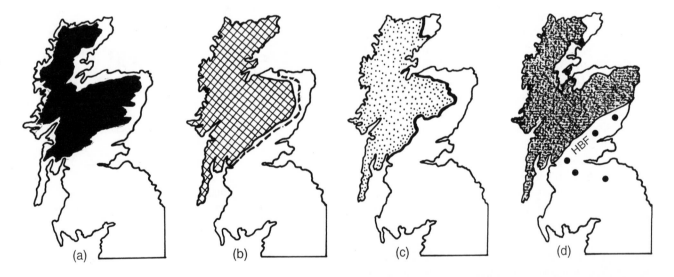

Figure 13.2 Maps showing the influence of geology on some aspects of the landscape of the Scottish Highlands. (a) Relief. The black area shows land higher than 300 metres. (b) The ruled pattern shows the area of the Highland clan system, and the dashed line is the 'Distillery line' that demarcates the area, to its NW, of important malt-whisky distilling (from Withers 1982). (c) The extent of Gaelic-speaking in Scotland (dotted area) until last century (from Withers 1982). (d) Much simplified geological map to show the extent of dominantly Precambrian rocks (schists and gneisses) in northern Britain. The southern boundary is sharply defined by the Highland Boundary Fault (HBF), often referred to as the 'Highland line'. The black filled circles indicate the locations of bottling plants for natural mineral water; see text.

	ROCK NAME	RESISTANCE	PHYSIOGRAPHIC FORMS
IGNEOUS FINE TEXTURED DARK (Basic)	**BASALT**	Usually resistant except when bearing olivine. Exfoliates readily	Columnar jointing, dykes and escarpments
MEDIUM	**ANDESITE**	Usually resistant	Not widespread enough to form typical landscapes
LIGHT (Siliceous)	**RHYOLITE**	Usually resistant, but sometimes decomposes badly	Bluffs and cliffs
COARSE TEXTURED DARK (Basic)	**GABBRO DIABASE**	Usually very resistant except when much jointed and when containing olivine	Dykes and escarpments
MEDIUM	**SYENITE**	Similar to granite but lack of quartz renders less resistant	Uplands
LIGHT (Siliceous)	**GRANITE**	Usually resistant. Disintegrates readily in arid regions	Exfoliation domes. Bosses and uplands
SEDIMENTARY FINE GRAINED: ARGILLACEOUS LOOSE	**CLAY**	Weak but usually coherent enough to form vertical walls	Bluffs and badlands
CONSOLIDATED	**SHALE**	Usually weak	Gentle slopes, valleys and lowlands
FINE GRAINED: LIMY LOOSE	**MARL**	Very weak	Low valleys
CONSOLIDATED	**LIMESTONE**	Weak in humid regions. Resistant in arid	Karstland, sink-holes, high escarpments
COARSE GRAINED LOOSE	**SAND**	Usually weak	Lowlands but sometimes caps uplands
CONSOLIDATED	**SANDSTONE**	Resistant if well consolidated	Cliffs and plateaus
VERY COARSE LOOSE	**GRAVEL**	Moderately resistant because of porosity	Often caps uplands
CONSOLIDATED	**CONGLOMERATE**	Very resistant	Ridges and mountains
METAMORPHIC CHANGED FROM SEDIMENTARY ROCKS from SHALE by moderate pressure	**SLATE**	Weak but more resistant than limestone	Lowlands
LIMESTONE by pressure	**MARBLE**	Weak	Lowlands unless associated with more resistant metamorphics
SANDSTONE by cementation	**QUARTZITE**	Very resistant. Perhaps the most resistant rock	Residual ridges, knobs and monadnocks
CHANGED FROM EITHER IGNEOUS OR SEDIMENTARY ROCKS BANDED	**GNEISS**	Usually very resistant	Uplands
SCHISTOSE	**SCHIST**	Usually resistant	Uplands and ridges

Figure 13.3 Lobeck's (1939) classification of common rock types and their resistance to weathering and erosion.

cuestas and flat-irons also reflect a close dependence of relief on bedrock. Faults may bring rocks of contrasting durability next to each other, with corresponding effects on the landscape. Figure 13.5 illustrates diagrammatically how differently disposed beds in different settings give rise to different physiographies.

13.3 Soils, Vegetation, and Habitat

The geology of an area affects the scenery in another important way: it influences the nature of the soil and hence the ecology. Bedrock directly affects the chemical make-up of the soil. For example, soils associated with scree slopes of sandstone or granite only allow

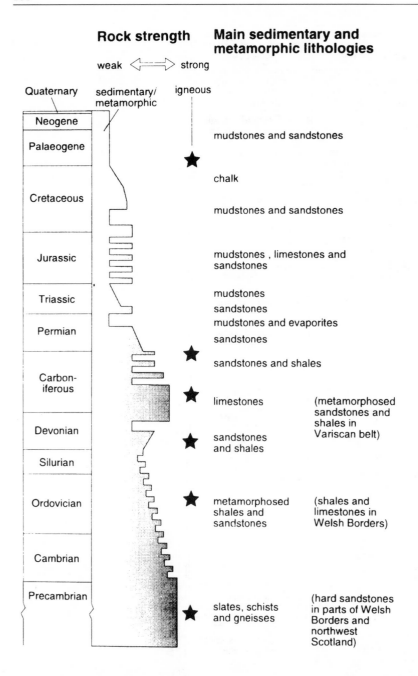

Rock strength **Main sedimentary and metamorphic lithologies**

weak ⟵⟶ strong

Figure 13.4 Diagram to show the resistance to weathering of the main rock types of the UK. Reproduced from Woodcock (1994), by permission of UCL Press.

the growth of certain mosses and mountain berries whereas if slates and volcanic rocks are involved, ferns and dwarf willows may appear in addition (Pearsall 1968); where basic igneous rocks are present, even flowers can flourish.

Bedrock also normally controls the proportions of sand, silt, and clay in the overlying soil, and these have a lot to do with the soil's water-holding and transmitting properties. Sandstone and granite tend to form porous,, well-drained, inert, sandy soils whereas mudstones and shales give richer, heavier, poorly drained material. Basic and ultrabasic igneous rocks can weather to give fertile soils, rich in lime (calcium

carbonate), potash, and magnesia. Mica-schists can give a similar result. Russian studies (e.g. Chikesa 1965) indicated that differences in vegetation could be used to map out such things as superficial deposits, subsurface salt and groundwater variations; higher trees and denser grass were noted to form on diamond-bearing rocks (kimberlites) in comparison with the surrounding sedimentary rocks.

Two particularly conspicuous relationships between bedrock and ecology are given by outcrops of serpentinite and limestone. Brooks (1987) has documented a range of influences of serpentinite, such as on the distribution of termite mounds in Zimbabwe, and of

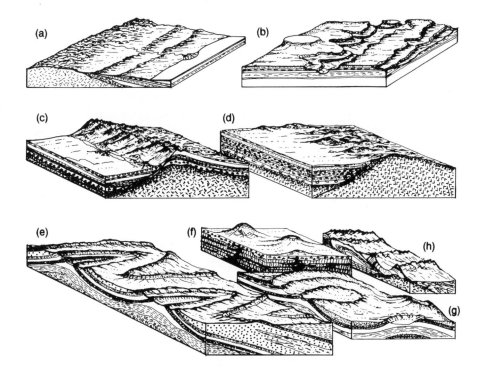

Figure 13.5 Block diagrams showing how landforms are influenced by different types of bedrock and their arrangement. (a) Coastal plains; approximately horizontal sedimentary strata. (b) Interior plains, developed on flat-lying strata. (c) Young fault-block mountains. (d) Old, eroded, fault-block mountains. (e) Fold mountains, with regularly plunging folds of differently resistant strata. (f) Volcanic landscape, with young volcanic cones on flat-lying lava plain. (g) Domal hills generated by periclinal folds, that is, the fold plunges vary in direction. (h) Young mountains in complex folded and thrust bedrock.

gopher holes on the campus of Stanford University! Limestone (including chalk) produces soils of markedly low acidity. This, together with its effect of unlocking nutrients, promotes the growth of particular plants.

Certain trees flourish in such soils – limes, box, and yews, for example. Although single yews are scattered across the British Isles, nearly all woods of yew are sited on chalk or limestone bedrock. This, in turn, has influenced place names. A geological map shows that places with names such as Yewdale, Yewbarrow and Yewcrag (even though there may be no yew trees growing there now!) are almost invariably sited on limestone. Similarly, although ash is a widespread tree, pure ashwoods are confined to soil derived from limestone. These trees even flourish on the elevated, windswept limestone pavements of the northern Pennines of England; the northernmost British ashwood is sited on an exposed hillside in Wester Ross in northern Scotland – where an anomalous strip of Cambrian limestone is brought amongst all the surrounding Precambrian rocks by a major thrust fault (section 9.7).

Many flowers illustrate the relationship. A glance at the local geological map will tell the gardener the chances of being able to grow particular flowers, as well as helping to explain their distribution in nature. As an example, consider where in the UK the wildflower called kidney-vetch can be seen growing, and where you would find bell-heather (figures 13.6a and 13.6b). In a general way, their distributions are mutually exclusive: the heather likes the Scottish Highlands, west Wales, southwest England and the heaths of southern England whereas the vetch does not. The reason is clear from a geological map. The vetch, unlike the heather, likes limestone soils (figure 13.6c).

The converse of looking at a geological map to help understand the vegetation is for the geologist to use what is growing to help him produce his map of bedrock. When Howell Williams, co-author of Greenly and Williams's treatise on geological surveying (see section 14.3.6 and chapter 15), was surveying the geology of the Snowdon district, North Wales, he noticed how green ferns and mosses abounded on one particular volcanic formation. The adjacent formation, in contrast, was covered by heather and bilberry. On this basis he was able to map in the outcrop of the formations, even where they were unexposed.

The nature of the vegetation influences the insect life that lives there, and, in turn, other life higher up the food chain. Thus animal life may to some extent be influenced by geology. Calcium is needed to build shells and bones and hence, in the UK, snails flourish in limestone areas whereas elsewhere slugs dominate. Red deer in Scotland living away from lime-bearing rocks have sometimes to eat discarded antlers in order to maintain sufficient calcium for growth. Fortey (1993) described the prized trout streams of England – cold, clear rivers flowing over chalk – where freshwater crayfish live. The creature lives there because it needs calcium, derived from the limestone, in order to construct its shell.

Bedrock has other effects on habitat. One is its influence on slope and drainage, and another is the control on exposure to wind and sunlight, in other words the microclimate. Figure 13.7 illustrates a variety of upland situations in the UK where the orientation of the beds affects the vegetation that grows there. Another role of geology is its influence on the surface water, both the course that the water takes and its chemical make-up. Frazer Darling and Morton Boyd (1970), in discussing northern Scotland, remarked how 'a pleasant hour could be spent comparing the freshwater systems of the Highlands with the geological map'. As an example, they described a loch in Wester Ross that abounded in crustacea and insects – fish food. As a result the loch is famous for the outstanding size and quality of its fish. Yet, just a few

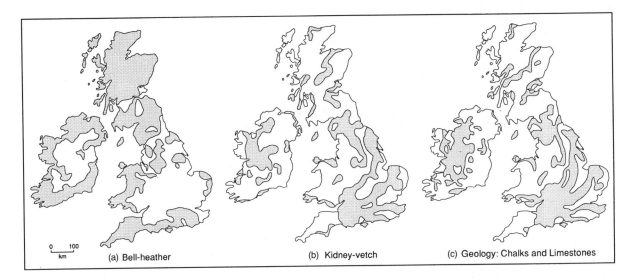

Figure 13.6 The influence of bedrock geology on the distribution of two wildflowers common in the UK. (a) Bell-heather. (b) Kidney-vetch. (c) Simplified geological map showing the major outcrop of limestones in the UK. Based on Perring and Walters (1976).

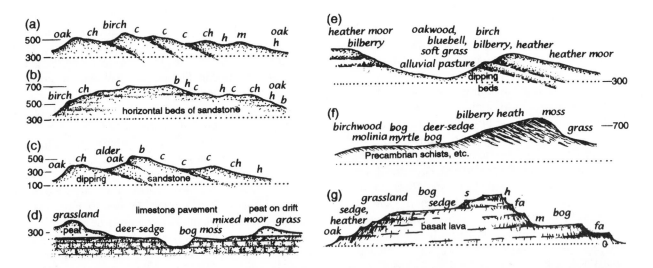

Figure 13.7 Diagrammatic cross-sections (vertical exaggeration × 2) to show the influence of rock type and differently dipping beds on plant ecology, based on Pearsall (1968). (a) Dipping sandstone beds, Pennines, England. c = cotton-sedge 'moss'; h = heather; ch – both together; m = moor-grass. (b) Horizontal sandstone, Bleaklow, Pennines, England. b = bilberry moor; otherwise as (a). (c) Dipping sandstone beds, Pennines, England. As (b). (d) Horizontal limestone beds, partly overlain by peat with, in places, glacial drift. Craven Uplands, west Yorkshire, England. (e) Sandstone beds with variable dip, northeast Yorkshire, England. (f) Part of Ben Wyvis, eastern Scottish Highlands, with foliation of Precambrian rocks affecting slope and exposure. (g) Terraces in poorly permeable basalt lava flows, Elgol, Isle of Skye, west Scotland. s = deer-sedge; h = heather; m = moor-grass; fa = fescue-bent grassland.

hundred metres away, another loch has but a few small trout. The reason is evident from the geological map – the first loch is located on a strip of metamorphosed limestone, which prompts alkalinity and fertility in its waters.

The calcium content of water can also affect the vegetation. One loch on the Scottish Isle of Lismore is cloudy in spring and summer while in winter it is perfectly clear. The change is at first puzzling, but the fundamental reason is apparent from the geological map. In the long days of spring and summer the vigorously growing plants need to extract large amounts of carbon dioxide from the water, so much that a fine dust of calcium carbonate is precipitated. The water is so rich in carbonate because – unusually for this region – the island is virtually wholly composed of a Precambrian limestone.

13.4 Human Activities

Humankind has been influenced by geology right from the start: the distribution of bedrock had a lot to do with where people chose to live, their tools, and their trading materials. Consequently, geologists and archaeologists, now working together in such new subjects as archaeological geology or geoarchaeology, find a number of uses for geological maps.

Neolithic man, for example, with only hand implements available to him, preferred to live on the loose, easily worked soils that develop on limestones. Hence, Neolithic settlements tend to be concentrated in regions of limestone outcrop. In areas of active tectonics the land can change so quickly that man has

continually to adapt, and the information on a geological map may be crucial to archaeological reconstructions (e.g. Bailey et al. 1993). In seeking to interpret lifestyles of the past, the archaeologist needs knowledge of the physical environment of the time, and evidence may be contained on maps that include superficial deposits. Moreover, there may be information on past natural disturbances of these materials, such as by flooding, erosion, and coastline change, and this information is needed in order to differentiate man's own activities. It may be important in planning excavation sites as well as in interpreting the results (Bell and Boardman 1992).

Archaeologists have long used geological maps for finding the natural distribution of materials used by ancient man and to help interpret trade patterns (e.g. Rapp et al. 1990). Many of the Bronze Age objects found in Britain originated in Ireland, which appears to lack an obvious source of tin. Budd et al. (1994) discussed the distribution of tin deposits in the remainder of the British Isles in an attempt to reconstruct the routes by which the raw and the worked material was transported in Bronze Age times. The World Wide Web is increasingly being used to archive maps that highlight geological sources of archaeological artefacts.

In more recent times, the local bedrock had a major influence on building – both the material and the style of construction (e.g. figure 13.8). Indeed one of the most striking aspects of many historic towns and villages in Europe is the material of which they are built: the red brick of much of clay-based eastern England,

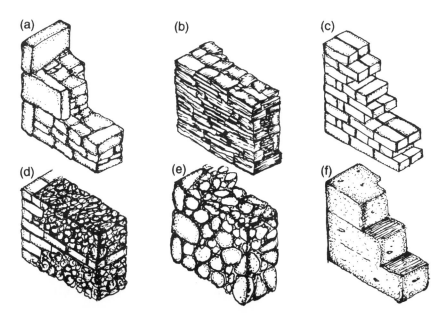

Figure 13.8 Diagrams to show how different kinds of rocks influence the construction and appearance of walls. (a) Stone wall typical of sandstone areas, with partly dressed blocks. Limestone walls tend to be similar but less regular. (b) Slate stone wall. Notice interior packing of finer, rubbly chippings. (c) Brick wall. In older walls the bricks are less regular and placed in a more traditional pattern, here the so-called English Garden wall bond. (d) Flint wall, typical of the chalk areas of western Europe. In this example the flint nodules are broken ('knapped') to give relatively smooth surfaces and bricks are incorporated. (e) Cobblestone wall, with split cobbles. (f) Clay wall, found in areas of heavy clay. Unlike bricks the clay is not baked; straw, etc. may be added for strength. A whitewashed rendered finish is normal.

for example, the honey-coloured limestone of the Cotswolds, and the grey limestones of the Pennines. Areas that lacked building stone or brick-clay relied more on timber. Older parts of the USA show similar features: the cream limestones of southern Indiana, the granite of eastern Maine, the basalt of San Francisco. In olden days, such materials could not be transported very easily, and hence a geological map provides the explanation of the local building materials.

The nature of the material influenced the style of construction – in field walls, barns, houses, and churches, including the roofs – and the building design (e.g. Penoyre and Penoyre 1978). The small, smooth slates quarried around Angers, France, for example, lend themselves to the delicate curving roofs of the Loire chateaux as well as lesser buildings of the region. In contrast, the rough-splitting, silty slates of Cantal are used in plain, small roofs of rustic appearance, heavily supported by thick timbers. To some extent the distribution of bedrock not only influences building design and construction but the spread of the people that know how to work the material. Just as, say, men of Cornish extraction are found today in a number of metalliferous mining areas of the world, the huge slate quarries that once operated in North Wales have led to Welshmen working slate in many areas abroad. It is no coincidence that the slate industry of Pennsylvania is centred on a town called Bangor.

The relationship between industry and geology is well known. The industrial revolution – spawned where coal, iron, and limestone occurred together – illustrates it well (Woodcock 1994) and geological

materials still hold the key to the siting of a number of industries. Evaporite deposits, an important source of chemicals such as salt, have governed the locations of major chemical factories: Detroit, Michigan; Trona, California; Basle, France; and Cheshire and Teeside in the UK. Cement plants tend to be located on outcrops of limestone.

Less well known is the relation between geology and the distribution of breweries. Besides hops and barley, the main influence on beer quality is the water, which many breweries extract from the local bedrock (Lloyd 1985). In the UK, important brewing centres became established in places such as Tadcaster in North Yorkshire, sited on Magnesian Limestone, and on the Chalk of eastern and southeastern England, because of the advantages of having calcium and bicarbonate in the brewing water. Other celebrated breweries are located on sandstone. Those of Burton-on-Trent, in the English Midlands, are sited on gypsum-rich Triassic sandstone and capitalise on the high sulphate concentration of the water, a major control on taste. The sulphate also promotes stability, and beer produced here last century was the first that could withstand the long journey to the colonies. The key to understanding Export Ale, India Pale Ale and their birthplace, therefore, lies in a geological map!

A remarkable relationship exists in some parts of the world between wine and the bedrock on which the vines grow. Consulting a geological map can help explain why certain kinds of grape vine thrive in particular spots (see, for example, figure 3.3). The vines of

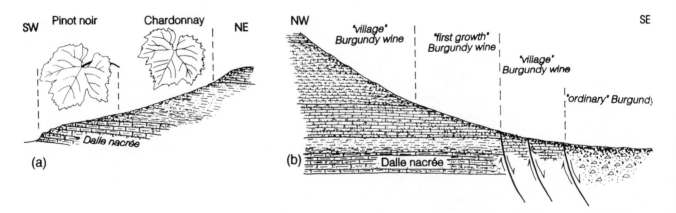

Figure 13.9 Geological cross-sections of two parts of the Burgundy area of France, illustrating the control of bedrock geology on the kind and quality of wine yielded by the grape-vines that grow there. (a) The Corton area, 5 km north of Beaune, where experience has shown that the red grape Pinot Noir grows best on a pearly looking limestone (Dalle nacrée) whereas the white-wine producing Chardonnay vines prefer a slightly sandy, highly calcareous shale. Both formations are Jurassic in age. (b) The Pommard–Volnay area, 5 km south of Beaune. Here, in the core of the Volnay syncline, the Dalle Nacrée has dipped well below ground, and the vines thrive on some overlying limestones. The best wine – the so-called 'first-growth' – comes from one particular limestone formation, and the next best – the 'village' wines – are produced from adjacent zones. Out on the down-faulted plains, where the limestones are deeply buried, 'ordinary' Burgundy wine is produced. Based on Vigneaux and Leneuf (1980), and Pomerol (1989).

Hermitage, which produce a rich, red wine that is among the greatest and most expensive in France, are restricted to granite soils, whereas Tuscany's best vineyards are on a complex of schist-like rock. And concentrated on a narrow strip of red soil leached from the underlying limestone – in places no more than 200 metres wide and known as Coonawarra – are some of the most famous vineyards of Australia. Figure 13.9 depicts another celebrated example, the geological setting of some of the best vineyards in Burgundy, France.

13.5 The England–Wales Borderlands (Plate 1)

This is the first of four examples selected to illustrate how a geological map can provide the key to understanding the landscape of an area. The area depicted on Plate 1 shows a variety of geological features, some of which were introduced in chapter 2. It is this diversity that accounts for the very varied scenery here. The striking change from the mountainous scenery of the NW of the area, where Cadair Idris [SH7515] – perhaps the best known Welsh mountain outside Snowdonia – is carved in the complex of resistant, intrusive and volcanic rocks coloured red, pink, and orange on the map, through the wild but rounded uplands around Plynlimon [SN8085], formed of tough Ordovician and Silurian mudstones and sandstones, to the mellow, verdant pastures around Hereford [SO5040], based on Old Red Sandstone, is all due to the overall change in bedrock. Further east, the plains of Permian and Triassic rocks show only a little topographic relief. There, marly horizons give lower land and fertile soils whereas sandstones give more elevated ground, with soils that sometimes dry out. The former were attractive to earlier settlers; the latter were the last areas to be cultivated.

The map also shows how the numerous intricacies of the bedrock geology account for the various local changes in relief, such as the limestone uplands of the Black Mountains [SN7518], and the scarp-and-vale topography of Wenlock Edge [SO5090], formed by a sequence of limestone units within Wenlock and Ludlow-age mudstones. The prominent N–S ridge of the Malvern Hills [SO7645] rises where Lower Palaeozoic rocks and a strip of tough Precambrian gneiss and schist (unit 30) protrude from the adjacent, soft, red sandstones (coloured orange on the map). The abrupt change in rock types gives rise to springs, from which the famous Malvern Water is bottled. Perhaps nowhere is the influence of bedrock on relief seen more sharply than in the odd little hills that jut out of the Shropshire Plain where older rocks outcrop, for example, at Caer Caradoc [SO4893], formed of rhyoli-

tic lava and tuff, and The Wrekin [SJ5904], of Upper Cambrian sandstone. The Longmynd [SO4295] is an airy, upstanding, heather-covered mass which Fortey (1993) has suggested has an ancient *feel* that magically transports one to the Scottish Highlands. It is no coincidence that both those areas are made of Precambrian rocks.

In the far NW of the map, a line of dashes and dots marks the trace of the Bala Fault. Its straightness is typical of major strike-slip faults (section 9.9), and the weakened rocks of the fault zone have been preferentially eroded to give lower land and, in places, steepsided valleys. Lake Bala is long and narrow in plan, elongate along the fault trace, and one of the deepest lakes in Wales. The setting is analogous to Loch Ness, a deep elongate lake lying in the Great Glen Fault zone of northern Scotland. Moreover, the deep, isolated waters of Bala Lake have allowed the evolution of a fish species unique to the lake – the gwyniad – a further analogy, some say, to the aquatic life of Loch Ness. Immediately NW of the fault line, centred on [SH7515], is the complex of volcanic and sedimentary rocks that makes the mountain of Cadair Idris, with a sparse vegetation on its windswept slopes. At one horizon, however, a band of lime-rich pillow lavas has produced alkaline soils which allow a variety of mountain plants to grow. Condry (1973) mentioned seven plants restricted to this unit and remarked that its outcrop could be 'traced as easily by botanists as by geologists'.

The hills of central Wales (labelled 'Cambrian Mou(ntains)' on the map) allow only scattered farms and isolated settlements. The fields are irregular, bounded by walls of the tough, local Silurian sandstone, often with blackthorn and mountain ash trees growing between the blocks. In contrast, the adjacent English lowlands have prompted centralised, compact villages to develop, with planned field systems outlined by hedgerows.

Similarly, the buildings vary across the area. Dolgellau [SH7018] is built from massive boulders of the local dark green dolerite; in the middle-west of the map area, unadorned dark-grey sandstone buildings abound. From the plains of Hereford [SO5040], blocks of Old Red Sandstone have long been carved to build cathedrals and castles, and a host of lesser buildings in the area. The 'New Red Sandstone', the Permian and Triassic rocks numbered 89 on the map, accounts for the vibrantly red walls and buildings encountered in the extreme east of the map area. The other striking buildings of these lowland areas are also influenced by the bedrock geology, but in an indirect way. The sandstones here tend to be soft and deeply weathered, giving fertile soils on which oak trees could flourish and farmers could do well. The exuberant, black and white 'half-timbered' buildings so typical of this area reflect this availability of first-class timber and an affluence which maximised its ornamental possibilities.

Much of the map area is agricultural, but of contrasting kinds. The acid, upland soils of the bleak Cambrian Mountains restrict farming to hardy sheep and the tough little Welsh Black cattle. Further east, the uplands are based on Upper Silurian siltstones (74 on the map) which are softer and make better pasture. It is still sheep country, though, and some famous breeds have originated here. Crossing eastwards onto the Devonian and younger sandstone plains, the soil becomes distinctly red and markedly more fertile. Hops and soft fruit are grown, and apples for cider. And the lush watermeadows of the sandstone plains around Hereford allow fattening of one of the world's great breeds of beef cattle.

There are, however, important industrial sites – where the bedrock geology has prompted it. The Lower Westphalian, unit 82 on the map, yields coal and accounts for the once great South Wales mines around Merthyr Tydfil [SO0208]. Here the E–W strip of grey on the map, with older rocks to the north and younger rocks to the south, represents the northern limb of a major syncline stretching off the map. The finer convolutions of the outcrop pattern, however, are topographic – the effect on the map pattern of the famous South Wales 'valleys'. The occurrence of the same rocks near Telford [SJ6608], together with iron deposits in the adjacent unit 83, helped spawn here the so-called 'industrial revolution'. The legacy of earlier coal mining is very evident in both the physical landscape and the folklore of the Forest of Dean, around Coleford [SO5811], where the coal-bearing rocks of unit 82 lie underground, beneath the outcrop of the younger, overlying unit 83. A further direct geological influence on the location of industry is provided by the outcrop of unit 89, which stretches NE off the map area towards the famous 'Potteries' of the Stoke-on-Trent region. The clay deposits within these and the underlying rocks were utilised at first for pots, and then whole ranges of bricks, tiles, and other ceramics. To the south, and based on the nearby coal and iron deposits, is the 'Black Country' of Wolverhampton.

All these contrasts in topography, history, vegetation, architecture, commerce, etc., between the industrial conurbations in the east of the map area and the empty uplands to the west, are striking, and the reasons are clear on the geological map.

13.6   Western USA (Plate 2)

Examples from the map reproduced here as Plate 2 have been mentioned earlier: the upfaulted block of resistant rocks forming the Wind River Mountains,

and the waterless, volcanic Snake River Plateau (section 2.2). The other structures mentioned in that section also have profound scenic effects. For example, the abrupt rise of the Teton range above Jackson Hole is the direct result of the Teton Fault, and the prominent N–S trend of the ranges to the south reflects exactly the thrust-faulted masses of resistant rocks. In fact, the striking physiography of this part of the USA is due essentially to upfolded and upfaulted masses of older, tougher rock alternating with basins filled with younger, soft materials. The NW–SE-trending massif in the NE of the map area, for example, the Big Horn Mountains, separates the Powder River Basin to its NE from the Big Horn Basin to its SW (figure 2.2), while the latter is divided from the Wind River Basin further SW by the Precambrian-cored mass of the Owl Creek range.

The Oregon Trail pioneers of the middle-1800s saw these differences in relief and used their features as landmarks. The trail itself, however, stuck to the easily traversed basins as far as possible. In the Wind River Basin, at lat. 42° 30', long. 107° 5', a very prominent lump of bare rock stands proud of the surrounding plain. It is a mass of Precambrian granite, far tougher than the surrounding Tertiary sands. Sighting it was crucial to the early emigrants. Proverbial wisdom had it that if they did not reach it by the 4th of July they could not tackle the high mountain ranges further west before the winter snow set in. It is still called Independence Rock.

The uplands form pine-covered slopes and clear aired, semi-alpine pastures enjoyed by tourists; the lowlands are dry and dusty but contain important economic deposits such as oil, uranium, the important industrial clay bentonite, and incredible amounts of near-surface coal. Some of the clays are used for bricks and as a primitive stucco but softwood timber is the main construction material if there are upland areas nearby. More up-market buildings may be faced with a local stone, for example the reddish Cambrian sandstones around Jackson (lat. 43° 30', long. 110° 45'), coloured pink on the map.

The geological map helps explain the scenery on a more local scale also. For example, the plains north of Sheridan, composed of Tertiary sands and gravels, are replaced to the NW by Cretaceous sandstones and limestones which give a complexly undulating, grassy country. The elevation differences between hilltop and valley are small, but in one of them, the valley of the Little Big Horn River, the differences were crucial in the 1876 US Cavalry–Indian battle there. A driver following Interstate 80 further west, between Billings and Bozeman, sees the sharp mass of Crazy Peak to the north, with ribs of resistant rock running from it. The map tells why. It is a patch of Tertiary granite, a prominent red patch on the map, with dykes radiating from it. Geology has had a very major impact on the development of these western states; upland, scenic areas catering for tourists, the semi-arid, poorly fertile plains for cattle.

The Tertiary granite mass in the far NW of the map area is the source of the rich copper mineralisation which spawned Butte, Montana; Casper became the largest community in Wyoming because of the oil reserves of the nearby Cretaceous sandstones. Young volcanic rocks give the high plateau of Yellowstone but are soft enough to have yielded dramatic river gorges in places; the National Park is framed to the east by the remote Absaroka range, a rugged expanse of Tertiary volcanic rocks still only traversed by one road. Even those visitors who marvel at this spectacular scenery without understanding its geological basis may take away a geological memento, say one of the celebrated fossil fish beautifully preserved in the Tertiary shales at Fossil Butte, near Kemmerer, or a carved piece of Wyoming jade, originating in the Precambrian granite and found as loose pieces in the nearby Sweetwater River.

13.7 Example: The Weald, SE England

The area of southeast England known as the Weald is an eroded, upright anticline (figures 13.10 and 13.11a). The oldest rocks are therefore in the core of the fold. They are dominantly alternating clays and sandstones, whereas the limbs are composed of tougher chalk. This information alone explains the main points of the physiography of the area, such as the asymmetric ridges known as the North and South Downs, underlain by chalk and with surfaces sloping away to the north and south, respectively. This relatively tough rock can form prominent cliffs and headlands where it meets the sea. But the geology explains much else about the landscape. The following section considers a few illustrations, in a traverse southwards across the region as far as the core of the anticline.

The northernmost parts of the area are the flat, poorly drained areas of south London, underlain by the London Clay. They are much built on, in brick, and some houses have subsidence problems due to the clay foundations alternately shrinking and expanding with the seasons. Small pastures still exist in places, bordered by oak woods and with rushes growing in the damp patches. Southwards, the land rises slightly as the sandier material of formations such as the Thanet Beds is reached. Better soil has formed in this area, and the land was favoured historically for settlement and agriculture. The oldest human remains in Britain are here – at Swanscombe – and at Oldbury

Hill a crude shelter made of sandstone slabs is one of the rare places preserved in western Europe where Palaeolithic man lived and worked. Older towns such as Croydon, Carshalton, and Epsom originated on these rocks, though they are now swallowed by the suburban sprawl of south London. Woolwich Arsenal was sited here partly because the then remote marshes of the Thames provided safety, but also because the bedrock outcrops provided the sands for moulding firearms.

To the south, the land rises to the chalk of the North Downs. The amount of rainfall increases (figure 13.11b) but most of the valleys are riverless. Villages rely on well water. The few streams are famous for their trout fishing and calcium-loving crayfish. Otherwise, surface water is restricted to temporary dewponds, still of mysterious origin. Open pastures of springy grass and wild herbs, loved by rabbits, mix with cornfields and beeches. The calcium-rich soils nurture a range of characteristic plants, which in turn encourage certain insects while discouraging others. Butterflies and moths have names such as Chalk-hill Blue and Chalk Carpet. Beechwoods flourish, together with groves of yews, but both lack undergrowth and so hole-nesting birds predominate.

The Chalk has a fairly uniform thickness; the variation in outcrop width seen on the geological map is a reflection of changes in dip angle (section 4.4). Neolithic settlements are concentrated here, as nodules of flint from the Chalk were a vital resource. Many hillforts are sited around the heights of the Chalk escarpment, while the prehistoric route called the Pilgrim's Way follows its foot. Early parish boundaries straddled either the dip-slope of the Chalk and the clay ground to the north, or the scarp slope and the land to the south; this oddly elongate shape is still apparent today. Buildings are simple in style here. Flint from the Chalk was used in the walls but, because it is difficult to lay squarely, brick was commonly used for corners and the surrounds of doors and windows (figures 13.11c and 13.11d). The bricks, made from the local chalky clay, tend to be yellow, grading into purple when they have been overfired.

To the south the ground falls away, slowly at first, and with heathland where sandy lenses of the Upper Greensand occur, and then more abruptly to the Gault Clay. Like the other clay formations of the region it forms low ground with wet, cold soils little suited to arable cultivation. These areas were late to be settled. Yet now this part of England is known for its orchards and market gardens: these tend to be sited on the better soil that forms over sandier beds within the clay units. The famous hopfields are commonly located where the clay is overlain by river gravels and loams. Construction of the M26 motorway across the

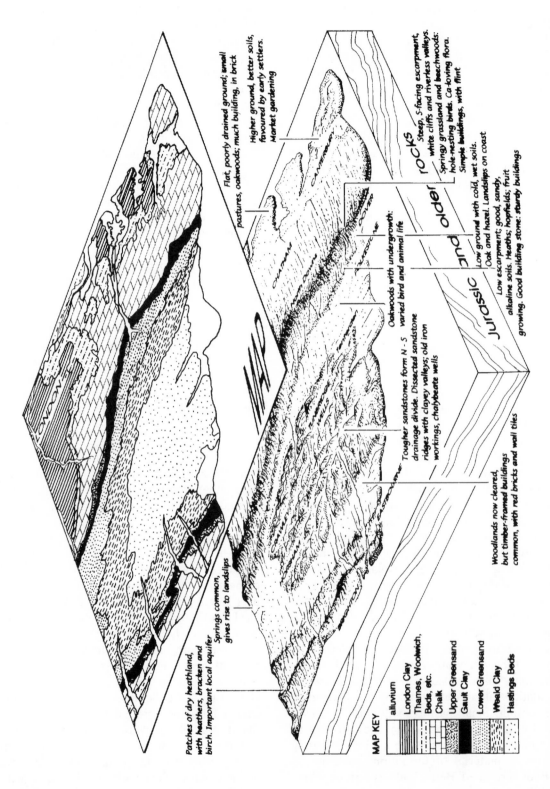

Flat, poorly drained ground; small
pastures, oakwoods; much building, in brick

Higher ground, better soils,
favoured by early settlers.
Market gardening

Steep, S-facing escarpment,
white cliffs and riverless valleys.
Springy grassland and beechwoods:
hole-nesting birds. Ca-loving flora.
Simple buildings, with flint

rocks

Oakwoods with undergrowth:
varied bird and animal life

and older

Low ground with cold, wet soils.
Oak and hazel. Landslips on coast

Tougher sandstones form N - S
drainage divide. Dissected sandstone
ridges with clayey valleys; old iron
workings, chalybeate wells

Jurassic

Low escarpment; good, sandy,
alkaline soils. Heaths; hopfields; fruit
growing. Good building stone: sturdy buildings

MAP

Woodlands now cleared,
but timber-framed buildings
common, with red bricks and wall tiles

Patches of dry heathland,
with heathers, bracken and
birch. Important local aquifer

Springs common,
gives rise to landslips

MAP KEY

alluvium
London Clay
Thames, Woolwich,
Beds, etc.
Chalk
Upper Greensand
Gault Clay
Lower Greensand
Weald Clay
Hastings Beds

Figure 13.10 Generalised block diagram to show the influence of bedrock, as portrayed on a geological map (upper surface of diagram), on aspects of the landscape of the Weald, southeast England. See text for discussion.

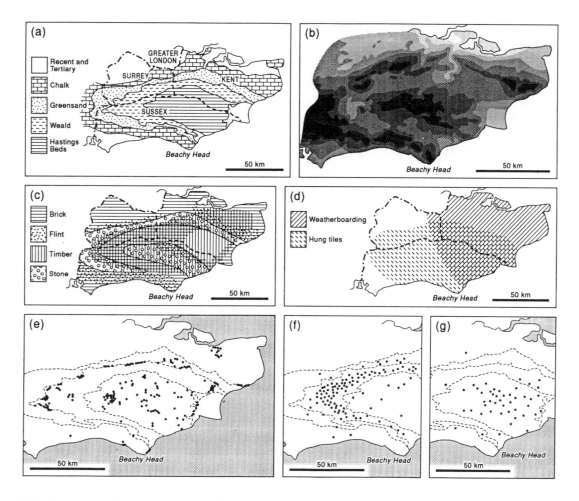

Figure 13.11 Examples of the influence of the bedrock of the Weald, southeast England, on aspects of landscape. (a) Simplified geological map. (b) Distribution of rainfall (densest shading represents areas of over 800 mm rainfall per year and the contour interval is 62.5 mm). Highest rainfall corresponds roughly with the outcrop of Chalk and Hastings Beds. (c) Distribution of principal walling materials, and (d) principal wall-cladding materials (based on Brunskill 1982). (e) Distribution of landslips and mudflows (based on Goudie 1993). (f) and (g) Distribution of place names: (f) shows the concentration of place names ending in -fold (a wood) and -hurst (a clearing) on bedrock of Greensand, and (g) place-names ending in -den (from *dene*, a woodland pasture) on Weald Clay and Hastings Beds. Based on Wooldridge and Goldring (1953).

sticky clay caused major problems – the black polythene that can be seen lining the drainage channels by the road is one very visible illustration.

The Lower Greensand, however, forms particularly good soils, as its calcium content is sufficient to give a low natural acidity. The densest population of England's new breed of vineyards is here. Some parts of this formation make an excellent building stone – it is seen in older buildings here but also in some of the grand buildings of London. Sandier layers give rise to stretches of dry heathland, where small birds such as buntings and pipits enjoy the bracken, birch, and expanses of heather. Pits mark where sand used to be extracted for foundries, now overgrown but excellent sites for wildlife.

Underlying the Greensand is the Weald Clay. Downward percolating water meets this impervious clay and forms springs, commonly the location of landslips (figure 13.11e). The Weald Clay contains a variety of lithologies, including limestones and sandstones that have been used as a building stone, and one sandstone bed which is capable of splitting into slabs to be used as a paving and roofing material. Localised deposits of iron and a relatively pure sandstone also occur, which once sustained a small-scale iron and glass industry. This clay area used to be heavily wooded and timber was much used as a construction material, but as it became more scarce brick was increasingly used. The bricks were often laid in random fashion between the structural timbers, and covered with wall tiles. The bright red colour of the bricks differs from the browns and brilliant oranges of the tiles, which were made from a different clay. The remaining woods are still dominated by oak, and have a rich undergrowth, home to warblers and wood-peckers, and carpeted by bluebells in the spring.

The core of the Weald anticline reveals the Hastings Beds, dominantly sandstones but with beds of siltstone and claystone. The sandstone layers are tough and form a complex series of ridges, much dissected by rivers flowing away from the central, E–W drainage divide. Iron nodules in the clay-rich parts stain the ditches and have led to place names such as Furnace Wood and Minepit Copse. Springs rich in iron salts spawned the spa town of Tunbridge Wells.

The localisation of certain place names has long been recognised in this district. Last century Topley (1872) reached an explanation with, as Wooldridge and Goldring (1953) put it, 'geological map in hand' (figures 13.11f and 13.11g). Names ending in -den, for example, originated between the seventh and thirteenth centuries when the clay ground, long unfavourable to settlement, became progressively cleared of woodland and settlements were created. Hence this particular name is concentrated around the outcrop of the Weald Clay (figure 13.11g). There are numerous detailed discussions of the landscape of the Weald: all are pervaded by the role of bedrock. The geological map is the key to understanding the Wealden landscape.

13.8   Example: the Pennsylvania Turnpike, Eastern USA

Figure 13.12 sketches the route of a much travelled highway in the eastern USA – Interstate 76, the Pennsylvania Turnpike – the road from Philadelphia to Pittsburgh. The 550 km drive crosses a very varied landscape, and a geological map of the region, simplified in figure 13.12, provides the explanation.

Philadelphia is sited on recent sands and gravels, overlying Tertiary bedrock, but right at the place where they meet the resistant metamorphic rocks that lie to the west. Such junctions arise all along the mid-eastern seaboard and are commonly points where streams and rivers tumble over low waterfalls and rapids before they meander eastwards across the coastal plain. The trace connecting these points has become known as the 'fall line', and along it are many of the region's earliest settlements. The falling water provided power, first for grain- and sawmills and then for electricity. Loaded ships could sail upstream this far. Just as at Philadelphia, some of these communities grew into major cities – Richmond, Baltimore, and Washington, for example.

The turnpike heads out from Philadelphia towards Pittsburgh, but not in a direct line; its route takes into account the geology. It veers SW at first, following the strike of the metamorphosed limestones and dolomites and then crossing schists and gneisses. Their structure is complex, but this largely Precambrian bedrock has by now been eroded to leave low, rolling hills and broad valleys, with open pastureland. The soil cover is thick but, away from the meta-limestones, somewhat acid. Nevertheless, arable farming is possible, to help satisfy the markets of the nearby major cities. In the wooded pockets hardwoods such as oak and hickory replace the conifers that dominate the sandy coastal areas. In some patches the trees are stunted, because of the unusual chemistry of the soil, sited on serpentinite bedrock. Little of the rock hereabouts was suitable for building, though sandstone was used in some houses for decorating the window openings and porches.

Further west, Triassic rocks underlie a lower, gently rolling area with sheets of intrusive dolerite (here called diabase), giving higher ground. Early settlers, such as the Quakers and various German sects, chose this gently sloping, fertile land for their homes, particularly where it was underlain by limestone (Lemon 1972). Sites with springs or year-round surface water were preferred, or situations where the soft surface rocks made digging for groundwater easy. Approaching Harrisburg, the road meets a belt of Ordovician slates and limestones which are thrust and folded, a precursor of what is to come further west. Here, erosion has planed these soft rocks and the area is often referred to as eastern America's 'Great Valley'. The slaty areas are slightly higher and more undulating than the gentle limestone slopes, where fertile, non-acid soils give good cropland and orchards. The limestone produces a karst surface in places, and springs where it overlies slate formations. The road from here, rather than continuing due west to Pittsburgh, swings towards the south, in order to follow lower land and more easily engineered bedrock. It follows the route laid out by British troops in 1785, during the French–Indian war.

The road meets the first of the tough sandstone ridges that begin to dominate the landscape 70 km SW of Harrisburg. The turnpike now has to cross the Appalachian Mountains and it does so through a series of spectacularly engineered tunnels. While not particularly high, the Appalachians are difficult to get around in. The reasons are geological and especially well shown on a geological map. The formations in this area are folded into numerous NE-plunging folds and some of them are very resistant to erosion. They make high ridges, which, because of the folds, every now and then double back on themselves. Hence, few roads run NW–SE, and even those running NE–SW tend to meet obstacles at the fold hinges. Early settlers called them the 'zig-zag' or the 'endless'

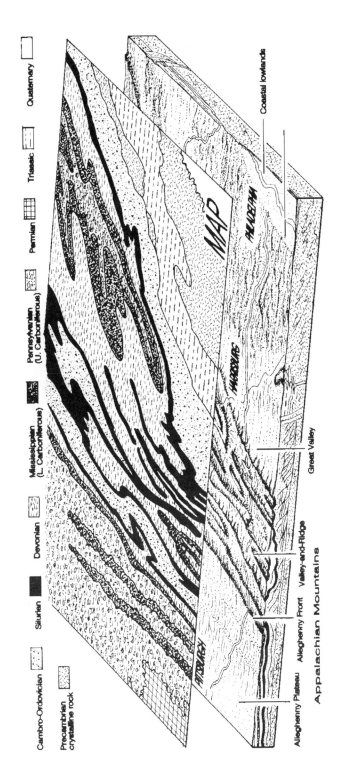

Figure 13.12 Generalised block diagram to show the influence of bedrock, as portrayed on a geological map (upper surface of diagram), on aspects of the landscape seen between Philadelphia and Pittsburgh, Pennsylvania, USA. See text for discussion.

mountains. Without involving elaborate engineering, the roads, like the rivers, have had to seek out gaps in the upstanding ridges. There are no railways just here.

The hillslopes are heavily forested, chiefly by oaks, but with other hardwoods such as chestnut and hickory, which form one of the few natural resources of the region. The other important resource is anthracite, and the lines of the old deep mines twist and turn on the map, recording where they followed the folded coal seams. But the economics of the local wood and coal have always been difficult; they always faced tough competition from elsewhere and have never generated much income for the local people. The geology here does not allow gracious living. Soils are thin and the hillslopes are steep. Forests on the SE-facing slopes are more heavily scarred than those facing NW, because they lie to the lee of the moist air swirling up from the Gulf of Mexico and are dry: they are more vulnerable to forest fires. In fact, where slaty rocks dip parallel to the steep slopes an effect like the tiles of a roof is created and rain simply runs off, leaving patches of what is effectively semi-desert. Salamanders and lizards dart between the dwarfed plants, which can include wild pinks, sedums and cacti – even the prickly pear cactus typical of the arid western USA. Hawks soar on the thermals focused by the sharp, rocky ridges.

Farming is possible in the valley bottoms but most of these are narrow. Some are floored by limestone and are a little more fertile. They were settled first, especially by those that sought seclusion: the Amish, for example. But most of the soil here is thin, and liable to flooding, leaching, and erosion. The markets are distant and travel difficult. Consequently, the farming is commonly of a subsistence type, supporting a family living in a hillside cabin constructed cheaply from the local timber – no stone houses here. In the house, old-time crafts, music, and speech might still survive. The family may have expectations and an outlook on life quite different from elsewhere in the USA. All this is a reflection of the limited resources and complex physical barriers formed by the folded Palaeozoic formations.

The change westwards from the folded valley and ridge country is abrupt. It gives way to very gently dipping Carboniferous age (here called Mississippian and Pennsylvanian) rocks and these begin with a prominent east-facing escarpment known as the Allegheny Front. It long formed a major barrier to westward migration and even today the routes across it and within the plateau region to the west are restricted. Hence, although the geology is here very different, its repercussions mean that culturally and economically this region is still within 'Appalachia'.

The oak-dominated forests to the east give way to mixtures of beech, maple, and tulip trees, with a thick undergrowth. The plateau is actually higher than the folded ridges to the east and it is heavily dissected by rivers. The shallow dips of the bedrock reverse direction from place to place, giving very open folds of formations with differing durabilities and drainage characteristics. For example, the cores of anticlines tend to comprise easily eroded, poorly permeable shales and siltstones, giving low-lying, marshy ground. In these glades, the water table is shallow, and insulated by thick layers of sphagnum moss; the ground is cold. A number of plants and animals reach their southernmost limit in these cold depressions within the plateau.

But following the road further westwards there is a much more conspicuous change in the human landscape and the geological map explains why. The bedrock is very gently dipping and contains whole series of coal seams. Not only is this good coking coal, ideal for steel making, it is very near-surface and it outcrops virtually horizontally – the easiest possible conditions for mining. The steep-sided valleys of this area, and the vulnerability of shale formations to landslip, are hardly ideal for locating major industry but the quality and the accessibility of the coal seams have led to it being the site of one of the world's greatest industrial conurbations. Limestone in the bedrock gave a source of crushed roadstone, and extensive deposits of alluvial sand and gravel provided construction material. Some of the sand is quite pure; there is now an important glass industry here. Alluvial clay spawned ceramics factories. So, as the road finally reaches Pittsburgh it crosses a scene still largely dominated by industry. The nature of the city and the scenery around it is still due ultimately to the region's geology. Thus, the basic key to understanding the landscape, here and all along this 550 km journey, lies in the geological map.

Selected Further Reading

Fortey, R., 1993. *The Hidden Landscape. A Journey into the Geological Past*. Pimlico, London, 310 p.
    An entertaining and highly readable essay on the influence of geology on the landscape of Great Britain.
Woodcock, N., 1994. *Geology and Environment in Britain and Ireland*. UCL Press, London, 164 p.
    Includes examples of the relationships between geology and landscape.

CHAPTER 14

The Heritage of Geological Maps

14.1 Introduction

There are facets of geological maps other than those discussed in the preceding chapters, such as the tremendous heritage they represent. The maps we use today have not always looked this way, and the earliest maps did not appear overnight. Geological maps have slowly evolved, in many ways reflecting the growth of geology itself. The first part of this chapter briefly outlines this history. Every geological map embodies parts of the story. The treatment here is brief, but references are provided which give much greater detail.

Then there is the human aspect to maps. Many a geological map represents the culmination of much individual labour, and its production may have involved personal conflicts, adventure, even tragedy. There have been individuals and institutions who have contributed influential advances in map methods, and who have bequeathed magnificent maps. The second half of this chapter sketches the lives of eight individuals, in order to glimpse the personal stories that lie behind their maps and, because they all made influential advances, behind all maps. The choice of people is a rather arbitrary one, aimed at giving some chronological and geographical spread. Some very interesting persons and some very significant maps have had to be passed by.

14.2 A Short History of Geological Maps

The striking thing about the birth of geological maps is that it was late and slow. People had always *thought* about the earth but only when religious dogma was escaped and careful field observations made was the scene set for the advent of maps.

First came the *idea*. John Aubrey, famous for his archaeological discoveries at Stonehenge and Ave-bury, but called by a contemporary 'a shiftless person, roving and magotie-headed', wrote that he 'often times wished for a mappe of England, coloured according to the colours of the earth, with marks of the fossils and minerals'. Martin Lister, a physician living in York, in 1683 presented a paper entitled 'An ingenious proposal for a new sort of map of countrys'. He described in some detail how the distribution of soils and rocks at the earth's surface could be displayed on maps, so that 'something more might be comprehended'.

Then came maps which actually plotted the locations of minerals and rocks. They were approximate maps, lacking good information and an accurate topographic base. Some were better than others, so that it is unrealistic to specify the 'first' geological map. None had any concept of a sequential or a three-dimensional arrangement of the units. Two early examples were Packe's (1745) depiction of the hills around Canterbury, and Guettard's (1746) map of France (section 14.3.1). Later, Desmarest (1725–75) mapped in detail the basaltic lava flows of the Auvergne, France, showing their sequential arrangement, and the varied geology of the Transbaikal region of the USSR was surveyed by Lebedev and Ivanov between 1788 and 1794 (Pavlinov 1984).

A tremendous advance in map-making was provided by William Smith (section 14.3.2). He first really grasped the sequential arrangement of sedimentary rocks, and depicted it on maps with reasonable accuracy. He advertised the practical applications of geological maps. In 1815 he published his momentous map of England and Wales, arguably the most significant geological document there has been, and among the most treasured of geological collector's pieces.

Following this, geological maps began to appear thick and fast. Maps of parts of western Europe were published, with ever-improving detail and accuracy, and maps of entire countries. Another map of England and Wales appeared in 1819, which in its own way was also a remarkable achievement. The Geological Society of London had been formed in 1807 and its first President, George Greenough, took on the job of compiling information from well-known geologists of the day, to prepare a national geological map. Although it is Smith's map which is today most widely remembered, many regard Greenough's map as the better, and it has been influential (Cook 1989). MacLure's 1809 map of the eastern USA does not aspire to the cartographic heights achieved by Smith and Greenough, but is an extremely important document in American geological history.

All these maps were the works of private individuals and societies. However, many of their authors stressed the value to applied geology of good maps, and it was inevitable that governments and public bodies would soon see the need for funding official geological surveys of their regions. There are competing claims as to which was the first official government survey. Guettard was receiving official money for mineralogical purposes in 1746 (section 14.3.1). MacCulloch carried out governmental geological work as early as 1811, and in 1826 was specifically instructed to make a geological map (section 14.3.3).

A geologist was attached to the Trigonometric Survey of India in 1818 (Eyles 1950), and in 1823 Denison Olmsted of the State University of North Carolina was given $250 a year for four years to undertake geological excursions – perhaps the first use of US public funds for geology. Some remarks on geology and mineral resources had been made in the earlier reports of government expeditions such as the epic Lewis and Clark mission of 1804–06, but the comments were sparse and incidental. The North Carolina effort did not continue, neither did the State Geologist appointed in South Carolina the following year. The first fully functioning State Survey, soon producing a seven miles to an inch coloured state map as well as a significant advance in knowledge, was inaugurated in Massachusetts in 1830 (Merrill 1924). A government-financed geological map of France was begun in 1825, completed in 1835, and published 1841 (Eyles 1950).

The Ordnance Survey in the UK took an increasing interest in geology, at first mainly in Ireland (Davies 1983, 1995). The Director apparently had the impression that if rock samples were collected from time to time by topographic map-makers, a geologist back at headquarters would be able to construct a geological map from the specimens (Wilson 1985). Parcels of samples were actually dispatched from the field for this purpose! However, the impracticality of the idea is shown by the issue in 1827 of a 70-page 'Directions for Geological and Mineralogical Observations', and

the appointment of a 'Superintendent of the Geological Sciences in Ireland'.

Shortly after this, in London, what is now the British Geological Survey was born. In 1832, de la Beche (pronounced Beech) offered to make available for £300 the results of his own geological surveying of Devon, to add to the recently published Ordnance Survey maps of that county. De la Beche had been 'dismissed with ignominy' from Military College in his youth, but had become an energetic and skilful geological observer. After some negotiation of conditions, a deal was agreed. At a rate of £37.50 for each of the eight sheets (Wilson 1985) the geological survey was conducted at a cost many orders of magnitude less than it would be today! It worked well, and in 1833 de la Beche was offered a full-time position. The Geological Survey was established, as a one-man department of the Ordnance Survey. Wilson (1985) describes what followed – a mixture of political manoeuvring, personality clashes, and shrewd appointments, as de la Beche entrenched and expanded his position. By 1845 the Geological Survey employed 24 scientists. Oldroyd (1996) has highlighted how establishing the BGS fitted in with the imperial thinking of the time, for geological surveying was seen, among other things, as a territory building exercise. It was almost military: the BGS surveyors were called 'officers' (a term still sometimes used) and wore an army-style uniform.

Geological mapping of the UK accelerated. By 1854, Wales had been surveyed at the one-inch scale, largely by the enthusiastic mapper A.C. Ramsay, who in the same year commenced the Survey's activities in Scotland. Mapping of superficial deposits began in about 1863–65; and the first Survey map of England and Wales was completed in 1883. By the turn of the century, most of Great Britain had been surveyed at the one-inch scale, apart from the Scottish Highlands. And it had become the aim, despite the altercations reported by Wilson (1985, chapter 10), to survey the nation at the six-inch scale, for eventual reduction. In general, this approach continues today. It has led to impressive accuracy, but it is very time consuming. In fact, the survey has never been completed for the whole country. Despite its flying start, the BGS is still pursuing its goal.

In the first half of the nineteenth century many parts of the world were geologically mapped, at the reconnaissance level, for the first time. Ireland (1943) gives a list. Detailed maps of small areas continued to proliferate. They were produced by learned natural historians, and by amateurs. The day when geological mapping would be limited to technical experts was still far distant.

All this took place in the era of hand-colouring. This produced maps of highly variable colouring standard, and restricted their distribution, despite whole teams of colourers being employed, often, apparently, rather poorly paid ladies (Wilson 1985). Colour printing of geological maps began in 1851, after some unsuccessful false starts (Ireland 1943). This made an immense difference to the availability of maps. The contrast is reflected in the prices these antique geological maps command today – hand-coloured maps fetch a considerable premium.

As the nineteenth century progressed, Geological Surveys were set up in more and more countries. A Geological Survey of Canada was established in 1842, under William Logan. He returned to Canada after work for the British Survey in South Wales (Bassett 1969a). Merrill (1924) recalls that when Logan began his work in Canada, 'a large part of the country was wilderness, without roads, and there were no maps. Little was known of the region besides the coast-line, of the geology practically nothing. From dawn to dusk he paced or paddled, and yet his work was not finished, for while his Indians (often his sole companions) smoked their pipes around their evening fire, he wrote his notes and plotted the day's measurements'. Since that time the Geological Survey of Canada has become a major organisation, with a prodigious output. Although Greenly and Williams (1930) felt that 'the whole system of publication of this survey appears to be complicated and difficult for an outside person to understand', the publications today are well catalogued and handsome maps have been produced.

In Australia various state surveys were established between 1856 (Victoria) and 1896 (Western Australia) (Darragh 1977), eventually leading to the formation of a coordinating body, the Bureau of Mineral Resources, now called the Australian Geological Survey Organisation. The Geological Survey of New Zealand was formed in 1865.

Geological work in the United States had steadily progressed for several decades, and by the later part of the nineteenth century, the west was rapidly being opened up, with several Territorial Surveys competing to reveal its geology (e.g. Bartlett 1980; see section 14.3.5). Coordination became imperative, and in 1879 the USGS was established (e.g. Rabbitt 1980a). From the start this organisation had the responsibility for both the topographical and geological surveying of that country. As Linton (1947) has pointed out, this coordinated effort gave an immediate advantage over the European counterparts, where often the geological surveys had to make do with the base maps produced for a different purpose by a separate institution which was somewhat insensitive to the geologists' needs. The growth of the USGS has been awesome, its output enormous. But its products have always been

well catalogued, its maps models of clarity and consistency.

Oldroyd (1996) catalogued the foundation of 126 governmental surveys. He noted that not a few of them were set up on the lines of the British model and staffed with British-trained mappers, in line, he argued, with the territorial ambitions of Victorian times. For surveyed ground, he suggests, was regarded as 'captured', brought under control, and added to the geological empire. Thus France and Britain saw that their colonies in Africa were mapped; the declining empires of Spain and Portugal did little in South America. Newly established states in the USA soon founded a geological survey to help bring the territory under proper bureaucratic control (e.g. Schwartz 1964; Bergstrom 1980). China and Japan started geological surveys as they flexed their military and economic muscles. Mapping represented power. Oldroyd (1996) remarks that the thinking continues today, 'with the power now projected under the oceans and towards other regions of the solar system'.

At the same time as the launching of government surveys, in many places learned societies were being founded, each commonly publishing its own journal. Until recently, these learned journals formed the main vehicle for individual geologists to publish geological maps. The *Journal of the Geological Society of London*, for example, contains numerous maps, of many parts of the world. Such journal maps have to be folded up if they are at all large, and few are in colour, but they represent tremendous advances in knowledge.

Many geological companies realised the need for systematic investigations of their geological properties. Maps of mineral prospects, detailed mine maps, quarry maps – all were produced, sometimes in great detail, but usually for the private use of the company. Exploration for oil spawned a host of new cartographic techniques, especially concerning the sub-surface geology.

Many of the principles that came into being in the early days of geological maps have carried through to today. The methods are greatly refined, but making a geological map is still basically the same endeavour, with the same goals. However, the increased specialisation of modern earth science has encouraged today's journals to print articles on more specialised aspects of geology. Similarly, the efforts of geological surveys have diversified such that only a portion of the staff are engaged in basic surveying. This is also a reflection of the increased efficiency of surveying methods (section 15.2), and the power of modern instrumental techniques. It does not imply any diminishment of the central importance of the geological map.

14.3 The Contributions of Some Individuals

14.3.1 Jean Etienne Guettard (1715–1786)

Jean Guettard was a great observer. At a time when thinking about the earth was still based on a mixture of religion and speculation, Guettard was a tireless collector of *facts*. And he was a pioneer in seeing the value of displaying observations on a map. His maps were a long way from the kinds of thing we expect today, but it was a bold start.

As a child, Guettard was interested in natural history, especially plants. He trained as a doctor, but continued to collect plant specimens. He became struck by how their occurrence appeared to depend on certain minerals and rocks. Eventually, it was the 'mineral substances' themselves, and *their* distribution, that came to dominate his interest. Guettard was a very energetic collector of information. He travelled widely in France, visiting mineral localities, observing all the time, and assiduously reading all the literature that was available to him. Unfortunately, as Rappaport (1969) put it: 'The talent he most conspicuously lacked was that of generalisation, of seeing the implications of his own observations'! He desperately needed some way of organising and presenting his vast amount of data.

With time, Guettard began to suspect that there may be some sort of pattern to the distribution of the mineral deposits. Rock classification was in those days a very hazy affair, but he began to think that rocks, too, were not distributed haphazardly. The three kinds of rock he recognised, which he called 'sandy', 'marly', and 'schistose', seemed to occur in broad bands. He thought that if he could establish the 'determinate trend' of these bands, it would be possible to *predict* the kinds of rock in unknown country. But how could he illustrate this?

Guettard hit upon the idea of marking the mineral deposits and rocks on a map. He had little to guide him on how to do it. What little there had been done on similar lines was probably not known to Guettard. So he took an existing map of France, and used symbols, mainly chemical ones, to mark the locations of mineral deposits, and added a few fossil localities. The distribution of his three rock types was indicated by an engraved shading. The rocks did, indeed, fall into bands. His 'Memoire et Carte Minéralogique' was presented in Paris in 1746, and published in 1751 (Geikie 1905).

At a stroke, the map enabled Guettard to organise, synthesise, and communicate his information, and to test his ideas. He had stumbled upon what were to become vital functions of geological maps. There was now no stopping him! On his first map, the three rock

bands stopped at the English Channel, and Guettard predicted that they would reappear in England. He searched the literature for supporting information – reading no English, he had to rely on what had been translated – and convinced himself that he was right. So, on to a map of England went the same system of symbols and engraving. He tackled next, still relying on literature, the whole of Western Europe, from Iceland to the Mediterranean, and, later on, the Middle East and North America!

His maps were well received. This 'new kind of map' was deemed to open up a new field for geographers and naturalists, providing a link between the two subjects. Guettard extolled the practical value of his maps, for example, their use in locating further supplies of good building stone and durable road material. He felt sure that the maps would lead to a unified understanding of the earth. He was, however, dissatisfied with the accuracy of his work, saying 'if you will only let me have a proper map of France, I will undertake to show on it the mineral formations underneath'. With the help of a young man called Lavoisier, destined to become the great chemist, and later on Monnet, Guettard went on to complete 16 detailed sheets of France and a large report. The 'Atlas et Description Minéralogique de la France' was published in 1780 (Rappaport 1969).

Clearly, Guettard had made a great step in the direction of geological mapping. But can his maps be called 'geological maps'? Probably not, in a modern sense. They were, at best, approximate, partly because the base maps of the time were insufficient for accurate work, and because much of the geological information, even in France, was second-hand. But there is a much more fundamental shortcoming. The maps were strictly two dimensional. There was no attempt to show the relationships between units, either geometrically or stratigraphically.

It is probably not that Guettard had no comprehension of geological relationships, more that he felt it involved too much inference. For Guettard, almost obsessed by the importance of observation, believed that if something were not factual it should not be shown on a map. As we have seen in this book, the modern geological map, in contrast, is intrinsically an interpretive document.

The point is well illustrated by a disagreement during the preparation of the 1780 Atlas. Lavoisier had grasped the importance of stratigraphic sequence, and realised that it may lead to an understanding of geological history. He had assembled some stratigraphic columns with this in mind, compound ones, drawn from various localities. This was all too speculative for Guettard – not the kind of thing he wanted on his maps. In addition Lavoisier would have liked to infer more about distributions : 'It is only necessary to link all similar symbols by lines which would show not only the size and extent [of each deposit] but also their various points of intersection' (Rappaport 1969, p. 283). However, Guettard, the senior man, had his way. Inference was not allowed on the map, although some of the sheets do have Lavoisier's compound sections in the map margin.

Before his death in 1786, Guettard published numerous, detailed papers, all based on his meticulous observations. But it is for his maps that he is most remembered. Besides illustrating geological distributions, he had pioneered the use of maps for data collection and synthesis, just as with modern maps. However, fundamental map problems were left untackled. The matters of dealing with the three-dimensional arrangement of rocks, of depicting stratigraphic sequence, and of combining inference with observation still lay ahead.

14.3.2   William Smith (1769–1839)

Smith is by any standards one of geology's greats. He first understood and communicated the idea of sequence in layered deposits, and recognised their characteristic fossil assemblages, two concepts fundamental to stratigraphy. He was a pioneer of applying his geological knowledge: to finding underground water; to drainage schemes; to coal mines and canals. And he produced his epoch-making geological map. Much has already been written about the man, and his maps, but they are too significant not to be included here. Also, it is very relevant to the theme of this book that Smith apparently found writing awkward and difficult – it was through his maps that his perceptive geological knowledge and understanding were communicated!

Smith was born in Churchill, Oxfordshire, in 1769. His father was a blacksmith and his ancestors ordinary farmers – facts which became increasingly relevant when Smith encroached on a geological community dominated by the well-to-do. By the time he was eighteen, Smith was an apprentice surveyor at Stow-on-the-Wold and taking on increasing responsibilities. In 1792, while surveying coal properties, he lodged at Rugborne Farm, near High Littleton, seven miles SW of Bath, and 'began to think about the succession of the strata' (Eyles 1969). This farmhouse he later called 'the birthplace of English Geology'!

According to Cox (1942), Smith was always 'ready to converse, with friend or stranger, upon many subjects, but usually reverting finally to his favourite topic – the strata of the earth and the application of their study to agriculture, mining and all works of public utility'. He began to realise that the geology of England is dominated by a regular succession of different strata, tilted towards the SE. His continuing work furnished him with more and more evidence. Later critics said that Smith became over-obsessed with this belief, but the important thing for us is that he hit upon the idea of showing the arrangement on a map.

He drew a map of the area '5 miles round the city of Bath', one and a half inches to a mile, which was 'coloured geologically in 1799'. In 1801 he produced two small maps of England and Wales to show the overall arrangement of the strata (Eyles 1969). In the same year he issued a prospectus for a large-scale and accurate map of the strata of the whole of England and Wales. He began putting this together while continuing his various applied geological activities – supervising sinkings for coal, devising sea defences, etc. – and all the while amassing a fossil collection. This was displayed on sloping shelves in order to correspond with the beds in which they were found (Eyles 1969)!

Here we glimpse some of the obstacles which Smith had to overcome in order to publish what was to become his celebrated map. To start with, there was no topographical base map sufficiently uncluttered and accurate for Smith to use. He had to spend time and money persuading a London map engraver and publisher, John Cary, that his project was worthwhile enough to justify the preparation of a base expressly for Smith's purpose. And Smith did not have much money (Eyles 1967). His nephew, John Phillips, later remarked that Smith's hearty laugh was a cover for his financial difficulties. Although making a reasonable living from his consulting work, with the Napoleonic wars in progress times were hard, and Smith was relying on his map to make him some money. After all, it took about 15 years collecting the map data, carefully, single-handedly, and without financial support. The cost of publishing the map became heavy. Each of the 16 sheets had to be hand-coloured, and several artists had to be employed merely to do the colouring. Smith began to see that profits from sales of his map would be thin. It was this situation which turned Smith to the idea of selling off his beloved fossil collection, a saga recalled in detail by Eyles (1967). It also meant that even when the map did appear, it was to have a publication life of only four years.

Then there was the matter of the rival map, being assembled by Greenough for the Geological Society of London, and representing many of the leading geological figures of the day. Why, in 1812, did Greenough even contemplate such a map, when Smith's project was nearing completion? The question has been much discussed (e.g. Bailey 1952). A friend of Smith's, John Farey, complained of the 'very unhandsome conduct of certain persons' in trying to strangle the sale of

Smith's map. Greenough said he thought that Smith had abandoned his project.

Why did the Geological Society not lend its support to Smith's endeavour, rather than launch a rival project? Perhaps Smith would not have wished it; he was an independent man. According to his nephew, John Phillips, Smith had said: '. . . I foresaw that the truth and practicability of my system must be tested far and wide before its uses could be generally known and its worth duly appreciated. I thought, of course, no one could do this so well as myself . . .' (Bassett 1969b). Also, Smith's working-class background had probably not gone unnoticed. Smith may have been a difficult fellow to work with, but one wonders if the stories behind some geological maps not only involve individuals and their personalities, but the circumstances of their birth! Smith had already stated his preference for 'Saxon and British words' rather than those contrived from 'dead languages', ruffling the feathers of the classically educated geologists of the time. Smith was not even invited to join the Geological Society until 1830, by which time recognition of his achievement was such that he could hardly be excluded.

Even 70 years later, Archibald Geikie, a later doyen of the geological establishment, was critical of Smith's uncultured approach. Geikie (1905) thought Smith's efforts were limited, and he saw them in the light of Smith's background:

> William Smith was tall and broadly built, like the English yeomen from whom he came. His face was that of an honest, sagacious farmer. His work, indeed bears out the impression conveyed by his portrait. His plain, solid, matter-of-fact intellect never branched into theory or speculation. His range of geological vision was as limited as his general acquirements.

Anyway, Smith's map did appear, in 1815. Not only was it the first geological map in any modern sense, it was of two entire countries, and the work entirely of one man. It embodied innovative principles, which are now fundamental to geology. There are 16 sheets in 20 colours. At a scale of about five miles to one inch, the entire map occupies six feet by eight-and-a-half feet. The units are shown in their correct overall stratigraphic order, with the bases distinguished from the tops by a heavier tint. The inclined units even 'V' across the valleys in the correct way. Of course, there were mistakes and shortcomings. Challinor (1970) gives a list. Many users felt it was soon superseded in accuracy and clarity by Greenough's map of 1819. Smith was aware of some of the inadequacies, and took the opportunity to improve the map at each printing (Eyles and Eyles 1938).

It has been said that with Smith's effort 'the geological map had arrived'. The colours Smith chose

had no little influence on succeeding maps, both in Britain and elsewhere (section 15.4.5). In fact, at first glance, Smith's 1815 map is not greatly different from the BGS 'Ten Mile' map of today! Smith continued to produce beautiful maps. Between 1819 and 1824 he published a *New Geological Atlas of England and Wales*, comprising maps of 21 counties. Some were recently reprinted by the British Museum of Natural History. In 1828, Smith moved to Hackness near Scarborough, in Yorkshire, and four years later published a remarkably detailed and accurate map of the Hackness Hills.

Of Smith's great 1815 map, the Eyleses were only able to trace in 1938 about 30 copies still in existence, distributed among only 12 owners. Will it ever be reprinted? No! Eyles and Eyles (1938) record that when in 1877 the owner of the 16 copper plates, Edward Stanford, offered to sell them 'at trifling cost' to the Geological Society of London, there was no support for the idea. The plates were melted down.

14.3.3 John MacCulloch (1773–1835)

MacCulloch was one of the first persons in the UK to receive government funds for geological purposes. He made numerous pioneering geological observations, wrote a travel book which is still in print, and made the earliest published geological map of Scotland.

Topographical surveying early last century was still in the hands of a government military body, called the Board of Ordnance. It had been set up in medieval times to survey artillery installations. John MacCulloch, trained as a surgeon but attached to the Board of Ordnance as a chemist, also had an interest in geology. In 1811 he was given an official geological job to do (Eyles 1950). He was sent to Scotland to find out what kinds of rock were safest for grinding gunpowder! In 1814, he was appointed Geologist to the newly formed Trigonometrical Survey unit of the Board of Ordnance, to examine certain aspects of Scottish geology. This unit subsequently became the Ordnance Survey of the UK. The appointment of MacCulloch arguably formed the earliest state geological survey.

In 1819 MacCulloch published what was to become a classic of travel literature and of Scottish geology, his book: *A Description of the Western Islands of Scotland*. An abridged version is still sold today. Challinor (1970) lists 27 original geological discoveries contained in the book, many of them now fundamental to Scottish geology. Ten coloured geological maps were included, covering many of the Hebridean islands.

Despite all this, the Head of the Board of Ordnance, the Duke of Wellington, terminated the geological appointment. MacCulloch returned to being official chemist, but continued his survey of Scottish geology *at his own expense*. Only after several years had passed did the Treasury restore Mac-Culloch's financial support. In 1826, he received government funds for the sole purpose of preparing a geological map. A new era in geology had begun. The need for official geological surveying had been realised, and funds earmarked for it. But even back then, patterns were being set of relationships with universities and of expense claims which are not unfamiliar today. For example, one of the best-known university geologists of the time, Robert Jameson, snubbed the official effort, saying that MacCulloch's Survey was 'utterly unknown to any public body in Scotland' (Eyles 1950).

Regarding expenses, MacCulloch was to receive £2 a day while in the field, £1 a day personal costs, and 2 shillings (10p) per mile travel expenses. In his first field season MacCulloch claimed over 7000 miles of travel costs, plus costs of boat hires, guides, stationery, and 'pedemeters' for measuring distances 'on foot, horse, and carriage'. The Treasury was embarrassed, having greatly underestimated what was involved in a geological survey. So MacCulloch had to account in detail for his movements and his costs: the auditor questioned his large mileages and the short time taken to cover them. MacCulloch replied that accountants failed to understand the nature of geological fieldwork.

MacCulloch pressed on, apparently working long days, including Sundays, and being paid considerably in arrears. Then doubts began to arise in official circles about whether or not the map should be published. Murchison was unsure about the quality of the field notes, though he thought the 'sketch' of a map might be useful. The topographic base map was known to be inaccurate – MacCulloch later spent much of the first 50 pages of his 'Memoirs' abusing it. In fact, if the Highland and Agricultural Society of Scotland had not taken an interest in the map (Boud 1985), it may have 'remained buried in the archive of a government department' (Eyles 1937). MacCulloch may not have helped the situation himself, being, rather like William Smith, fiercely independent. Flett (1937) says that MacCulloch had the reputation of being short-tempered and difficult to agree with. He became increasingly bitter that his work was not being properly recognised.

However, the map eventually did appear, in 1836. It comprised four hand-coloured sheets, each 86 × 69 cm, at a scale of four miles to an inch (1:250 000), and used 18 different colours. The main elements of Scottish geology show clearly. It is now a famous document, but it somehow typifies MacCulloch's dogged career that he did not live to see it published. He married, for

the first time, when he was 62 – the year before the map came out. Whilst on his honeymoon, MacCulloch was killed by a runaway carriage.

As a final irony, Eyles (1950) draws attention to the review of the published map in the *Scotsman* newspaper. The price of £5 for the map was thought high by the reviewer. Yet, the topographic base map sold for the equivalent of £4.50. MacCulloch's life work, a classic document on Scottish geology, was being deemed to be overpriced at 50p!

14.3.4 John Phillips (1800–1874)

John Phillips saw and helped engineer the change from the restrictions of hand-coloured maps to the increasingly accurate and more widely available colour printed maps.

He was born on Christmas Day, 1800, at Marden, in Wiltshire. His father was from near Carmarthen, in Wales, but migrated to Oxfordshire, where he married the sister of no less a person than William Smith! John Phillips lost his father when he was seven, and his mother soon after, and therefore came under the care of his Uncle William. When Phillips's schooling was finished, he went to live with Smith, in a house 'full of maps, sections, models, and collections of fossils'. He began to accompany his uncle on long walks, and on what would today be called consulting work. In this way, Phillips learned his geology.

In 1824, Phillips accompanied his uncle to York, where Smith was giving a series of lectures, and became interested in the fossil collection of the Yorkshire Museum. Phillips was eventually appointed Keeper to the Museum, and lived in the old museum buildings now known as St Mary's Lodge, until 1853. During this time, Phillips was a busy man. He was instrumental in setting up the British Association for the Advancement of Science, had periods as Professor of Geology at King's College London, and Trinity College Dublin, and published much on the geology of Yorkshire, including several maps. He coined the names for the stratigraphic eras – Palaeozoic, Mesozoic, and Cenozoic. He advertised the pleasures of geology, encouraging people to ramble in the hills and valleys of Yorkshire 'for a better knowledge of its natural beauties'. Phillips seems to have been an amiable and easy-going fellow. Perhaps this is why he could get along with his somewhat prickly uncle. And during his time at York, Phillips introduced to Britain a major advance in map production.

He had witnessed the delays his uncle had experienced before publication of the 1815 map (section 14.3.2). The hand-colouring of each copy of that map took about a week, and added greatly to the cost. John Phillips learned of the new German technique of lithography, where, through the antipathy of grease and water, a picture drawn in ink on fine stone can be pressed on to a sheet of paper. He learned the art of drawing on stone, and lithographed pictures for his uncle (Butcher 1983). The plates in his 1841 work on SW England are lithographed by William Monkhouse, who operated in Lendal, York, close to Phillips's residence. Butcher (1983) remarks that the two seem to have struck up a friendship, so that when in 1849 Phillips began compiling a general work on the geology of Yorkshire, it was Monkhouse who lithographed the 35 plates. The book, the *Geology of Yorkshire*, appeared in 1853, and one of the plates was a geological map of Yorkshire *lithographed in colour*. This is the earliest known colour-printed map in Britain (Butcher 1983).

Colour printing of maps had been attempted in Europe for over a decade, using various combinations of tin foil and cardboard masks, but there were always problems where the colours met (Ireland 1943). Different stones were tried for the different colours, with varying degrees of success. Phillips's map employed four stones, for the black, blue, red, and yellow inks, with which he was able to generate 14 shades. The results were acceptable, although Butcher (1983) comments on the 'widespread and patchy discolouration' of the first edition of the map. The second edition, published in 1855, was an improvement.

In those infant days of colour printing, the results were variable. Ireland (1943) cites an English 1853 map of North America as being 'poorly executed and much inferior to the excellent hand-coloured maps typical of the English publication of that date'. But the advantages of colour printing were obvious, and as soon as the practical snags had been ironed out, the days of hand colouring were numbered. John Phillips went on to become Professor of Geology at Oxford, organising the geological museum there, and to publish numerous works – on geology, on other planets, even on the weather. But in many ways his main contribution to geology had been made while at York; this is certainly true of his contribution to geological maps. It is fitting that he was returned there to be buried. There is a memorial window to him in St Olave's Church, adjacent to where he lived.

14.3.5 John Wesley Powell (1834–1902)

John Wesley Powell – US Civil War soldier, Grand Canyon explorer, and government scientist – directed the USGS during its formative years. He was energetic in promoting the importance of geological maps to the Government at a vital time, and was instrumental in establishing methods and standards which are apparent in today's maps of the USA and elsewhere.

Powell was born in New York state in 1834, four years after the arrival from England of his Methodist minister father. He took an early interest in natural history – studying it, and eventually teaching it. By 1860, civil war was looming in the USA, and Powell, with his strong Methodist anti-slavery background, was one of the first to volunteer for Lincoln's army. He quickly rose in rank, already showing his leadership talents. At the Battle of Shiloh, while raising his arm to signal 'fire', he was badly shot. His arm had to be amputated. Within months, however, he was back in the action, with his love of geology intact. According to a colleague (Rabbit 1980b), while marching his regiment through new areas, Powell would 'familiarise himself with the geology'. He was in the trenches at the Battle of Vicksburg, at the same time collecting rocks and fossils for the Illinois State Museum!

At the end of the war, Powell, tough and war-hardened, and with administrative and scientific experience, looked for employment – and excitement. He took a geology teaching job in Illinois, which provided the former but not the latter. The unexplored West was beckoning. The great rivers attracted Powell, and especially that last largely unknown place – the Grand Canyon of the Colorado. And so, while retaining his teaching position, he organised an expedition there. It was largely self-financed, and it was ill-thought out. Faul and Faul (1983) remarked that 'at the time it could have been reasonably viewed as a reckless and pig-headed adventure', although 'in retrospect we know it was the beginning of a great survey'.

Excitement it certainly provided. Boats were lost, and the food turned rancid. The crew got sick. One day, what was left of their clothing caught fire at the campground, and the explorers had to continue their journey half-naked. At one point, Powell fell while climbing the canyon wall, but managed to grab a jutting ledge. He dangled over the canyon, hanging by his single arm, while a colleague scrambled to the only accessible point, above Powell, but out of reach. A barometer case was stretched down, but it also failed to reach Powell. Finally, the colleague took off his 'drawers', and lowered them down. Powell let go of his vital handhold, being momentarily suspended over the canyon before grasping the rescuing underpants!

The crew grew increasingly despondent. They were exasperated at Powell's being continually captivated by the geology and oblivious to their dire situation. One member wrote about Powell that 'if only he can study geology he will be happy without food or shelter, but the rest of us are not afflicted with it' (Bartlett 1980). Eventually, three of the group decided to leave the expedition. They managed to climb out of the canyon, whereupon they were shot by Indians.

Information accumulated about the natural history and Indian culture of the canyon, though not a great deal about the geology. The individual mainly responsible for the geological surveying was one of the three that had been shot and, in any case, the surveying equipment had been lost in the river! However, beautiful ink drawings were produced (see figure 6.1), with sufficient care and detail to bring out the main configurations of the rocks (Powell 1895).

A rumour spread that the entire party had drowned in one of the rapids. The story appeared in several national newspapers, with an obituary to Powell included. As a result, when the bedraggled expedition reappeared having navigated the canyon, Powell got national publicity beyond his dreams. Fame came instantly – together with funding for more exploration. A second expedition brought back considerably more geological information, and led to a complete topographic map of the region, and also tremendous popular interest. Powell was a national hero.

At this same time, other government surveys were operating in the West, and conflicts of interest began to develop. Congress saw fit to consolidate activities, and in 1879 established the USGS (Rabbit 1980a). The first director did not enjoy the politics involved in such a position, and after only two years resigned, in favour of Powell. He had the right administrative and political qualities, as well as common support. He assembled round him a group of exceedingly able scientists, on whose advice he could draw and on whose work he could rely. Powell, himself, did little more in the way of surveying, indeed geology remained as only one of his interests. He was, however, committed to the fundamental importance of geology and geological maps, and was adamant about the necessary standards of accuracy and presentation.

Powell established a series of committees to consider just how geological mapping should be executed, and how the map is best produced. The results were radical. For example, accuracy of the topographic base map was deemed paramount, and it had to be contoured. Hachuring and shading, so much in vogue elsewhere, were out. The typeface, symbols, colours, etc., of the topographic map were carefully designed, not as an end in themselves, but to produce a base on which other data, particularly geological, could be shown effectively. Such a coordinated approach was unknown in Europe, where the geological surveyors had to make do with the base maps they were given.

Powell's committees gave a lot of thought to the nature and status of the map units, and designed a colour scheme based on several criteria. At first, the approach was 'to follow common usage', which was essentially the William Smith approach of mimicking the colour of the natural rock. After four years of attempts, this usage was deemed to be 'vicious and bad' (Powell 1881). An 'entirely new colour scheme' was to be employed. For example, by a logical system of printing the spectrum colours, plus brown and grey, in spaced lines of various combinations of orientations, 'several hundred distinctions for the clastic rocks' were demonstrated to be feasible. Colour printing of hieroglyphic-like symbols, with various weights, sizes, and closeness, gave several hundred distinctions for the Archaean alone. A scheme of letter symbols was devised, to facilitate further distinction of the units, and in case the colours should fade with map use. The 'explanation of colours' conventionally employed was replaced by a new kind of map legend, designed to help each map be self-explanatory and of maximum value to all kinds of users.

These ideas were immediately put into operation by the USGS. Under Powell's directorship, these innovative, high standard maps began to pour out. At the same time, new subdivisions of the Survey were founded. The USGS burgeoned into the world's premier geological organisation.

Meanwhile, Powell's strong opinions were brought to bear on broader concerns, such as organisation of the settling of the West and irrigation of the lands. These were emotive issues, in a time of political delicacy. Powell had 'made warm friends and strong enemies' (Merrill 1924), and eventually the latter won out. Powell resigned the directorship in 1894, although continuing his interests in Indian affairs. His health deteriorated, further surgery was required on his arm, and in 1902 he died, to be buried in Arlington National Cemetery.

Today, the significance of his scientific ideas are being rediscovered (Rabbit 1980a), and the romance of his military and exploring achievements glows brighter than ever. But his geological legacy lies mainly in maps. It is seen in every quadrangle produced by the USGS, and all the other national surveys which have followed the lead of that institution.

14.3.6   Edward Greenly (1861–1951)

On the staff of the British Geological Survey around a hundred years ago were some of its most illustrious surveyors ever. And pre-eminent among these was the trio working in Scotland: Benjamin Peach, John Horne, and Charles Clough. These were heroic times, as the complicated geological structure of Scotland was being unravelled, including the remarkable thrusts being discovered in the NW Highlands, and the complexities were being recorded on unprecedentedly fine geological maps. The maps of Clough, especially, were legendary. Even the great Sir Edward Bailey (next section) believed that 'for thoroughness, minute and accurate detail, Clough's maps probably have never been equalled'. Into this coterie came, in 1888, a new appointee to the Survey, an intensely cerebral young man named Edward Greenly. This man was already obsessed by his childhood sweetheart and wife-to-be, Annie, and now a second devotion entered his life – the science and beauty of geological mapping. These two things provided Greenly with an almost mystical enchantment throughout his life. Greenly's maps were to become very widely admired but he always acknowledged that he 'learnt under the inspiration of the greatest master of the art of geological surveying, C.T. Clough'. Unlike Clough, however, Greenly spoke and wrote extensively about geological maps and mapping, and through this has had direct influence on generations of surveyors.

Greenly was born in Bristol, but, because his father who was a doctor changed practices, soon moved to nearby Bath. There he met Annie, the girl who was to be 'the love of his life from early youth to grey hairs'. Annie seems to have been a jolly girl, playing and picking flowers by the banks of the local canal – the same canal constructed just a few decades earlier by William Smith. However, she suffered from glaucoma and was to become blind in one eye before she was sixteen, a matter that assumed pivotal importance later. The intense emotions of the fervid courtship are recalled in detail by Greenly in his autobiography, cleverly entitled *A Hand through Time*. By the time he was nineteen, Edward was planning a career in law and took up as an articled clerk. However, the following year he took a brief hiking holiday in Wales and there he experienced a St Paul-like revelation. In the early morning sun of a hot summer's day, Edward and a friend climbed the Welsh mountain called Cadair Idris and noticed large cracks (columnar joints) in its igneous rocks. After 'marvelling at what it might mean, we deplored our ignorance aloud' and as soon as he could Edward began reading books about geology. The notion of a law career was forgotten. He frequently recalled that 'from that day I never wavered. That early morning on Cadair Idris made me a geologist'.

Greenly enrolled at the University of London to study geology but took classes erratically. He never graduated; his geological training came from avid reading, from specimens in the London museums, and from travelling, at his father's expense, to do his own fieldwork. He pored over maps at the Geological Survey, in awe. It seemed to him that 'to be able to use a geological map was no small achievement, but a being who could make one must surely be a superman'. He began attempting geological mapping, first in the Bristol area, without any training and always on his own. Returning by train from fieldwork one day, he got

talking to a fellow traveller, bearded and wearing knee-breeches. Greenly, as he jumped off the train, accepted the stranger's scribbled name and address. Only then did he realise that he had been talking with the Professor of Geology at Leeds, who was greatly amused as he watched Greenly read the note and in his excitement spill his knapsack full of fossils all over the station platform.

As it turned out, this encounter was to become another turning point in Greenly's career. The two travellers established contact again, and carried out some joint fieldwork, which led to a professorial letter of recommendation for employment being sent to the BGS Director. Providentially, a vacancy arose not long after. Greenly applied for the position and at interview mentioned that he hankered to meet the celebrated Peach and Horne. To this the Director replied, 'It is to them I am about to send you!' And so the young Greenly found himself in the job of his dreams, despite little formal geological training and having to head north to the 'land of rain and rheumatism'. 'No recruit to the Survey could have had a more felicitous introduction or nobler mentors', as one writer later remarked, and Greenly revelled in it. Forever he was grateful that he had climbed Cadair Idris in that morning sunlight and caught that particular train home from fieldwork.

It was from John Horne that Greenly obtained his training in geological surveying. From him, the importance of three-dimensional thinking was emphasised and thoroughly absorbed. Greenly later recalled a story of Horne's. Talking of apprentice surveyors, Horne said 'Some cannot think in three-dimensions. One was being taken up a hill where the beds dipped at a higher angle than the slope of the ground and we remarked:- "Here, you see, the higher you go, the lower you get", upon which we heard him muttering:- "The higher you go the lower you get; the higher you go, the lower you get . . . damn it all, how can that be" '. Apparently the man decided he had better become a museum curator.

It was the hallowed maps of C.T. Clough that were to provide the model and inspiration for Greenly's mapping. Although Horne spoke of Clough's maps 'with a solemnity that seemed to indicate something unapproachable' and felt that his mapping 'was on a higher plane so lofty that emulation was not to be thought of', Greenly had other ideas. Whenever he got the chance to be with Clough, Greenly tried his best to absorb everything he could from the acknowledged master, about accuracy, fineness of penmanship, diligence, and detail. Above all, he was struck by Clough's insistence on distinguishing between observation and inference. One day Greenly asked Clough why he plotted the boundaries of superficial deposits with such care, as these seemed to be of subsidiary importance to the bedrock. 'Well', Clough replied, 'is it satisfactory to be careless about anything? But I have a special reason. These lines delimit exposures, and thus show how much of any line has been inferred'. This, the beginning of the exposure method of mapping, was another revelation to Greenly. From then on the distinction between actual exposure and interpreted boundaries, which underpins much large-scale surveying today, formed the basis of his mapping technique.

Greenly was entranced with his work. He threw his heart into it and soon Annie, now his wife, joined him in the Highlands. She was happy enough to endure the 'coarse lodgings and coarse food' and the 'houseless, pathless, fenceless, treeless, bushless, boggy moorland' in order to be with Edward, but a snag began to appear. Edward grew acutely aware that there was nothing to occupy Annie's time and mind, and although some employment for the Survey might be possible it would demand close visual work and Edward would not tolerate any risk to Annie's remaining eyesight. Thus it seemed to Edward that his two passions – his geological mapping and the well-being of his beloved wife – were becoming incompatible. Then he came to an amazing decision. He would resign his BGS position and move to somewhere more amenable to Annie – and continue geological mapping *at his own expense*! Being from reasonably well-heeled backgrounds, the two of them reckoned they had the means to allow this astonishing course of action. After only five years in his cherished and much envied position with the BGS, Greenly resigned.

The question then became: where to move to? They sought a place with a substantial area of challenging geology in need of modern mapping, a reasonably 'cultured society', and not too far from Edward's ailing father, now in London. Eleven areas were considered. Parts of Ireland were geologically suitable but, along with the Lizard and Channel Islands, were considered too remote. Start Point, the Malverns, Charnwood Forest, the Precambrian parts of Shropshire, and the Lleyn Peninsula were too small in area; Pembrokeshire, they felt, 'contained no cultural centre'. Central Wales was considered, because of the university at Aberystwyth, but Ben Peach suggested that apart from the graptolite zones, the geology would be too monotonous. All this left just one place – Anglesey. So, in 1895, with very mixed feelings, the two Greenlys moved there and Edward embarked on his project to survey the entire island to his own newly developed, exacting standards.

What Greenly set out to do was no less than to emulate, conceivably even surpass, the mapping methods of Clough, on an island of profound geological complexity. Right from the start the mapping would be carried out on at least the 1:10 560 scale (six inches to a mile) in order to record meticulous detail, and even larger scales would be called on where appropriate. There would be no compromise on three-dimensional accuracy and the separation of observation from inference – every individual exposure of bedrock would be precisely delimited. Greenly aimed at a map which, although possibly leaving some 'knotty problems' of interpretation, would be 'thoroughly sound and reliable and would never go out of date'. So, day by day Greenly would go out, sometimes walking, other times travelling by bike, pony trap or train, to apply his principles and systematically pursue his goal. When, eventually, completion neared, the BGS agreed to publish the map and an accompanying memoir, after appropriate thorough inspections on the ground. By 1919 the work was complete and was published. Producing the Anglesey map and the two-volume explanatory book took 24 years of Greenly's life. The end product is one of the world's monumental geological maps and the memoir has been called 'a veritable masterpiece of geological literature'. And all of it was done for love. Sir Edward Bailey remarked that 'it is doubtful whether so methodical and sustained a research has been carried out by any other individual working without pay as incentive or excuse'.

Greenly then went on to survey further areas and to write an influential treatise on geological mapping, co-authored with the young Snowdonia vulcanologist Howell Williams who was to become Professor at Berkeley, California. The work is referred to a number of times in chapter 15 of the present book. Then in 1927 the dreaded blow occurred – Annie passed away. Greenly lived a further 24 years but his life was never again complete. He studied Italian, in order to absorb the full beauty of Dante's words about love and loss, and he learned to play the piano in an attempt to recapture Annie's music. He became progressively immersed in spiritual matters, and his thinking explored ever more obscure planes. He became engrossed by cats and convinced about immortality, eventually arriving at the belief that his favourite cat was Annie reincarnated. And because he knew Annie wished it, he continued to busy himself with geological mapping, even after the amputation of one of his legs in 1939. It has to be said though, that this later work lacked impact. Without doubt, it is the Anglesey map that represents the apotheosis of Greenly's geological career.

Greenly had endeavoured with his Anglesey surveying to produce 'a truthful, beautiful map as an end in itself' and felt that because he always 'distinguished very clearly between observation and inference' it would stand the test of time. He was right; his work still remains the definitive map of the region. Although understanding of Anglesey geology has undergone quantum leaps since Greenly's day the map still holds up to close use, because it only records features that can actually be seen. Greenly said that 'to separate inference from observation and to part the probable from the certain was the life and soul of my methods in surveying', and the Anglesey map continues to prove him right.

14.3.7 Sir Edward Bailey (1881–1965)

Edward Bailey had many parallels with J.W. Powell. For example, both were infused with the importance of geological maps, both sustained serious war wounds, and both rose to become Director of their nation's Geological Survey. Unlike Powell, however, Bailey disliked the administrative aspects of Survey work, and was perfectly happy making maps – including some of the most outstanding maps ever produced by the British Geological Survey.

Edward Bailey was born weak, and suffered a sickly childhood. At school he was picked on by bullies, and took some cruel beatings. Bailey's reaction to this was to set about 'toughening himself up'. He did this in no small way, until he built up a mental and physical fearlessness which was to become a hallmark for the rest of his life, to the point of eccentricity. As an undergraduate student he would go for marathon walks at night, he took up boxing, and he would sleep all year with his bedroom window wide open and only a cotton sheet for a cover. He graduated in 1902 and applied to join the Geological Survey. Only two people were appointed that year, but Bailey was one of them – despite having broken the nose of the Director's nephew in an undergraduate boxing match.

To his great delight, Bailey was assigned to the West Highland unit. There, the complex geology and the difficult terrain were to be his delight for many years. Perhaps more importantly, he joined a team comprising three of the finest geologists the BGS has ever employed. Ben Peach and John Horne were fresh from their brilliant unravelling of the Moine Thrust region of northwest Scotland, and the field maps of the equally gifted C.T. Clough have often been cited as the standard against which others should be judged. The mapping training that Bailey received, and his sudden exposure to 'a jostling crowd of problems awaiting solution', exhilarated Bailey for the rest of his life. By his twenty-first birthday he was mapping independently for the BGS, and Bailey went on to be involved with many of the superb geological maps of the Scottish western seaboard.

Even so, for Bailey the maps alone were not sufficient. The accompanying memoirs were vital channels for Bailey to expound his imaginative interpretations of the geological histories of the surveyed areas. The memoir accompanying the map of Glencoe, for example, allowed the first detailed three-dimensional analysis of a volcanic caldera. It was a characteristic of Bailey's writing that terms were used exactly, and if no suitable terms existed, he would invent them – antiform and synform, for example. In addition, Bailey was beginning to publish the results of geological studies carried out in his own time – evenings, weekends, holidays – in journals other than official Survey publications. As we shall see shortly, the liberty to do this was to become to Bailey a principle of profound importance.

All was not plain sailing in these early days, though – Bailey's eccentricities were already raising eyebrows. Why was some of his fieldwork being done barefoot? And was he really conducting official Survey work *wearing shorts*? The hierarchy reeled, and informed Bailey that he was not wearing 'sufficiently formal attire'. Hearing this, Bailey promptly resigned, which rattled his senior supervisors even more. A special dispensation was produced which allowed Bailey to continue his work in shorts.

At the start of World War I, Bailey volunteered for duty, which took him away from his beloved Highlands – he was posted to an island off Plymouth. This did not part him from geology, however, for he immediately began surveying the island, the results of which he published after the war. He was eventually sent to the front line, and was badly wounded during the Battle of the Somme and later at Ypres. He sustained a badly damaged arm and lost an eye. He was later knighted for gallantry in battle.

After the war, Bailey was able to return to the survey of western Scotland, and was put in charge. Most significantly, in the present context, he was put in a position to complete a project he had been involved with off and on since 1907 – the mapping of the island of Mull and nearby areas. The finished one-inch map appeared in 1924, and the following year the accompanying two-volume explanatory memoir was published. The map is superb. It has been called a 'landmark in the history of geological cartography'.

The intricate geology of the island of Mull, a deeply eroded Tertiary volcanic complex, had presented a formidable scientific challenge, and it was tough terrain. Bailey was entranced, both with the routine surveying, and the gradual deciphering of the volcanic story that was hidden in the rugged mountains. His commitment to geology and oblivion of danger is reminiscent of J.W. Powell. On one occasion of prolonged heavy rain, the survey team was camped in a meadow which became flooded. The team was forced to leave during the night, only to realise the next morning that Bailey had not joined their evacuation. One of them waded back to Bailey's tent, to find the water lapping just inches below the level of the canvas camp bed. On the bed lay Bailey, happily reading a textbook on physics.

The mapping of Mull has been called 'one of the most wonderful chapters of the geology of Britain'. When the map was published 'new lustre had been added to the Geological Survey' (Flett 1937). The precision with which the intricate geology is portrayed, including over 40 kinds of igneous rocks, is a joy to the eye. It is the kind of map that draws gasps when first seen. One can find criticisms – the density of the colours tending to obscure the topographic base, the oddly slanting annotations – but these are quibbles. Without doubt the Mull sheet represents one of the highest pinnacles of achievement of the BGS.

Nevertheless, Bailey was continuing to annoy his superiors with his machismo activities. He would soak his boots at the start of a day in the field, so that further wettings would go unnoticed. He was renowned for eating nothing more than a chocolate bar for lunch, or, if he had been supplied with more, eating it first thing to 'get it over with'. On one occasion Bailey received an official reprimand for leading his new Assistant Director through a chin-deep river, when he knew there was a perfectly good bridge just out of sight upstream!

But much more significantly than all this, in 1920 a Director was appointed to the BGS who felt that all his staff's scientific writings, whether or not they were related to survey work, had to be vetted by him. This was completely contrary to the principle of scientific liberty cherished by Bailey. He became increasingly uneasy with the new regime. In turn, his Director grew paranoid about enforcing his new rule. Things eventually came to a head. When Bailey was banned from looking at any of the Survey's field maps, even though they were available for consultation by the general public, it was the last straw. Bailey knew he had to move. He resigned, taking up the life of a university professor, at Glasgow.

There, Bailey clearly enjoyed himself, travelling far and wide, coming up with innovative interpretations, and letting his geological imagination roam. He was one of the first in the UK to embrace a wild new theory which involved the continents actually drifting apart. He seized on the significance in rocks of 'way-up' indicators, was one of the first to recognise submarine slumping in rocks, and anticipated the concept of growth faulting.

Bailey had an unrivalled grasp of the three-dimensional aspects of mapwork. He began to apply his skills to geometrically difficult regions such as the Swiss Alps. The convoluted forms of the sections he

constructed became nicknamed 'Bailey's bicycles'. He, himself, enjoyed a play on his first name in his paper entitled 'Eddies in mountain structure'. He formulated some of the methods for dealing with the outcrop patterns of structurally complex areas, and, following a visit to the Appalachians, helped promote the 'down-structure' viewing technique (section 8.3).

But all the time, despite the treasured scientific liberty which a university offered, it seems that the Geological Survey was still his first love. In 1936 he was invited to return – as Director. Bailey accepted, but on his own conditions regarding research. As it turned out, he had little time to put his ideals into action before once again he and his Survey were plunged into supporting the national war effort.

In World War II, BGS geologists were not involved in the fighting, but in all kinds of strategic support work. Geological surveying was required for underground vaults, new airbases, temporary water supplies, etc., as well as for sources of raw materials which could no longer be imported. These ranged from coal and iron to less obvious but vital needs – sand for sandbags, mica for electrical equipment, silica for optical glass, even sapphires for aircraft compass pivots. Bailey threw himself into his new responsibilities. He was later to be proud of the BGS war effort, and was eager to point out that its duties had been greatly facilitated by the fact that high-standard geological maps of the UK already existed.

During the war, the BGS office was evacuated, becoming the Civil Defence Headquarters. The geological maps and materials were shipped to North Wales for safety; the geologists, too, were moved. Except Bailey, that is. He found space sufficient for a temporary office on the topmost gallery of the Survey building, and worked there throughout the London blitz. True to form, he seemed oblivious of the danger, even after he was 'temporarily buried by a V_1 or Doodlebug' (Bailey 1952).

At the end of the war, Bailey saw the need for the Survey to have a fresh look. Reorganisation was necessary, and Bailey felt that it was time for him to make way for new blood. He retired, although until his death 20 years later he continued actively in geology, publishing research papers, and attending geological meetings – invariably wearing shorts! And in 1948, there came to fruition one of his pet projects.

Back in 1942 Bailey had been appointed to look after the geological aspects of a 'National Atlas'. He had arranged for the BGS to produce a map at the innovative scale of 1:625 000, or 'about ten miles to an inch'. The scale was selected in order that the land mass could be accommodated on 'two not over-large sheets'. The map was to carry a new kind of 10-km grid. This was to help unify the two sheets and give easy correlation with larger-scale maps which from then on were to carry a kilometre grid. This 'National Grid' would also provide a convenient and precise reference system for locations 'much more easy to employ than that given by longitudes and latitudes' (Bailey 1952).

Most of the detailed survey data were already available on the one-inch sheets; it was largely a matter of compilation and reorganisation to fit the new scale. Bailey took the project to heart. The north sheet covered the Scottish Highlands and Islands and he saw this part, especially, as his own. We can imagine him during those long dark evenings of the War, perhaps even with the sound of air-raids around him, drafting his map.

On its publication, the map was hailed a great success. Since then it has remained the standard geological map of Great Britain, and a model for many other national maps. Without a great deal of alteration, a second edition was published in 1957, and a third in 1979. Part of the south sheet of the map is, of course, reproduced here as Plate 1. The north and south sheets are widely available today for purchase. Also still sold are the one-inch maps of the Scottish western seaboard which Bailey helped produce. They remain the definitive maps of those areas (some now photographically enlarged to 1:50 000), especially the celebrated Mull map. The legacy to geological maps of Sir Edward Bailey remains very tangible.

14.3.8 J. David Love

The individuals discussed in the last two sections had in their mapping difficulties of scale – how to squeeze their detailed observations onto even the largest scale maps of their area. David Love also has had scale problems – how to recognise equivalent mapping units across an area considerably larger than the entire British Isles. For he is a man of the vast, high plains of the western USA; he has mapped more of the Rocky Mountain states, and Wyoming in particular, than anyone.

David is the son of pioneer parents (his father was a nephew of the seminal conservationist John Muir), whose colourful lives and correspondence are carefully recalled by Fowler-Billings (1993). McPhee's (1986) *Rising from the Plains* is a fascinating account of their profound influence on David's early years, and on the manner of his subsequent engagements with Rocky Mountain geology. David's birthplace was a log cabin, 12 miles from the geographic centre of Wyoming, that sat alone under a vast blue arch of warm summer sky and that seemed even more remote in the piercing winds of the long winters. The nearest neighbours were more than a dozen miles away;

David and his brother were the only children in a thousand square miles. 'There was not a lot to do', he later recalled, 'but look at the landscape'. And David was the best in the family at it, at noticing details in the shapes of hills and, even at the age of five, being the first to spot Indian arrowheads and chippings. David's education was from his mother's knowledge of the classics, delivered at the kitchen table (which had been bought second-hand from a defunct gambling saloon).

One day young David watched in awe as a two-horse wagon emerged from the lonely Owl Creek Mountains and attempted to cross the Wind River, only to be toppled and swept vigorously downstream. His father ran to help and, after several attempts, managed to rope the lead horse and haul the ensemble back to the bank. This heroism and skill impressed David, but what really intrigued him was the mysterious individual who had been driving the wagon. It turned out that this man was a geologist – a new word to the young boy – who was returning from a season's fieldwork aimed at understanding the mountains. His name was Nelson Horatio Darton, the first to map geologically wide areas of the American west, and with a quality that has ensured he has remained one of David Love's heroes. In that river mishap all the season's field documents were lost, so that Darton returned the following year, vanishing into the mountains to start his surveying all over again. David became haunted by the mystery of the Owl Creek Mountains and what this geologist was doing there. Geology began to take on more meaning to the young boy. 'Day after day you had nothing but the terrain around you – you had nothing to do but think about why the shale had stripes on it, why the boggy places were boggy, why the vegetation grew where it did, why the trees grew only on certain types of rock, why the water was good in some places and bad in others. Everything depended on geology'.

Later, other geologists began to appear in the area, some of them staying in the Love household. But these men were interested not in the mountain ranges but in the sedimentary basins in between. They were oil geologists, most of them, and they soon found plentiful oil. One well, only six miles from the Love Ranch, gushed with such force that it blew the wooden drilling derrick to pieces. Some of the geologists, though, were from the USGS. To David Love, they 'raised a magic curtain'. When he won a place at the University of Wyoming, in Laramie, it was natural that he went to study geology. There, new vistas opened up, and when on graduation he accepted a doctoral position at Yale University, he was to encounter the very different world of the east coast of America. Even so, when the question arose of which region of the world David should go to for his PhD field studies, the answer was immediate: the Owl Creek Mountains in Wyoming.

No topographic base maps were available for the area. His rigorous undergraduate training with a plane-table therefore became invaluable in allowing David to make his own base maps where appropriate, otherwise he had to rely on aerial photographs lent by the Soil Conservation agency. His mapping training also underpinned a series of summer jobs, surveying lakes and prospective dam sites, in order to help finance his continuing studies. The region he surveyed for his postgraduate work eventually stretched far beyond the Owl Creek Mountains into the Absaroka range, covering over 500 square miles of virtually unknown geology. This was wild, roadless country – frequented by the likes of Butch Cassidy – and David tackled it alone, often on foot. The terrain included, as McPhee (1986) put it, 'folded mountains and dissected plateaus, sedimentary basins and alpine peaks, desert sageland and evergreen forest'. After six years of field surveying, the work earned David his doctorate degree, justified a special publication of the Geological Society of America, and remains the basis of modern knowledge of that region.

Love first took a job as an oil geologist but he soon became disenchanted. 'Looking for oil for some damn fool to burn up on the road', seeing the sky lit at night by gas burning in thousands of flames because it seemed to lack economic value, and attempting to stay in hotels which barred 'dogs and oilmen', all conspired to make him take up employment with the USGS. The Director assigned his new recruit, because of World War II armaments, to explore for vanadium – in Wyoming. Love soon acquired, on the campus of the University of Wyoming, his own office, a field outpost of the USGS. From that base, Love was to carry out his mapping of the Rocky Mountain area for the USGS for almost 50 years. It was his own mapping station and only now, as David Love enters his eighties, is it being wound down!

During this time, Love has produced numerous maps for the USGS, which deal with all aspects of the region's very varied geology. Herein lies one of his strengths in mapping – he is equally at home working with any kind of geology. He has surveyed Precambrian gneisses and Pleistocene alluvium, intricately deformed strata and 2-m thick beds that stretch horizontally over $50\,000\,\text{km}^2$. Some of his maps emphasise particular commodities, such as oil, gas, coal, uranium, phosphate, and gold. He has been, after all, working in one of the world's mineral-rich areas. Wyoming has coal seams 70 m thick, 90% of the world's reserves of trona (used in glass, ceramics, textiles, etc.), and, with parts of adjacent states, shale that contains more oil than Saudi Arabia. When mapping, each evening he would copy up his work, invariably by the night's camp fire. On the fire his dinner would be cooking, perhaps a sage chicken or rattlesnake obtained by a throw of his geological hammer. (His leather USGS saddlebags – ex-US Cavalry – had waterproof liners to hold wild game.)

By the early 1950s, it was time for the accumulating geological knowledge to be brought together, and Wyoming contemplated producing its first state map. The tradition in the US is for such maps to be compiled by a team and published with a long list of authors, perhaps under the charge of an overall coordinator. In this case, David Love was the obvious person for such a directing role. The task was a difficult one because of the sheer expanse of the region to be portrayed. Although some of the potential map units of this region are spectacularly continuous, others undergo dramatic lateral variations or are tectonically displaced to such a marked extent that correlation is a major obstacle. Love's first-hand experience of the rocks became critical, and eventually a practicable correlation scheme was devised. In 1955 an impressively detailed map of the entire state was published, with Love as chief author. It summarised the state-of-play of geological understanding of Wyoming, for 'a geologic map should serve as an epitome of what is known and what is not known about a region, up to date' (McPhee 1986).

Love pressed on with his exploration of the region's geology, producing maps at all scales. Some of the efforts of which Love is most proud are the maps that cover the stunning but geologically complex valley of Jackson Hole, Wyoming. The Precambrian rocks that make the spectacular Teton Mountains, back-drop to many a movie; the complicated, still active, tectonics; and the intricate terraces of the Snake River, have all been subjected to Love's unravelling abilities and beautifully recorded in a series of 1:24 000 sheets. Love also put together picture books and field guides to help explain the geological story for visitors, and the USGS published his large, carefully designed, block diagram of the area's geology.

After a further 30 years of gathering knowledge, a new state map was needed. Love was still very active, had by far the greatest experience with the region's geology, and was again the obvious person to lead the project. The challenge was to produce an even more comprehensive, detailed map, even though the region remained vast and complex. Despite Love's encyclopaedic knowledge, as well as the long-standing efforts of the USGS and the Wyoming State Geological Survey, for information on some areas he had to go back to the 1906 work of N.H. Darton, and in some cases even earlier maps. At the same time, the most modern fossil and radiometric age-dating techniques were brought to bear in order to aid correlation of equivalent units. Hence it now became possible for this map to show, for example, subdivisions of the Precambrian rocks. In some areas of intensive mineral exploitation, there was more information available subsurface than for the land surface. In such cases, Love took the well data, structure contours, isopachs, etc., in order to help him construct more accurately the boundaries on the map (see section 3.4).

Such detail had to be balanced with clarity. The general tradition is for each state to devise its own colour scheme, rather than attempting to integrate its system with that of adjacent states. Love went for a pleasing but essentially functional scheme, using pale colours for those units that outcrop over wide areas, and bold colours to emphasise thinner units of distinctive nature. The result was published in 1985. Love was again lead author and thus became only the second person ever to direct a US state map twice. The map was widely acclaimed. As one commentator put it, 'the 1955 map set a new standard for state geological maps in the detail of its coverage and in its fossil dating, a standard set anew in the 1985 edition'. Love's personal input shines out. One of Love's colleagues remarked that 'most regional maps are patched together from various papers and reports. Dave has looked at all the rock. It's all in one mind. To compete with Dave you'd have to do a lot of walking!'

Love is a modest man but clearly proud of his maps, whether of an entire state or a small quadrangle. In this part of the world, geological maps are routinely used for all manner of practical purposes. One reason for the success of the maps is that the map units work. Section 1.3.1 introduced the fundamental role of map units and section 15.2 emphasises that the division of an area into such formations is somewhat arbitrary and depends on the surveyor. A geological map can only be as good as its map units, and maps of a region the nature and size of Wyoming depend on meaningful units being devised. This fundamental aspect of the nature of geological maps is well illustrated by David Love when he is asked of his most valuable contribution over the years to mapping. He points not at a map at all, but at the coloured columns of his chart which correlates in detail the stratigraphic units between different mountain ranges and sedimentary basins (Love et al. 1993). It is Love's success at correlation across this huge region that has made the mapping possible; it is the key to the geological maps of the region in more ways than one.

CHAPTER 15

The Production of Geological Maps

15.1 Introduction

Although this book is about working with completed geological maps, an understanding and appreciation of them would be incomplete without some knowledge of what goes into their production. Many stages are involved, from the detailed field surveying required for large-scale maps through the preparation of 'fair copy maps' to the design and execution of a final printed version. Greenly and Williams (1930, p.99) list 23 steps. This chapter merely outlines what is involved.

In many ways geological maps have changed little during the couple of hundred years of their existence. The previous chapter mentioned improvements such as the advent of colour printing, refinement of ornament and increase in detail, but a geological map produced today would probably be quite intelligible to, say, William Smith. Suddenly, however, the rate of change has begun to accelerate, not least because of the rise of computer methods. This chapter also touches on some aspects of these developments.

There is a further purpose to this chapter. Often, field surveying is seen as an endeavour which is completely separate from interpreting existing maps. Part of the reason is that the two procedures are commonly taught to students, for perfectly good logistical reasons, in quite separate courses, and the connection is not made. However, a thorough understanding of the three-dimensional behaviour of map units, and the genetic implications of the way they are presented, is as essential to field surveying as dealing with a published map. Moreover, a field map has to be compiled with the needs of the eventual user in mind. Attention will be drawn at several points in this chapter to the need for the finished map to be clear and visually attractive, as well as scientifically sound. Such qualities cannot arise through cosmetic adjustment to the final document; the need has to be appreciated throughout the production of the map.

15.2 The Field Survey

Surveying to produce a geological map, or 'geological mapping' as it is often called, usually begins with the acquisition of a topographic base document on which geological data will be plotted. Very large-scale work, such as the drawing of mining maps, may well require the creation of a sufficiently large-scale base map at the same time as the geological survey, using the plane-table methods described, for example, by Compton (1985). Where a topographic map of appropriate scale is already available, the geologist plots his information directly on to this, or on to air photographs (e.g. Barnes 1995, chapter 3).

One of the first important judgements the geologist has to make in his survey is the basis on which he will divide the rocks of the area into map units. Much of his geological mapping will be concerned with marking on the base map the course of these units, and particularly the boundaries between them. These will form the basic framework of the map. In more detailed work, it is common to mark on the map the actual exposures of bedrock, so that the distinction between what has been observed and what has been inferred is clearly recorded (see also section 14.3.6). This can only be done realistically at scales of 1:10 000 and larger. In all methods, the sketching of cross-sections as the survey is progressing is necessary, to keep clear the understanding of the three-dimensional geology. The various techniques are discussed in detail by the standard works on field mapping (e.g. Ahmed and Almond 1981, Moseley 1981, Compton 1985, Barnes 1995). Much of the summary given by Ramsay and Huber (1987, appendix F) is relevant to general field mapping. Two points are emphasised here.

First, except in the rare case of perfectly exposed ground, the geologist will have to infer the course of the outcrops. He will have to project traces between exposures, bearing in mind the *three-dimensional nature* of the boundaries. He will need to visualise how the boundary will interact with the ground surface. Commonly, the shortest route is unlikely. It is a matter of applying the principles of earlier chapters in reverse, e.g. drawing the trace to 'V' across valleys in the appropriate way (chapter 6). In large-scale mapping it may be possible to apply the principles of chapter 4 in reverse, and sketch some structure contours from parts of a boundary that are well known, in order to predict where the unexposed boundary is likely to run (figure 15.1).

Second, as already emphasised in section 1.3.4, the geological map is the result of inferences of many kinds by the surveyor. There may well have been some arbitrariness in the definition of the map units, and the surveyor will be aware that implications for the geological history will arise from the way he judges the outcrop relationships. Although much of

this book is about interpreting completed geological maps, it is important to remember that the map itself is an interpretive representation of an area.

As the boundaries are plotted on the map, the outcrop patterns will begin to take shape, reflecting the geology of the area. The presence of structures such as folds and faults may be revealed. The surveyor may colour his field map, to accentuate the outcrops, and he may add supplementary information, such as orientation data, relevant topographic features, and perhaps some brief annotations. Many of these supplementary data will be recorded in the field notebook (e.g. Barnes 1995, pp. 91–94), which becomes an essential adjunct to the field map, or, increasingly, on computer-readable data sheets. All this wealth of information will somehow have to be streamlined if it is to be stored and made more generally available, and here computers are playing an increasing role.

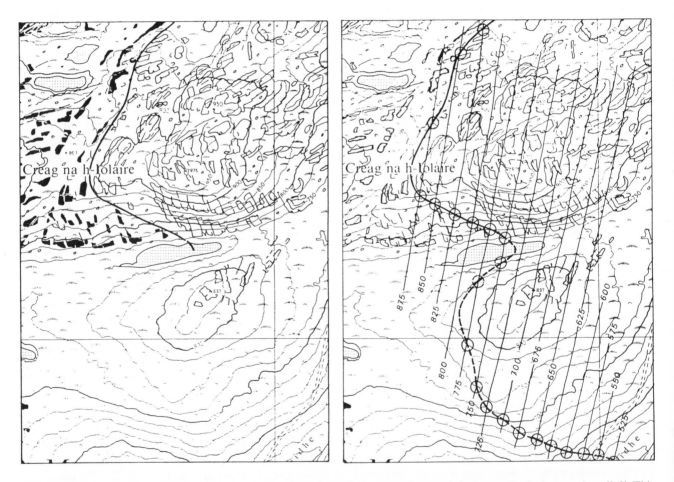

Figure 15.1 Using structure contours to trace a surface through unexposed ground during geological surveying. (*left*) This shows exposures of Torridonian sandstone, in black, and the overlying Cambrian quartzite, unornamented. Numerous exposures in the northern half of the area control the course of the junction between the two units. In the southern half, however, sparsity of exposure leaves the course unclear. (*right*) Structure contours, drawn from intersections (circled) of the trace known from (*left*) with topographic contours, are straight and evenly spaced, indicating a uniformly dipping surface. Extrapolating the structure contours southwards gives intersections with the topographic contours (circled) which enable a reasonable tracing of the junction in the southern area.

15.3 Storage and Manipulation of Map Data

Traditionally, the field surveyor next produces a 'fair copy' map: a simplified version of the field map that gives emphasis to the interpreted outcrop pattern, and gives selected representative supplementary information. Barnes (1995, chapters 8 and 9) discusses the nature and preparation of fair copy maps and cross-sections. The detailed field maps and notebooks are archived, for consultation as required. However, it can be difficult to retrieve specific information from such documents, even the most assiduously organised ones, and here computers are providing attractive new alternatives.

Storing the field data in digital form is proving convenient in a variety of ways, and as it becomes linked with the burgeoning use of Geographical Information Systems (e.g. Bonham-Carter 1994; Giles 1994), it is greatly enhancing the power and flexibility of geological maps. Nickless and Jackson (1994) summarise the efforts of the BGS in this direction. Vast amounts of data can be stored in little room, with relatively easy access, providing the information is well organised. Special aspects, say geochemical, geophysical, tectonic or borehole data, can be stored and integrated as required. Maps can be examined and experimented with on-screen and, if required, printed to particular requirements, say a specific scale, projection, form, or resolution. The traditional sheet boundaries become irrelevant and particular aspects of the geology can be omitted or emphasised. Items can be kept up to date without the need for complete resurveying. And one major advantage is that the three-dimensional methods developed in the early chapters of this book no longer have to be carried out by laborious manual construction!

Computer programs are now available that allow cross-sections to be constructed rapidly from digitised maps, and vice versa (e.g. Ichoku et al. 1994). Section lines and vertical scales can be varied; structural and stratigraphic sections can be interchanged; multiple sections can be transformed to fence diagrams and block diagrams, with various projections and perspectives (chapter 5). Other constructions such as structure contour or sub-crop maps can be carried out. The manipulations are executed extremely swiftly. You can use trial and error: if a particular form of map or cross-section is displayed on a screen and seen to be unsatisfactory, you can modify it before printing. Additional underground information such as seismic sections and well logs can be readily incorporated.

Not surprisingly, it is the oil companies that are making most use of these facilities. The kinds of data they work with are very amenable to this treatment, and their needs are specific and easy to define. Consider, for example, a geologist investigating a petroleum reservoir and requiring the standard isopach maps (section 4.4) of the relevant formations. Working manually, this would involve consulting the well and seismic data, calculating and interpolating apparent and true thicknesses, and the construction of structure contour maps before visually overlaying and subtracting them in various permutations to obtain the thicknesses. And some oilfields have scores of formations. Seismic and well information, however, is readily handled by a computer, which can store, assemble and interpolate the data. As required, relevant information can be instantly retrieved and any permutation of thicknesses, thickness ratios, thickness differences, etc., calculated and plotted for any specified stratigraphic intervals.

In fact, so rapid is the rise in power of these manipulative programs and the sophistication of the output, that it is reasonable to ask if the content of the early chapters of this book will soon be redundant! This is extremely unlikely, for at least two reasons. First, the methods still remain expensive. The programs are costly to purchase and to run. Realistic operations require substantial computing power and high-quality graphic output facilities. Second, it is only sensible that such expensive procedures, and on which much may depend, are operated effectively. You still have to instruct the equipment which manipulations to carry out. You have to tell it which line to draw the cross-section along, on which surface to draw structure contours, etc. The rapid manipulation is merely an efficient means to an end. So the operator has to know what it is the computer is doing for him, and its limitations. The technology is removing time-consuming tedium, but it is making geological understanding more important than ever.

15.4 Preparation of Maps for Publication

Despite the rise of digital methods, the vast majority of geological maps you are likely to consult in the near future are likely to have been produced by traditional means. Indeed, most new geological maps are still published in conventional, hard-copy form. So it is worth while glimpsing what goes into completing a map for publication, to help give you a greater understanding and appreciation of the final document.

15.4.1 Scale

The scale at which you see the published maps is not necessarily the scale at which the area was surveyed. For many purposes the field maps are considerably reduced. For example, it has been standard practice for the BGS to survey at 1:10 000, a scale highly convenient for the kind of accuracy and detail appropriate to a country the size of the UK. These maps are available for public inspection, but are reduced for publication to 1:50 000 (or, in pre-metric days, six inches to a mile field mapping for reduction to published one-inch maps). Plates 5 and 8, for example, were surveyed at a scale of six inches to a mile (1:10 650).

For the published document, the geologist has to decide what detail to retain. Mapping units may have to be grouped, and features excluded. Edward Greenly mapped parts of the island of Anglesey for the BGS at a scale of 25 inches to a mile. He later recalled that although this made it possible to record 170 different beds of jasper at one place, when the map was reduced to six inches to a mile for archiving, the number of jasper horizons had to be cut to 100, while the one inch to a mile map that was published (today at 1:50 000) could show only 45 jasper beds (Greenly and Williams 1930, p. 382).

15.4.2   Boundaries

The geologist will have indicated on the reduced map which boundaries were observed and which inferred, by using some system of dashed lines explained on the key (figure 4.2). He will also have ensured that his boundaries have continuity with adjacent maps. Stories abound of the need to insert geological faults at the margins of a map in order to explain offset between the boundaries on adjacent sheets.

Some simplification of the courses of the junctions may have been made to promote the clarity of the map; different weights of line may have been used to distinguish between depositional and faulted junctions – the final effect depends as much on the cartographic draftsman as the surveyor. Normal stratigraphic boundaries are best shown as a firm, but narrow line.

15.4.3   Key, and other information

Careful attention will have been given to the map key. The geologist will have decided how much information can be incorporated while maintaining clarity. A stratigraphic column or vertical section showing thicknesses and relationships may have been added. Most maps include cross-sections on the map sheet. Maps published by official surveys often include a statement on authorship of the map, which can provide an interesting insight into the map's history. Some modern maps, such as the current BGS 1:50 000 series, include short descriptions and explanations of the geology next to the maps; some USGS maps have lengthy descriptions on the reverse side of the sheet.

The design and layout of all these items should make a well-balanced whole. Modern BGS maps even have coloured frames! Returning to the map key, its main function, of course, is to explain the use of black and white ornament or colour on the map. The decision of how to represent each unit may seem a simple one, but it needs care to produce pleasing and effective results. There is, in fact, a fascinating and instructive history to the matter.

15.4.4   Ornament

The map units of an uncoloured map are normally ornamented to facilitate distinction between them. There have been attempts to standardise the patterns used. For example, Evans (1921) proposed that Precambrian rock be given NW–SE ruled lines, Lower Palaeozoic rocks NE–SW lines, Upper Palaeozoic N–S lines, etc. Such systems have never caught on, except for certain patterns becoming associated with particular lithologies. It is common, for example, for limestones to have a 'brick-wall' pattern, sandstones to be shown by various dots, and shales by dashes. Compton (1985, appendix 8) gives examples of some common ornaments.

15.4.5   Colour

Widely published maps have traditionally been coloured. The earliest maps, including those published in journals, were water-coloured by hand. The translucence of water-colours was ideal for allowing the other elements of the map to be easily examined, and in many cases, extremely pleasing effects were obtained. Some of them were very skilfully executed (e.g. Ireland 1943). Many are now valuable collector's pieces.

However, there were serious snags. Such labour-intensive methods greatly increased the cost of publication and restricted the number of copies. Some colours looked too dense and muddy, and broke the child's first rule of colouring – to 'not cross the line'. Mistakes were difficult to correct. The best colourists used a soaked map laid on wet newspaper in order to get even colours, but this led to considerable distortion (Wilson 1985). All this changed in the middle of the nineteenth century (section 14.3.4), when colour printing became used for geological maps.

Colours are so central to the effectiveness of a geological map that it does seem extraordinary that after about two hundred years of map-making there should still be such disparate colour systems. There seem to be four main approaches to choosing a colour scheme.

The first approach is largely aesthetic. The intention is to convey the outcrop patterns clearly, but in colours that are pleasing to the eye. The approach is often used in 'one-off' maps, which are not part of a wider effort such as those of the national surveys.

The second approach relates colour to the stratigraphic age of the rocks. The principles go back to a meeting of the International Geological Congress in 1881, but although a number of European maps followed the guidelines that were agreed at the meeting, there has never been universal acceptance. In that scheme, each stratigraphic system was allotted its own colour, which, apart from orange and red being reserved for igneous rocks and pink for metamorphic rocks, spanned the spectrum from the pale colours for younger systems to dark colours for older rocks. Although fine in theory, this did mean that whereas Tertiary rocks, for example, were assigned nice translucent yellows and Eocene deposits a pale yellow-green, Lower Palaeozoic and older rocks had to suffer dull, somewhat opaque colours. An analogous approach, though differing in detail, has traditionally been adopted by the US (section 14.3.5) and Canadian Surveys.

A third approach is based on the actual colours of the rocks themselves. Early map-makers such as William Smith followed this idea, and it has been influential on the colours used by the BGS ever since. For example, the browns of the Torridonian and Old Red Sandstone, the oranges of the New Red Sandstone, and the dark grey of the Coal Measures seem to be based on this principle. The BGS maps of the region around Oban, West Scotland, portray the purples of the dark-red weathering volcanic rocks, the dark blue-greys of the graphitic slates, dark greens of basic intrusives, and browns of the Old Red Sandstone deposits. Linton (1947) felt that such 'imitation of rock colours gave to the colour scheme a harmony which resulted in maps of real beauty . . . but that . . . it also imparted to many sheets an obscurity and gloom which made them almost impossible to read'. The same could be said of many of the French BRGM 1:80 000 sheets.

Fourth, some colours have come to be associated with certain lithologies. Blue has been used in many countries for limestone and calcareous deposits, and red for granites. Basic and ultrabasic igneous masses are commonly shown in dark greens or purples, and alluvium and other superficial deposits in pale cream. In addition to the colour, use can be made of various overprints, such as stipples and rulings but this has to be done judiciously in order to avoid clumsy and opaque effects.

Whatever approach to colour and overprinting is adopted, it is imperative that the topographic base remains clearly legible. It is unfortunate that the BGS, perhaps because it has no control over the final production and printing of its documents, tends to issue maps with indistinct topography. Even on otherwise superb geological maps, such as the 'one-inch' Bristol sheet and the 1:25 000 Snowdonia sheets, it can be difficult to read topographic information such as contour lines. Linton (1947) criticised the maps being produced then for being 'quite opaque so that it is impossible to follow the contours beneath', and some modern examples seem hardly to have improved in this respect. The 1:50 000 map of Denbigh, for example, published in 1975, has such dark colours coupled with a heavily ruled overprint that it is impossible even to read some of the place names!

15.5 Availability of Maps

The basic source of maps is the official geological survey of the relevant area. Usually this is a national institution, but some of the states of the USA, for example, have very active surveys. The easiest way of obtaining a particular published geological map that you know about is through a good book shop. It may hold stocks of current official geological maps of the local area; otherwise your map will have to be ordered. Besides the name or number of the sheet you require be sure to specify the scale. Some maps are available either folded, perhaps in a protective plastic wallet, or flat, normally delivered rolled up in a tube. With BGS maps you may have to specify 'solid' or 'drift' editions (section 1.3.2). Flat maps obviously avoid unsightly creases and are excellent for display purposes, but in any quantity create a major storage problem. You can make maps more durable by mounting them on plastic or cloth (Groves 1980).

The main difficulty with obtaining geological maps has always been: 'I don't know what is available for the area I am interested in'. Fortunately, things have suddenly become very much easier, because of the World Wide Web. Using one of the Web search engines to scan, say, 'geologic maps' (which will probably include 'geological maps' in the search) will reward you with listings of the products of all the major national geological surveys (updated monthly for the USGS), state and provincial surveys. Many give prices and addresses from where you can order maps directly. The Web listings also give a host of commercial producers, map libraries, courses on geological maps, and home pages of companies and individuals who deal with maps. In some cases maps can be downloaded directly, free of charge, from the Web, providing you have sufficiently powerful facilities to handle this. All these advances are a real boon and hold even more promise for the future. But even so, it is still nice to end up with at least one traditional hard copy, properly printed to give greatest accuracy and a pleasing appearance.

15.6   Justification of Geological Map Costs

The cost of 'official' geological maps, produced by national or state surveys, will have been underwritten by public funds. Therefore, in these days of scrutinising the use of public monies, the question is increasingly asked 'Are geological maps worth it?' The answer is entangled with politics and opinions on exactly what the state should be responsible for, but a number of studies show that on purely cost grounds geological maps are a bargain.

For example, the state of Nevada, USA, costed a detailed survey of the state to meet its new environmental geology needs (Fracolli 1985). It concluded that one geological map alone would cost $500 000 to produce, and the entire programme would total $31 million – a large amount of public money. However, these sums shrink when put against potential losses if the information were not available. As one example, the state was recently sued for $12 million property damage (due to land subsidence focused along Quaternary faults, and in a small area of one geological map) which could have been avoided if the geological information was available at the time of development. The Illinois State Geological Survey estimated that the ratio of savings to mapping costs varies between 5 and 55 to one, depending upon such things as aquifer depth and pollution potential in the mapped area (Bhagwat and Berg 1991).

The USGS estimates that the cost of remapping Loudoun County, Virginia, to modern standards was about $1 160 000, distributed over 6 years (Bernknopf et al. 1993). It made a series of statistical calculations on the chances: (1) of one new landfill-waste site in the county being sited on inappropriate geology and developing a leak, and (2) roadcuts for a new highway being routed through unstable material. It then went on to estimate the cost to the public purse of reparation of such failures. Taking property damage alone (i.e. ignoring repercussions for health, insurance, economic disruption, etc.) the cost arising from the former hazard was calculated at $1 504 432 and the second at between $935 000 and $3 157 000. That is, even on the basis of only these two hazards, the net benefit of producing an adequate modern geological map of Loudoun County (avoided charges minus cost of map production) was somewhere between $1.28 million and $3.5 million. Similar cost–benefit analyses were urged for the Netherlands by de Mulder (1990).

In 1992 the US Congress decided to deliberate on this matter of cost. It eventually declared that 'geologic maps are the primary database for virtually all applied and basic earth-science investigations' and are 'tangible products that can be immediately useful to a variety of non-scientists'. Consequently, recognising this 'societal importance of geologic maps', Congress duly passed into public law what it called the 'National Geologic Mapping Act' to authorise the continuing use of public funds for geological mapping of the USA.

15.7   Conclusions

A geological map shows so much more than just the distribution of rocks at the earth's surface. Besides all the projections into three dimensions and back into geological time that we have been discussing in earlier chapters, together with all the various practical uses, we have glimpsed some of the human stories and the endeavours that lie behind geological maps. A good geological map not only represents the summation of geological understanding of that area at the time, but conveys it pleasingly to the reader. To quote from Edward Greenly's treatise on geological surveying (Greenly and Williams 1930): 'Nature is beautiful; we are attempting a representation of her. A geological map is made to be looked at . . . let the process be a pleasant one. The surveyor should have kept beauty in view throughout his work. In its colour, as well as in its line, a geological map should be a thing of beauty'.

The view 70 years later has changed little. In 1996 a Congress Science Fellow in the US remarked that 'the geologic map can be considered a meshing of science and art, replete with valuable information, yet retaining that valuable imprint of the individual mapper, along with the individual's insight, creativity, and occasional oversights. As such, the geologic map should be considered invaluable to most earth-science investigations' (Folger 1996).

We have also glimpsed the future. Maps emphasising particular aspects of the geology of an area are likely to continue to proliferate – this is, after all, an age of geological specialisation. However, that specialised work will normally have to be set in its broader context and for this the geological map will continue to provide the central, coordinating thread. Maps will be increasingly produced and manipulated using the technologies and methods mentioned in this chapter. The results may not in some ways look like traditional maps but geological maps they remain. The principles are just the same and it is more important than ever that the user understands them as he takes advantage of the sophisticated techniques now possible. If you are familiar with how these maps work then digital methods are opening up yet new opportunities. As the power and flexibility of geological maps increase still further, all the signs are that the geological map will remain a fundamental tool.

Summary of Chapter

1. Field surveying involves subjective judgements, on dividing areas into formations and on tracing boundaries, which are normally covered. It utilises the same principles as interpreting completed geological maps.
2. Computers are revolutionising the storage and manipulation of map data, increasing the scope and reducing routine tedium, but making imperative operator understanding of the principles involved.
3. Published geological maps have involved numerous steps in their conversion from field documents, such as decisions on scale, key, and colouring.
4. Specialised and thematic maps continue to grow in range and usefulness, linked by the central thread of the conventional geological map.
5. Searching the World Wide Web gives access to up-to-date listings of available geological maps.

Selected Further Reading

Barnes, J.W., 1995. *Basic Geological Mapping*, 3rd edition. John Wiley, Chichester, 133 p.

A student-level outline of the basics of field mapping.

North, F.K., 1985. *Petroleum Geology*. Allen & Unwin, London, 607 p.

Chapter 22 discusses the kinds of maps used in petroleum geology.

Warren, W.P. and Horton, A., 1991. Mapping glacial deposits in Britain and Ireland. In: Ehlers, J., Gibbard, P.L. and Rose, J. (editors), *Glacial Deposits in Britain and Ireland*. Balkema, Rotterdam.

Pages 383–387 outline field and map production methods used for Quaternary deposits.

Map 24   Near Tywyn, Gwynedd, North Wales

The topographic map opposite is reproduced from the 1981 Ordnance Survey 1:10 000 map SH60SW, with the permission of the Controller of Her Majesty's Stationery Office, Crown copyright reserved. It is slightly enlarged here to 1:8000. The area is about 2 km east of Tywyn, North Wales, just off the region covered in Plate 1. The area comprises rocks of the Ashgill series of the Silurian; the continuation of the rocks can be seen on Plate 1 around [SH6603].

Within the Ashgill, a unit of slate known as the Narrow Vein is poorly exposed but has been quarried sporadically for roofing material. The courses of the top and bottom surfaces of the slate are marked on the map where they are exposed in the old quarries at a and b. The course of the top surface is marked where it is exposed at locations x and y. There are no further exposures of the slate, but in quarries nearby, the unit can be seen to be of reasonably constant thickness and persistently oriented at 062/13°S.

Complete the outcrop of the Narrow Vein in the east of the area. *(Using structure contours, locate the unexposed course of the Narrow Vein. Begin with the two quarries and, say, the top surface of the Vein. Establish the location and route of the 160 m contour, bearing in mind the known strike direction. Add other structure contours whose location can be derived from topography, say the 150 and 170 m values. Having established the spacing, add further structure contours, from about 200 m to 130 m, checking their spacing against the known dip of the Vein. Project the outcrop course between the two quarries. Repeat for the base of the Vein.)*

Repeat the procedure independently in the west of the area, using the two exposures of the top surface of the Vein. You should find that the structure contours derived here fail to link readily with those drawn from the quarry information. Because the Narrow Vein is thought not to be folded in this district, the explanation may be the presence of a fault.

Suggest on the map where the trace of the fault may run to best explain the structure contours, complete the Narrow Vein outcrop, and comment on the possible nature and displacement of the fault. State the stratigraphic throw.

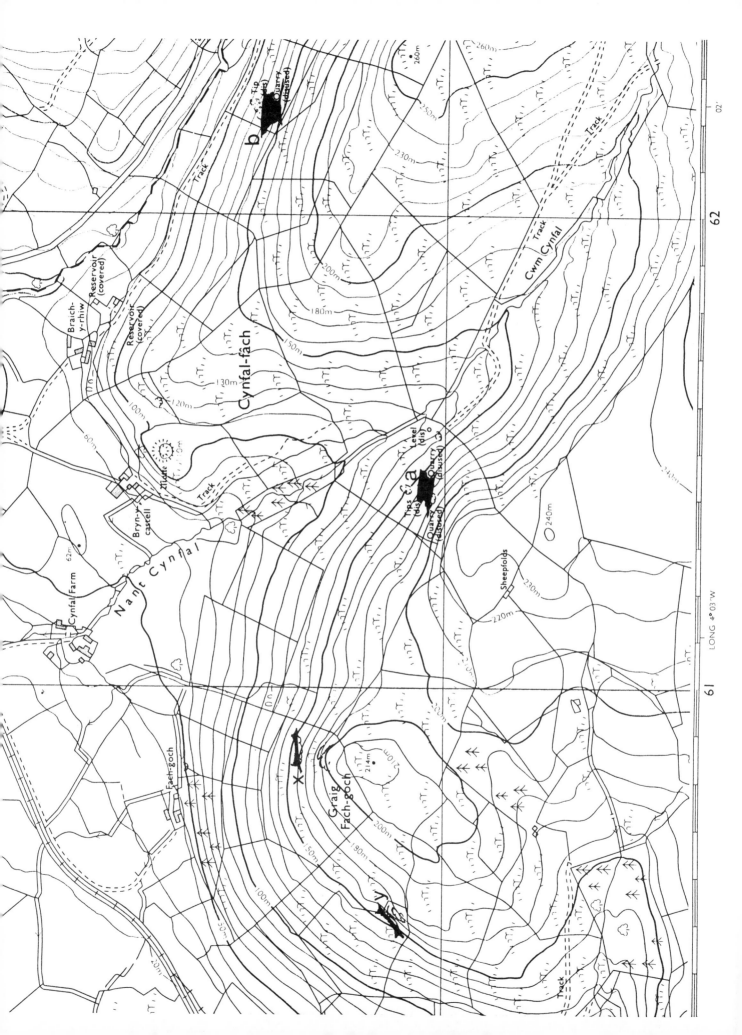

Solutions to Map Exercises

Solution to Map 1   Raton, New Mexico, USA

The thickest accumulation is likely to be where the surface of the substrate, the top of the Precambrian rocks, is lowest. This is in the pronounced basin centred 40 km south of Cimarron, in the NW part of township 22 North, range 19 East. The altitude there at the lowest point is just lower than –750 metres so that the total thickness of overlying sediments must be in excess of 2750 m. The Precambrian surface forms a trough elongate in a NE–SW direction and with smoothly sloping sides, but with a pronounced extension towards the NW, where the sides at higher altitudes have steep gradients. The next thickest accumulation is the much shallower, more open trough west of the town of Raton. The two troughs are separated by an undulating ridge with an approximate E–W trend.

The highest point on the Precambrian surface is at >2700 m, about 20 km WSW of Cimarron. The surface is steepest where the structure contours are most closely spaced, in the NW extension of the major trough – the area around 35 km SW of Cimarron. It is least steep west of Raton, where the absence of structure contours indicates that it is not significantly varying in altitude.

Solution to Map 2   Lacq gas field, Aquitaine, France

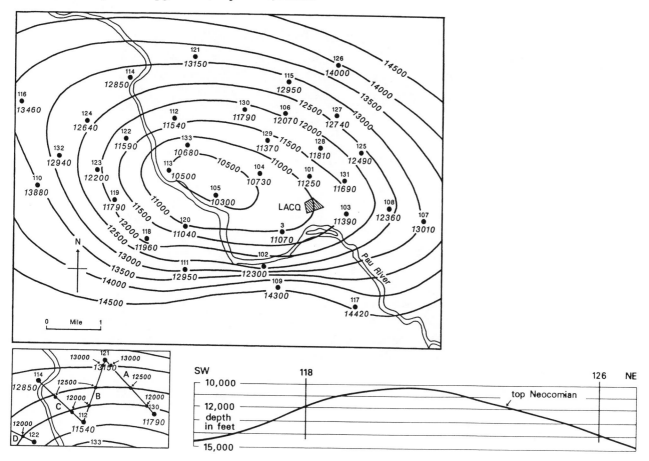

The Lacq gas field problem is solved by interpolating to obtain whole-number values for the structure contours, and linking these derived points into a reasonable pattern. Given the range of depths, a 500 foot contour interval is practicable.

Begin by selecting two adjacent locations of known depths that will give several interpolated values. The inset figure, reproduced here at a smaller scale than the map exercise itself, shows how locations 121 and 130 have been treated. The depth range between these two localities is 1360 feet (13150 – 11790 feet) and this occurs on the map over a distance of 39 mm, the length of line A. If 39 mm represents a depth interval of 1360 feet, then dividing 39 by 136 gives 0.28 mm as representing 10 feet of depth change. 12000 feet is 210 feet deeper than the 11790 feet at locality 130, therefore 21 × 0.28 mm along line A from point 130 gives the location of the 12000 feet contour. Similarly, 12500 – 11790 = 710 feet, so the 12500 contour is located at 71 × 0.28 mm along line A from point 130. The 13000 feet point is 121 × 0.28 mm from point A, or 15 (from 13150 – 13000 feet) × 0.28 mm from point 121.

Another adjacent series of interpolations is needed to gauge the course of the contours. Interpolation

along line B, between points 121 and 112, has yielded the locations of the 12000, 12500, and 13000 points, and so a roughly E–W course for the contours begins to appear in this district. Interpolation along lines C and D reveals a progressive arcing towards the SW. Continuing with other selected lines enables the pattern over the entire region to be derived. The structure turns out to be a dome, elongate in an E–W direction and asymmetrical, with its southern flank showing a steeper dip together with a slight undulation. A sketch cross-section is given.

Note that because of the profusion of depth data these are not strictly the 'three-point methods' outlined in figure 4.12. Linking localities 112 and 130, for example, to complete a triangle from lines A and B, would yield no additional information. Note also that the interpolated locations are guides only, and assume an even angle of dip between the control points. A more realistic contour pattern may result from adjusting the localities a little, hence the 12000 and 12500 contours do not, in the final interpretation (large figure) intersect line C at the exact 12000 and 12500 points derived by interpolation.

Solution to Map 3   Bear Hole, Montana, USA

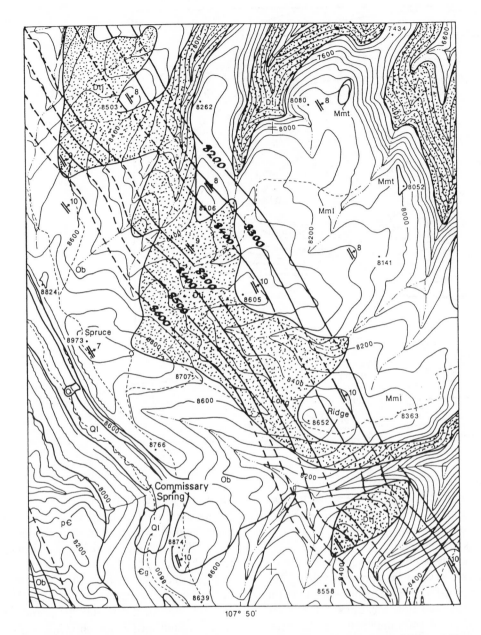

The outcrop of the Devonian is stippled on the figure. The 8200, 8300, and 8400 structure contours are shown for the top of the formation, and the 8300, 8400, 8500, and 8600 for the base. The contours are dashed where the surface no longer is present, through having been eroded away.

The structure contours indicate a shallow dip towards the NE, in common with the strike and dip symbols. The slight curves and variations in spacing reflect slight changes in strike direction and dip angle, respectively. The variations are not exactly the same for the two sets of contours, indicating that the top and bottom surfaces of the formation are not truly parallel: the thickness is not consistent.

The broad outcrop of Upper Devonian across the centre of the area is followed to the NE, down-dip in the normal way, by the overlying, younger, Mississippian rocks. Further NE, however, there is an abrupt and rapid loss of altitude of the land surface, indicated by the closeness of the topographic contours, and this governs the outcrop here. This is why Upper Devonian rocks reappear at the surface, despite the continuing NE dip. The steep slope continues to the NE corner of the map area, and hence it is the underlying unit that is outcropping there, namely the Upper Ordovician.

Solution to Map 4   Maccoyella Ridge, Koranga, New Zealand

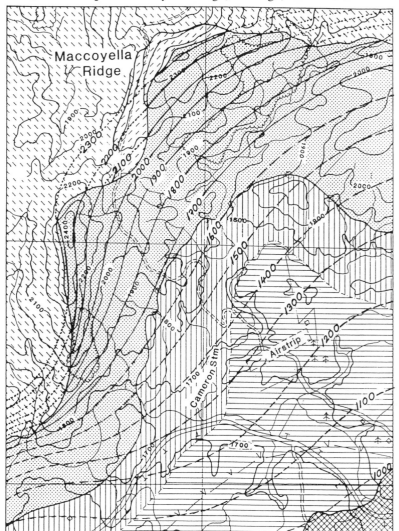

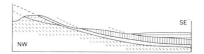

The oldest rocks outcrop in the NW of the area, and the formations become progressively younger towards the SE. The overall dip direction is therefore towards the SE. Similarly, the boundary of the Te Were Sandstone outcropping further towards the SE must be the top surface. It is the same as the basal surface of the overlying, next youngest unit.

Structure contours for the base of the Te Were Sandstone are shown in the figure. The continuous lines are those drawn from direct topographic control, i.e. intersections of the outcrop of the base with topographic contours. The contours portray the southeastwards dip of the Te Were unit, with its base getting deeper and deeper subsurface towards the SE. The dotted portions of lines, towards the NW of the area, also represent structure contours for the base of the Te Were Sandstone, but where the surface no longer exists. The surface is here 'up in the air', so to speak, having been eroded away to leave the underlying Koranga Sandstone to form the present-day land

surface. These lines are shown here to aid the explanation but such structure contours, of a surface that no longer exists at that place, would not normally be drawn on a map. The dashed lines are merely those structure contours for which there is no direct control. Their course and spacing is gauged from the structure contours which are directly constrained by topographic intersections, those shown as continuous lines.

The southeastwards dip of the surface is, however, not uniform. It is greatest in the southwest of the area, where the structure contours are closely spaced, and least in the southeast, where the surface appears to be levelling out. The strike direction, parallel to the trend of the contours, varies from almost E–W in the east of the area round to N–S in parts towards the west, before returning to a WSW course in the SW. This form is difficult to describe succinctly in words, thus highlighting a value of the structure contours.

Solution to Map 5   Coalbrookdale Coalfield, Shropshire, England

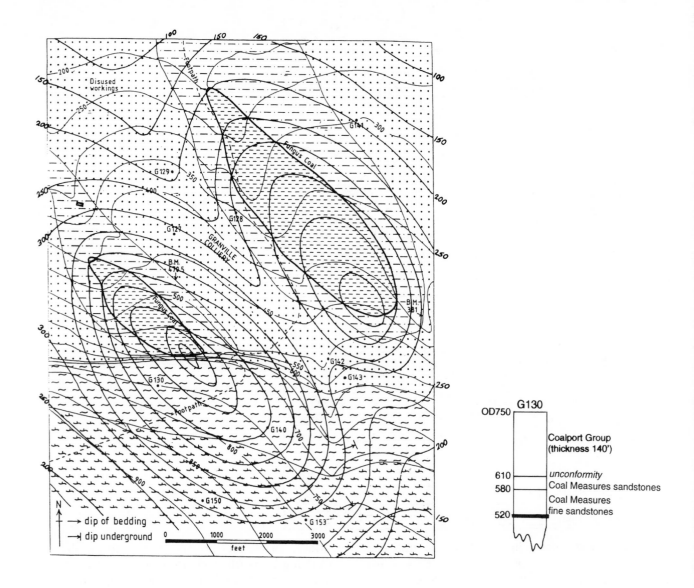

Lower Westphalian rocks, the grey-coloured unit 82 on Plate 1, are in contact with rocks of Wenlock, Middle Silurian age (pink-coloured unit 73), and various younger formations.

The accompanying figure brings out the structure of the Fungus Coal by showing all its structure contours, including those derived by extrapolation and those for areas where the coal has now been eroded away (i.e. the contoured surface is now above ground level). The structure consists of two elongate domes. (Described properly, see section 8.2, the domes are upright, gentle, periclinal anticlines, with axes trending NW–SE.)

Also illustrated is a vertical column of the predicted borehole geology at G130. At this site the land surface is at an altitude of 750 feet; from interpolation between the structure contours, the Fungus Coal is predicted to occur at an altitude of 520 feet. The coal will therefore be encountered at a depth below ground of 750 − 520 = 230 feet. It is also possible to construct structure contours for other surfaces in the map area. The other data shown in the vertical column were derived from contours for the top of the Coal Measures fine sandstones, and for the base of the Coalport Group. (Note that because the latter surface is an angular unconformity, see chapter 7, its structure contours bear no relation to those for the Fungus Coal. They run approximately E–W, losing altitude southwards.)

Solution to Map 6 Boyd Volcanics, New South Wales, Australia

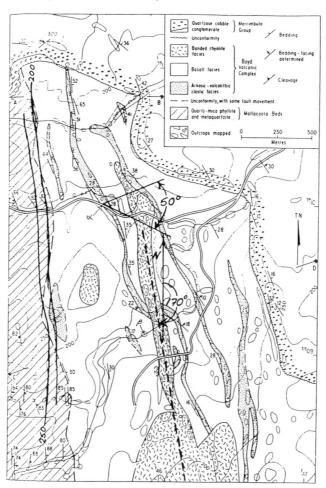

The outcrop of the Banded rhyolite facies is straight-looking and narrow. In addition to cross-cutting the other units, it completely ignores topography, crossing topographic contours at right angles. The only reasonable explanation is that it forms a vertical sheet (see section 6.4). The trend of the outcrop is not exactly N–S but approximately 170° from north, as indicated on the accompanying figure. Taking the land surface to be approximately level, the outcrop trend (the line where the igneous outcrop intersects with the land surface) represents the strike of the three-dimensional sheet. Its strike and dip is therefore 170/90°.

The strike and dip symbols immediately north of the road at the point marked α show a 28° dip towards the NE. The road runs oblique to this direction, therefore any road-cuts would display an apparent dip: the beds would appear to dip at some shallower angle. The angle (in the horizontal map plane) between the true dip direction given by the dip symbol and the road-cuts, which follow the route of the road, is 50° (see figure). Multiplying the cosine of this angle by the tangent of 28°, the true dip angle gives, from the lower equation in figure 4.7, the required apparent dip angle: 19°.

The outcrop width immediately north of the road, measured in the true dip direction, is 45 metres. There is no change of altitude across the outcrop here; the land is horizontal. Therefore, from the equation in figure 4.8, the outcrop width (horizontal apparent thickness) multiplied by the sine of the dip angle, 28°, gives the true thickness, 21 m.

The point β is 620 m from the nearest outcrop of the top of the Mallacoota Beds, which is the same surface as the base of the Basalt facies. The strike and dip symbol shows this distance to be in the true dip direction. The dip angle of the surface is 30°; the direction is NE. The land surface loses altitude from this point towards β. Thus the ground slopes in the dip direction, and the middle equation of the three in figure 4.10 is the relevant one here. The outcrop of the Basalt facies base, at the strike and dip symbol, is about halfway between the two nearest topographic contours, at 200 and 150 m, and is therefore at an altitude of 175 m, 50 m higher than β. From the equation in figure 4.10, multiplying 620 m by the tangent of 30° and then subtracting 50 m gives the required value: 308 m.

Solution to Map 7   'Northcrop' of the South Wales Coalfield, UK

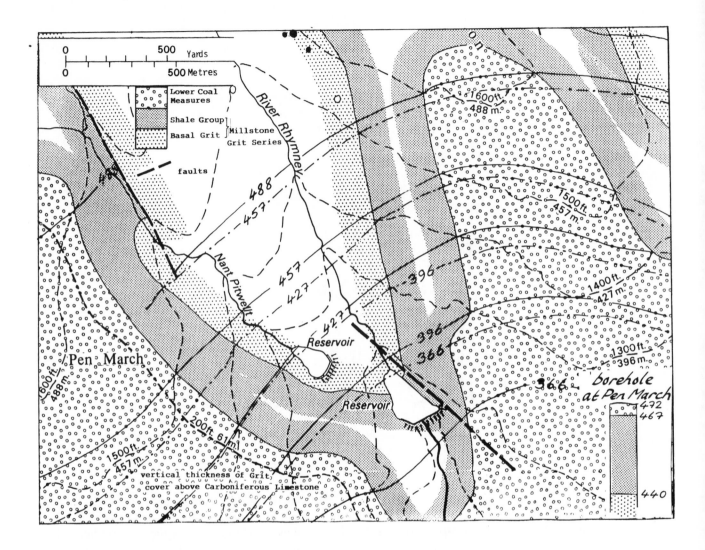

The figure shows possible structure contours (solid lines) for the top of the Shale Group. They show a very general southeastward dip direction, as suggested by the major V-shaped outcrop made by the Shale Group as it crosses the valley of the River Rhymney. The dip direction is far from constant, however. In the western half of the area the strike direction is approximately 055 but swings gradually round to about 105 at the eastern margin of the area. From the spacing of the structure contours, the dip angle varies a little but everywhere is gentle. It reaches 7° in places (where the contours are most closely spaced) and is as little as 4° in the NE of the area; a typical value is 6°.

The true thickness of the Shale Group also varies, but a typical value is 27 metres. This can be derived trigonometrically from the outcrop width or from interpolation between structure contours for the top and bottom surfaces of the Shale Group. Structure contours for the base are shown in the figure as dashed and dotted lines. The thickness encountered in a borehole, the vertical apparent thickness, will be a greater value but given the shallow dip angle the difference will be very slight.

Solution to Map 8   Long Mountain, Powys, Wales

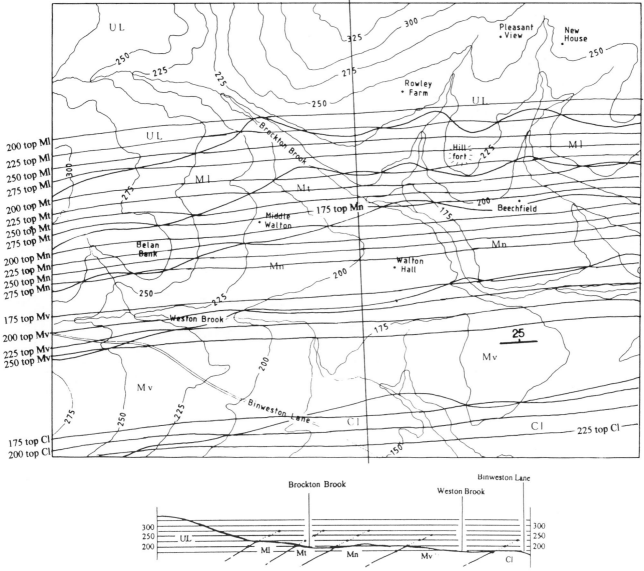

Although it is possible that Long Mountain is a plateau of flat-lying strata, there are many outcrop patterns of the Ordovician and Silurian rocks nearby that indicate folding (see section 2.3.3). The structure of Long Mountain, with older rocks curving around younger ones, is therefore more likely to be a downwarp or basin. The rocks, unit 74 on Plate 1, occur in the uppermost division (Ludlow) of the Silurian system.

Structure contours are shown on the accompanying figure. From these, using the equation in figure 4.3 to derive the dip angle, a representative strike and dip of the units is 085/25°N. Walton Hall, at an altitude of 185 metres, is 520 m from the outcrop (measured in the true dip direction) of the *Cyrtograptus* unit, which is at an altitude of 150 m. From the equation in figure 4.10c for ground slopes opposite to the dip direction, the depth of the unit below Walton Hall is 520 × tan 25

+ (185 −150) = 277 m. Rowley Farm, at 240 m altitude, is 1240 m from the 145 m altitude *Cyrtograptus* unit outcrop. Therefore the depth of the unit below Rowley Farm is 1240 × tan 25 + 95 = 673 m.

Discrepancies between depths obtained from structure contours and from trigonometric methods arise, besides measurement and cartographic inaccuracies, from the appropriateness and consistency of the dip angle. The structure contours for the *Cyrtograptus* unit indicate a dip angle less than the 25° typical of the district, though the even spacing shows that this area does have reasonably consistent dip values. A cross-section drawn from structure contours is shown.

Binweston Lane trends at about 295°, 60° from the true dip direction of 355°. The apparent dip of rocks alongside the lane would therefore be 13°, from the lower equation in figure 4.7 (tan apparent dip = tan 25 × cos 60).

Solution to Map 9   Silverband orefield, North Pennines, England

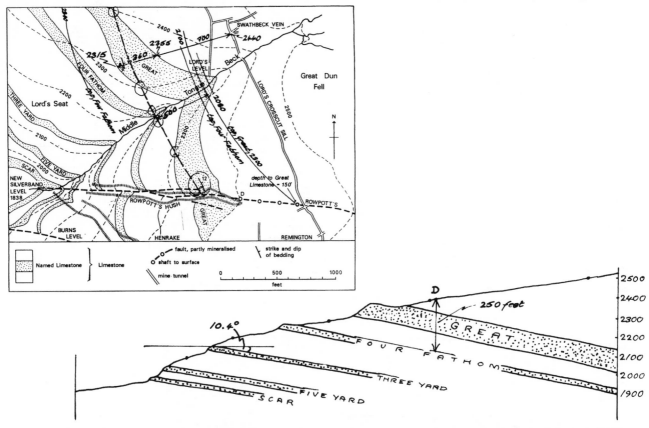

The outcrop width of the Great Limestone at the strike and dip symbol (towards the north of the outcrop), in the true dip direction, is 360 feet. The land surface slopes in the opposite direction to the dip direction (figure 4.8a) giving an elevation difference across the outcrop, interpolating between contours, of $2355 - 2315 = 40$ feet. Hence,

$$\text{true thickness} = (360 \times \sin 12) + (40 \times \cos 12)$$
$$= 114 \text{ feet.}$$

Swathbeck Vein is 700 feet in the true dip direction from the outcrop of the top of the Great Limestone and 85 feet higher ($2440 - 2355$ feet, by interpolation). Again, the land surface is sloping in the opposite direction to the dip direction, so from figure 4.9c,

$$\text{Great Limestone depth at Swathbeck Vein} =$$
$$(700 \times \tan 12) + 85 = 234 \text{ feet.}$$

The cross-section uses an apparent dip angle of 10.4° in the line of section, derived (figure 4.6) from the (true) dip angle given on the map:

$$\tan 10.4 = \tan \text{true dip angle} \times \cos 30°,$$

where 30° is obtained by measuring *on the map* the angle between the line of section and the true dip

direction. The depth on the cross-section of the Four Fathom Limestone in the shaft at D is 250 feet.

On the figure, the thick, dashed construction line, based on the control points circled on the figure, happens to represent three different structure contours that virtually coincide on the map. This one line serves as the 2200 feet structure contour for the top of the Four Fathom Limestone, the 2300 feet structure contour for the base of the Great Limestone, and the 2400 feet structure contour for the top of the Great Limestone. Their coincidence implies that the vertical thickness of the Great Limestone is 100 feet as is that of the underlying, unnamed, limestone. Note that the known depth to the Great Limestone at the place labelled Rowpott's, in the SE of the area, provides an important control point for helping establish the overall structure contour pattern.

At Middle Tongue Beck, the horizontal spacing between the 2400 and the 2300 structure contours for the top of the Great Limestone is 500 feet. Dividing this into the 100 feet altitude difference between the contours (figure 4.3b) gives the tangent of the dip angle, 11.3°. The location of the 2100 contour for the top of the Four Fathom Limestone is added at the same spacing as the 2300 and 2200 structure contours. At Lord's Level, by interpolation, the altitude of the limestone will be at 2080 feet.

Solution to Map 10   Zambian copper belt

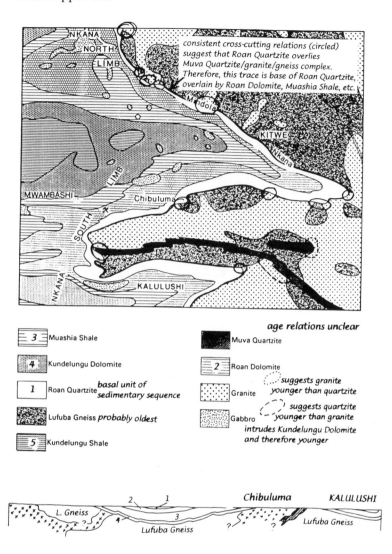

The areas circled on the figure (solid lines) show consistent cross-cutting relationships between the boundary of the Roan Quartzite and units that appear to be older, namely the Muva Quartzite, Lufuba Gneiss, and granite. The other boundary of the Roan Quartzite, the more westerly one, is in contact with a dolomite unit which itself is adjacent to a sequence of sedimentary units of similar outcrop pattern. It appears, therefore, that a sedimentary succession, beginning with the Roan Quartzite and numbered in sequence on the accompanying key and cross-section, was deposited on top of a complex of older, igneous and metamorphic rocks.

The age relations within this complex are unclear from the map. The Lufuba Gneiss may well be the oldest unit, intruded by masses of younger granite. But whether the Muva Quartzite was deposited on the gneiss before the granite intrusion or vice versa is ambiguous. Cross-cutting relationships in the area within

the dotted outline on the figure would suggest the former whereas those within the dashed outline support the latter interpretation. The complex is unconformably overlain by the Roan Quartzite, and the sequence of progressively younger, overlying units up to and including the Kundelungu Shale, is clear.

The one remaining unit of which the relative age is not exactly clear is the gabbro. It cross-cuts and intrudes the Kundelungu Dolomite, for example, and so must be younger than this, the second youngest sedimentary unit. It is not actually seen in contact with the youngest sedimentary unit, the Kundelungu Shale. Therefore the gabbro could be the youngest rock in the area or it could be the second youngest, having been emplaced before deposition of the shale.

The overall age relations mean that the sedimentary sequence has been folded into two, W-plunging synclines (see figure 8.4) with an intervening anticline, as depicted on the cross-section.

Solution to Map 11   Builth Wells, Powys, Wales

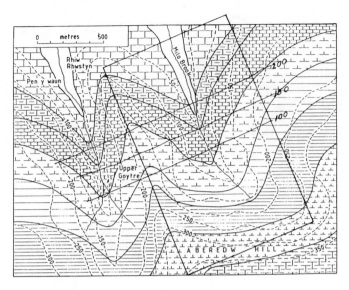

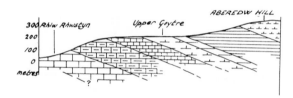

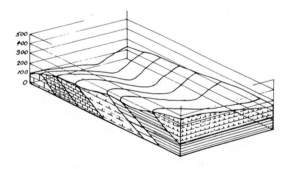

	vertical thickness, from section, metres	true thickness, from section, metres
Transition beds		
Chonetoidea beds	>180	>170
Lingula beds	250	235
Atrypina beds	50	45
Pterina beds	110	100
Oriostoma beds	125	115
Gypidula beds	200	190
Chonetes beds	200	190

On Plate 1 the rocks of this area are shown as unit 74, the uppermost division of the Silurian system. They are of Ludlow age.

The formations become younger towards the SE. Together with the SE-pointing V-shapes made by the boundaries as they cross both stream valleys, this suggests a dip towards the SE. This is confirmed by structure contours for any of the boundaries, which show that the units have a consistent orientation, at about 067/20°S. Three such contours are shown in the figure, for the top of the *Oriostoma* beds. The figure also shows

the outline of the base of the block diagram, selected so that the diagram is drawn with its sides in the true dip direction and the front parallel to the strike direction. The visible side of the block therefore corresponds closely to the cross-section. The angles of tilt and rotation were chosen to give representation of the front and side panels of the block, and of the top surface, while not introducing significant distortion to the dip angles.

The vertical apparent thicknesses of the units and the true thicknesses are shown, as measured from a cross-section in the true dip direction.

Solution to Map 12   The Helderberg, South Africa

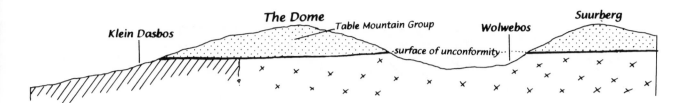

The base of the Table Mountain Group cuts discordantly across boundaries in the underlying rocks and therefore is an unconformable junction. Its trace approximately parallels the topographic contours and so it is horizontal. The Table Mountain Group at the Dome is an outlier, presumably having been eroded away from the surrounding areas. The cross-section shows the arrangement; the structure of the older rocks below the unconformity is not clear from the map.

Solution to Map 13   A sub-Permian unconformity and inlier

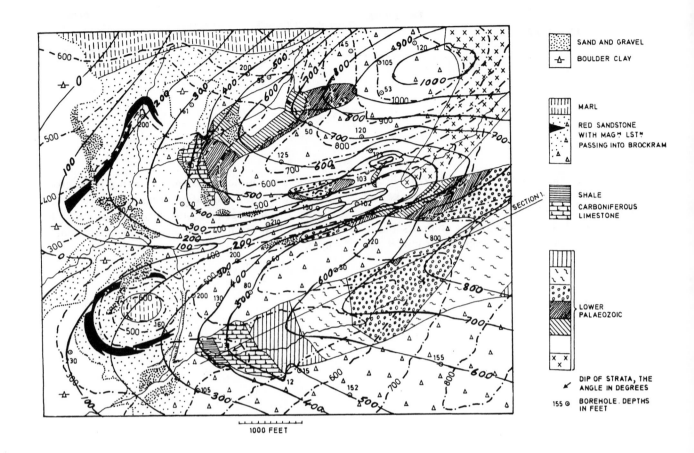

1000 FEET

The Carboniferous rocks overlie different parts of the Lower Palaeozoic sequence in different places and, in places such as the NW part of the large, southern inlier, with an angular discordance. The base of the Carboniferous sequence is therefore an unconformity.

The patterns of repetition in the Lower Palaeozoic rocks indicate the presence of folds. A SW-plunging anticline in the west of the northernmost inlier is flanked to the east by a S-plunging syncline which appears to reverse plunge direction as it appears in the central, oval-shaped inlier. The anticline might conceivably correspond to the S-plunging anticline apparent in the southern inlier, having undergone fault displacement. The extent to which the Carboniferous rocks are folded is unclear from the map. They are probably not folded at all, as nowhere is there evidence of repetition around the structures mentioned above. There is outcrop repetition in the westernmost part of the major, southern inlier, but this is probably a topographic effect, because the V-shapes of the outcrops as they cross the stream suggest a persistent westwards dip.

Structure contours for the sub-Permian unconformity are shown in the accompanying figure. The form of the early Permian landscape shows remarkable similarities with the present-day topography, consisting chiefly of two elongate hills separated by a narrow, approximately E–W, gorge that opens westwards. This gorge parallels the course of a fault apparent in the outcrops of Lower Palaeozoic rocks in the west of the area. The fault appears not to affect the Permian rocks and so cannot have had any weakening effect on them; the course of the valley may be inherited from an earlier landscape where the faulted older rocks were being preferentially eroded.

The superficial deposits (sand and gravel, and boulder clay) lie with angular discordance upon all the bedrock sequences, which are very much older. The base of the superficial deposits is therefore effectively an unconformity.

Solution to Map 14 Llandovery, Dyfed, Wales

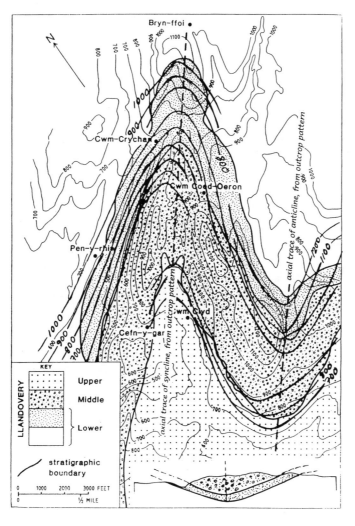

The repeating, converging outcrop patterns (compare with the middle part of figure 8.4) indicate a pair of plunging folds. The fold in the centre of the area contains younger rocks in its inner part and is therefore a syncline; the fold to its SE has older rocks in its inner part and is therefore an anticline. Both plunge towards the SW. Structure contours for the three stratigraphic boundaries are shown in the figure but only partially in order to retain clarity here. In the Cwm Crychan–Bryn-ffoi area, the hinge zone of the syncline, the structure contours show that the stratigraphic boundary within the Lower Llandovery loses 200 feet altitude in a horizontal distance of 1130 feet so that from the trigonometric relationship given in figure 8.7 the fold plunge is 10°. In the Cwm Clyd–Cefn-y-gareg area, the structure contours for the top of the Middle Llandovery/base of the Upper Llandovery surface show that it takes a slightly greater horizontal distance (1300 feet) for a 200 feet loss of altitude, giving a plunge of 8.7°. The difference in plunge between the two places is small; the fold is showing no real sign of being periclinal.

Although the form of the structure contours for the anticline is slightly different to that for the syncline, their spacing is similar, indicating that the two folds have similar plunge angles. The contour spacing is similar on each limb of the anticline, indicating similar angles of dip, i.e. an upright, symmetrical structure. The inset to the figure shows the appearance in cross-section.

In fact, the style of the folds can be judged well by down-plunge viewing of the map (figure 8.5). This reveals a style for both folds that is less tight than normal viewing of the map pattern suggests, and the anticline is open in style (see figure 8.3 for terminology). The locations of axial traces assessed from the outcrop pattern are not likely to differ much from those derived from cross-sections *in this case*, because the folds are upright. If the structures were inclined (figure 8.2), however, there could be a significant discrepancy.

Solution to Map 15   Millstone Grit of northern England

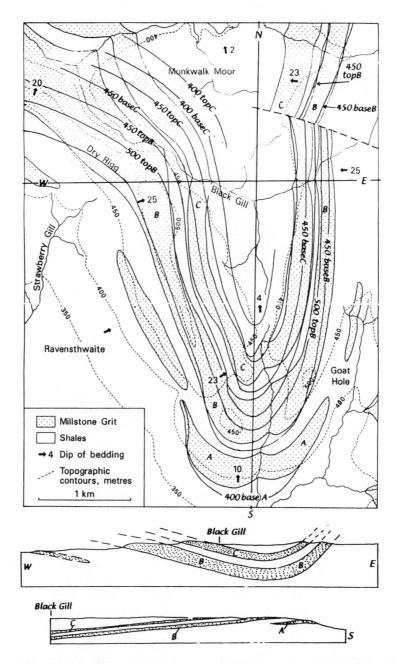

The fold is a fairly upright, synformal syncline that plunges gently northwards. Although the style of the structure appears tight in map view, this horizontal section is so oblique to the fold axis that the appearance is likely to be wholly misleading. In profile the style is likely to be very much less tight.

The outcrop width may reflect natural changes in formation thickness, especially with a lithology like coarse sandstone and the obvious changes in thickness in some of the grit units. Topographic slope cannot explain the difference here, and it is not obvious from the dip values provided on the map that variations in dip play a part.

However, on constructing structure contours (see figure), their spacing is seen to be closer in the east of the area (south of the fault) than in the west, so that dip variation is influencing outcrop width after all. For the purposes of constructing the cross-sections the Grit units have been labelled A, B and C. Clearly unit A is highly discontinuous.

To some extent the spacing of the topographic contours tends to decrease where they are located on sandstone, suggesting those units give steeper ground. The highest ground occurs within Grit units.

Solution to Map 16   Wenlock Edge, Shropshire, England

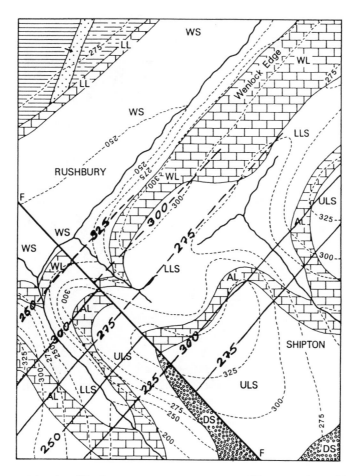

The structure contours on this simplified map are remarkably straightforward, representing an evenly dipping sequence displaced a consistent amount by a single, presumably vertical, fault. Several contours for which there is good control are shown in the accompanying figure, as solid lines, for the base of the Aymestry Limestone. The 300 metre contour NE of the fault abuts immediately across the fault with the 225 m structure contour. The amount of purely vertical displacement of the stratigraphic boundary, the stratigraphic throw, is simply the difference between these two elevations, that is, 75 m (see figure 9.9c). The same procedure could be carried out with the top of the Wenlock Limestone, for which a few structure contours are shown as dashed lines on the figure. In this case the 250 m structure contour SW of the fault abuts against the 325 m structure contour NE of the fault, thus also leading to a 75 m stratigraphic throw.

The stratigraphic throw can also be found trigonometrically (see figure 9.9b), but because no dip measurements are provided on the map, some structure contours would have to be drawn so that the dip angle can be derived from their spacing and elevation difference (figure 4.3). The dip angle in this example

averages 10°. A stratigraphic surface is selected which has known altitude immediately next to the fault, such as, in this example, the top of the Wenlock Limestone or the base of the Aymestry Limestone SW of the fault. Knowledge of the dip angle allows the calculation of the altitude at which that same surface will occur at the point immediately on the other side of the fault. The difference between the two elevations is the stratigraphic throw.

The structure contours for the base of the Aymestry Limestone coincide with those for the top surface. The 225 m structure contour, for example, for the base of the unit coincides with the 250 m contour for the upper surface. This indicates immediately that the vertical thickness of the Aymestry Limestone is 25 m. Similarly, the 300 m structure contour for the top of the unit, shown as a dashed line on the figure, coincides with the 275 m structure contour for the base, yielding a 25 m vertical thickness for this unit also. Thickness values derived in this way are always vertical thicknesses, i.e. the vertical apparent thickness (figure 4.8), and they have to be multiplied by the cosine of the dip angle in order to find the true thickness of the units.

Solution to Map 17   Aspen, Colorado, USA

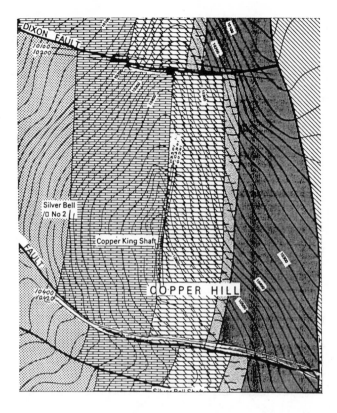

The traces of the Copper Fault, in the east of the area, and of the Justice fault in the west, trend very roughly N–S whereas the traces of the other faults run very approximately E–W. The two N–S faults appear to have larger displacements and to be younger than the E–W set of faults. The various boundaries displaced by the Good Thunder Fault all have different amounts of separations in the horizontal. Therefore, while oblique-slip may be involved to some extent, this fault cannot be a strike-slip fault.

The sedimentary formations become younger towards the west and therefore dip in that direction. If the Burro Fault is a dip-slip fault, the pattern of displacement it shows, acting on W-dipping units, can only be the result of downthrow to the south. Two structure contours, for 10 400 and 10 420 feet, are shown on the accompanying figure for the Silver Bell Fault. They lose altitude northwards and therefore the faults dips north. Its trace is slightly irregular, indicating localised changes in orientation of the fault surface, but in general the pattern is analogous to the lower right-hand corner diagram of figure 6.3, where a somewhat similar hill is crossed by a surface dipping steeply north.

The strike separation shown by the Dixon Fault, the distance indicated by the arrowed line in the figure, is 260 feet. The stratigraphic throw can be derived by drawing structure contours for the top of the Leadville Dolomite, but they are extremely close together, and sensitive to error. This readily generates a large error in the depth estimation. The trigonometric approach is perhaps easier. Although it depends on deriving the dip angle from closely spaced structure contours, this is less sensitive as the dip angle is clearly very steep. A difference between, say, 82° and 84° will make much less difference to the depth values than compounding errors in spacing over contours spanning several hundred feet. Three contours are shown (10 600, 10 580, and 10 560 feet) around Copper King Shaft, and their spacing yields a dip of 83°. Applying this angle to the Dolomite immediately south of the Dixon Fault means that at the point adjacent to the outcrop of the Dolomite north of the fault, it has an elevation of 8243 feet. The Dolomite outcrops, north of the fault, at 10 190 feet. Therefore the stratigraphic throw is 1947 feet.

Two structure contours (10 100 and 10 200 feet) are shown for the N-dipping Dixon Fault. They, too, are very closely spaced, reflecting the very steep dip of the fault surface. In this situation, unlike a shallower dipping fault, the difference between the stratigraphic throw and the true throw measured in the dip direction of the fault is insignificant. The Justice Fault is a relatively young, steep, E-dipping normal fault (younger formations to the east are brought against older formations to the west).

Solution to Map 18   Glen Creek, Montana, USA

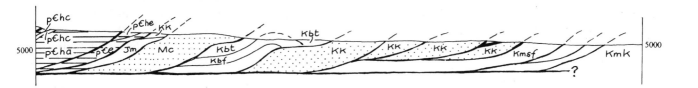

The interpretive cross-section shows the kinds of structures, typical of fold-and-thrust belts (see figure 9.10), that can be interpreted from the map. Note that most of the thrust traces on the section have been interpreted to have a listric form. The dashed line in the central part of unit Kbf on the map represents the hangingwall anticline seen in the cross-section and this also explains the symmetrical repetition of unit Kbt.

The trace of this thrust on the section has been interpreted to have a staircase form, to account for the anticline.

Several thrust traces on the map meet, suggesting that the thrust planes also coalesce at depth. This has also been incorporated into the section, together with a basal thrust into which all the thrusts root. Compare also figure 5.7d.

Solution to Map 19   Hamblin Bay Fault, Nevada, USA

The rocks of the volcano, which is essentially a vertical structure, have been displaced laterally to an extent not possible by dip-slip faulting. The NE boundary of the volcano shows a left-lateral (sinistral) strike separation of 19 km.

Faults splaying from the Hamblin Fault, such as that immediately NW of Pinto Ridge and that running along the Bitter Spring Valley, have little in the way of matchable marker horizons either side and are also likely to be strike-slip faults. At Saddle Mountain, the southernmost fault, which truncates the volcanic rocks, may well be strike-slip in nature. Its splay faults, though, which displace the shallow-dipping northern margin of the volcanic rocks, may well be dip-slip faults, downthrowing to the east. Such arrays of dip-slip faults are commonly associated with major strike-slip faults. In this example, although the dip of the minor faults is not known from the map, they may be reverse faults, resulting from localised contraction arising on the concave side of the curving strike-slip fault.

The faulting occurred after the Tertiary Muddy Creek Formation had been produced, but before deposition of the Quaternary alluvium and colluvium as this covers the fault. Approaching the major faults, the strike of bedding swings into parallelism with the fault, in response to the strike-slip movement. There is also some tendency for bedding to become steepened in proximity to a fault, reflecting a component of dip-slip movement.

Solution to Map 20   Honister Slate Mine, Cumbria, England

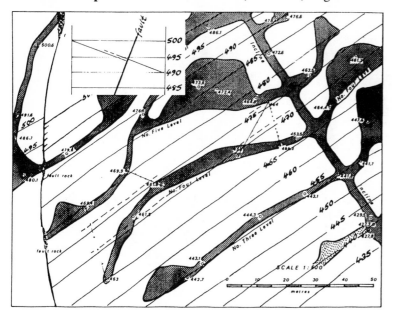

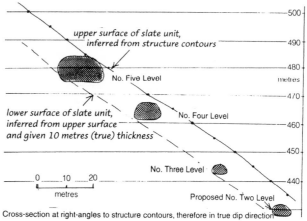

Cross-section at right-angles to structure contours, therefore in true dip direction

Note that the elevations provided on the map are for the floors of the levels, and these have been worked along the *base* of the slate unit. Structure contours are required for the *top* of the unit. Interpolation (as in figure 4.12) between the known elevations enables a few structure contours to be constructed for the base of the formation (the dashed lines in the figure around the No. Four Level are examples). From these the dip of the unit can be derived (figure 4.3). It is 36° towards the SE. From this the vertical apparent thickness can be calculated (true thickness divided by the cosine of the dip angle, 10 m/cos 36°, from figure 4.8); it is 12.3 metres. Assuming these values hold constant across the map area, 12.3 m can be added to each of the known elevations in order to derive the elevations of the top of the slate. From these the structure contours shown in the figure have been constructed.

The overall strike and dip of the slate unit is therefore 055/36°SE. The stippled zone in the SE corner of the map is actually part of a No. 2 level that was

opened in the mine. It successfully exploited the slate unit, as the structure contours predict, but encountered zones of intense jointing and was abandoned.

Level 4 was deviating from the strike of the unit, and needed to be driven towards the miner's right-hand to regain the full width before continuing along strike. Because the fault downthrows to the west, continuing Level 5 past the fault zone would cause it to enter rocks above the Honister Slate. The level would have to be driven just over a metre northwards in order to retrieve the slate unit. The fault trace on the map shows a stratigraphic throw, by interpolation of the structure contours (figure 9.9c), of 1.4 m. A cross-section in the dip direction of the fault, see inset, gives a vertical separation, the fault throw, of 1 m. The values differ because the strike of the fault, and therefore its dip direction, is oblique to the strike of the slate unit. Although the differences are small in this example, they are nevertheless of significance in the planning of the mining operations.

Solution to Map 21   Gwynfynydd gold mine, North Wales

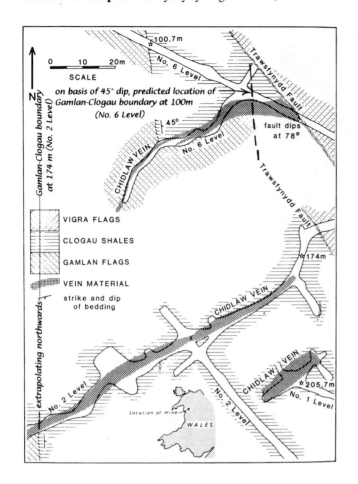

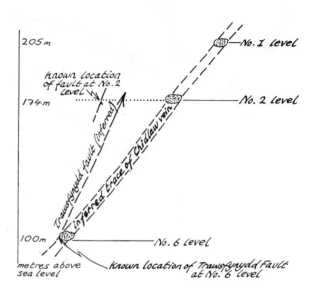

The figure shows a N–S cross-section which incorporates the Chidlaw vein in all three levels. In this section, the vein dips at 40°. From the map, the strike direction of the vein is at 060; the true dip direction is therefore 30° west from north. From the upper equation in figure 4.7, the tangent of the true dip angle is the tangent of the apparent dip (tan 40 = 0.84) divided by the cosine of the angle between the true and apparent dip directions (cos 30° = 0.87) = 0.96. The true dip angle is therefore 44°. The Chidlaw vein is oriented at 060/44° NW.

The location of the Trawsfynydd fault at the No. 6 level is shown on the figure, together with its projected trace in this cross-section, using an apparent dip of 68°. The fault is losing altitude roughly northeastwards. The true dip direction is 58° (clockwise) from the N–S orientation of the cross-section, and the true dip angle is given on the map as 78°. From the lower equation in figure 4.7, the tangent of the apparent dip in the line of the cross-section is the tangent of the true dip (tan 78 = 4.70) multiplied by the cosine of the angle between the true and apparent dip directions

(cos 58 = 0.53) = 8.87, giving the apparent dip of 68°. This projected trace does not, however, meet the known location of the fault at the No. 2 level, suggesting that the fault surface is not truly planar but steepens dip upwards.

Projecting the Gamlan–Clogau junction northwards while maintaining the 174 metre altitude of Level 2 takes the trace close to the north arrow of the map. At the 100.7 m altitude of Level 6, however, on the basis of the given 45° dip angle, the trace should be 73 m further to the east. (The depth difference, 73.3 m, divided by the tangent of the dip angle, tan 45 = 1, gives the horizontal distance, = 73 m.) In fact this position is 28 m further east than the actual location shown on the map, because of its displacement by the fault along which the Chidlaw vein formed.

The actual Gamlan–Clogau boundary in the No. 6 level is 45 m east of the 174 m trace projected from Level 2. At this point the 45° dip should have dropped the elevation to 129 m (174 – 45 m). Because it actually occurs in the No. 6 level at an altitude of 100.7 m, the stratigraphic throw of the fault is 28 m (129 – 100.7 m).

Solution to Map 22 Woore Moraine, Cheshire, England

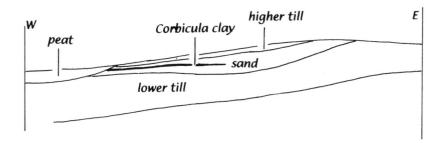

The depth to bedrock appears to increase northwards and more especially towards the west, suggesting a surface similar to the present-day topography. (Bedrock outcrops in the Old Clay Pit at about 80 metres altitude, but is 30 m lower in borehole B, in the north. It is lower still in the west: at 30 m in borehole C and 23 m in A.) The peat, the youngest deposit, occupies valleys, and these may be eroded in bedrock.

Borehole D suggests there are two till deposits, but borehole A gives the critical information. The *Corbicula* clay separates a sand–till succession from an underlying till–sand sequence, which rests on bedrock. That is, there must be two separate tills, from different glacial events. The till–sand of borehole C, also resting on bedrock, is likely to be the same as the lower sequence at A, though somewhat thicker. The Sand Pit where borehole C was drilled is therefore working the lower of the two sand deposits seen in A. However, the ground immediately to the south of the borehole C sand-pit is topographically higher than the sand in the pit, and is therefore composed of the higher of the two tills seen in A. The Old Sand Pit in the centre of the area exposes the lower till in its floor, the overlying sand in its walls, and the upper till to its north.

The till recorded at borehole B is thicker than any seen elsewhere in the area, and is actually thicker than both the tills at A taken together. The most likely explanation is that it represents a substantial development of both tills, the upper and lower not having been distinguished in the data from B. This means that the *Corbicula* clay was either not recognised at B or is not present in the north of the area, and that the sand also has thinned out completely to the north. The superficial deposits also tend to thin southwards, thus giving bedrock exposure in places.

The sand therefore is best developed in the central part of the area, and towards the west. Southwest of the Old Sand Pit, and just north of the marked topographic valley, below the peat it reaches a thickness of over 17 m, its greatest remaining thickness.

Solution to Map 23   Baraboo, Wisconsin, USA

The geological history of the area recorded on the map begins in early Precambrian times with the formation of the rhyolite. It outcrops only in the NE of the area, so it is unclear from the map whether the rock is part of an extensive, largely concealed 'basement' to the area, or just a localised igneous sheet. The cross-section, however (which incorporates information from boreholes) shows that it is the former. Being a fine-grained igneous rock, it is more likely to be a series of lava flows than one large intrusion. Rhyolite is rarely erupted in submarine situations; this area may well have been land at least at this period within the Early Precambrian.

In Middle Precambrian times, quartz-rich sands were deposited, destined to become, after some burial, the Baraboo Quartzite. The sediments may represent a transgression of a shallow sea across the land surface. They have been folded, presumably after burial and lithification, into an asymmetric, periclinal synformal syncline with E–W elongation. The cross-section indicates some complexities to the basic synformal shape of the structure. The limbs of the syncline, following erosion of the overlying material and exhumation, now make upstanding ridges that dominate the topography of the area (the 'North Range' and 'South Range'). The amount of erosion during later Precambrian times may have been great. The cross-section shows projections of how the Quartzite may have appeared after folding but before this erosion took place. This reconstruction implies removal of several kilometres of Quartzite before the younger (Cambrian, etc.), flat-lying materials were deposited at or around the level of the present-day land surface.

In Cambrian times, sands and coarse gravels, later to be lithified to produce sandstones and conglomerates, were deposited on top of the folded Quartzite with marked unconformity. The presence of an unconformable relationship is shown by: (1) the map, where, for example, the outcrop of Precambrian Quartzite is flanked on all sides, both down-dip and up-dip, by the Cambrian unit; (2) the key, which shows rocks of Late Precambrian times to be missing; and (3) the cross-section, which shows a marked angular discordance between the Precambrian and Cambrian units. The Cambrian deposits may represent another marine transgression across the area, with the conglomerates representing the shallower water sediments. It is possible that these coarser deposits were banked up against the ridges of Baraboo Quartzite which even in Cambrian times was governing the surface relief. Cambrian-age rocks are noticeably absent today from the ridge crests. This may be because they have been eroded away from such places, perhaps because the deposits were thinner, but it is also possible that there was no sedimentation there – that the ridges rose above wave-base as linear islands.

There are two places in the west of the map area where the Ordovician unit directly overlies Precambrian rocks, indicating an unconformable relationship at these places. Elsewhere, however, neither the key nor the cross-section suggests this, nor those parts of the map where the sediments are in contact with Cambrian rocks. The base of the Ordovician over the map area as a whole is therefore more like a disconformity. Dolomite tends to form as an extensive, shallow-water deposit. Because outcrops of dolomite are scattered all over the map area, including some quite high ground in the SW, it may be that in Ordovician times the entire area was under water.

The map and section record nothing further of the geological history until the Pleistocene. The Precambrian rocks still governed the relief then, as they do today. The map shows the position of a major, Pleistocene, terminal moraine. The morainal debris now partially blocks the gorge through the South Range at its north and south ends, leaving Devil's Lake trapped in between. In Pleistocene times such glacial blocking was even more effective, and this may have forced the Wisconsin River to leave its traditional course and find new valleys further east.

There may also be a link between the Ordovician rocks and the present-day topography. The drainage system is strikingly anomalous in that some of the rivers transect the quartzite ridges rather than flowing round them, and the major gorges at Lower Narrows and Devil's Lake are thought to have been cut by the Wisconsin River which now flows far to the east. The explanation may lie in (1) superimposed drainage, and (2) river diversion. The drainage pattern may have become established shortly after the deposition of the Ordovician materials, which blanketed the area and subdued the relief. This same system was maintained even when the rivers eventually began to cut down into the underlying, differently oriented quartzite. In fact, this superimposition may have begun very much earlier, using some now vanished Late Precambrian deposit, as the presence of Cambrian rocks in the present-day gorges (for example at Lower Narrows) may indicate that such valleys across the ridges were already in existence in Cambrian times.

Solution to Map 24   Near Tywyn, Gwynedd, North Wales

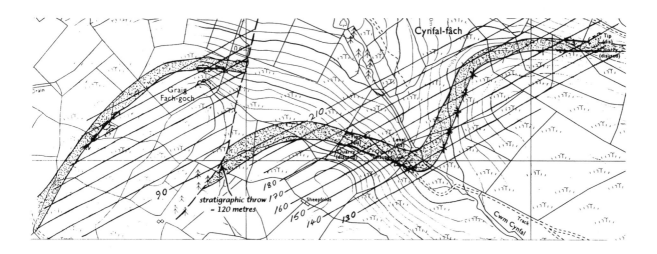

Shown in the figure are some relevant structure contours for the top surface of the Narrow Vein. Two points, one in each of the disused quarries in the eastern half of the area, where the outcrop of the top surface of the vein intersects with a 160 metre topographic contour, give good control for drawing the 160 m structure contour. Additional control points at 180, 170, 160, 150, and 140 m for the exposure labelled 'a' on the map, and at 150 and 140 m for that labelled 'b', together with the knowledge provided of a persistent 062 strike, enable further structure contours to be drawn in. The crosses show some of the points that were useful in deriving the outcrop from these structure contours. Note that the outcrop is tracing a 'V'-shape across the river valley that is consistent with the gentle SE dip of the Narrow Vein.

Structure contours for the lower surface of the Vein are not shown in the figure, to preserve clarity; they are exactly parallel to those for the top surface. One possible course of a fault that explains the mismatch of the structure contours is shown. The trace curves in order to help explain breaks in slope indicated by the topographic contours but really its location and course are very little constrained. With the interpretation shown, the stratigraphic throw of the fault, derived from elevation difference between the 210 and 90 m structure contours contiguous at the point of the arrow, is 120 m. A small, isolated fault like this one is unlikely to be a strike-slip fault. Assuming that it is a dip-slip fault, then the downthrow side is to the west. The interpretation of the fault trace is far too poorly constrained to be able to say anything about the orientation of the fault, and hence if it shows normal or reverse movement.

References

Ahmed, F. and Almond, D.C., 1981. *Field Mapping for Geology Students*. Allen & Unwin, London, 314 p.

Anderson, R.E., 1973. Large-magnitude Late Tertiary strike-slip faulting north of Lake Mead, Nevada. *US Geological Survey Professional Paper* 794, 18 p.

Appleton, J.D., Fuge, R. and McCall, G.J.H. (editors), 1996. Environmental geochemistry and health. *Geological Society of London Special Publication* 113, 260 p.

Archer, A., Luttig, G.W. and Snezhko, I.I., 1987. *Man's Dependence on the Earth: The Role of the Geosciences in the Environment*. Schweizerbart'sche Verlagsbuchhandlung, Unesco, Stuttgart, 216 p.

Aust, H. and Sustrac, G., 1994. Impact of development on the geological environment. In: Lumsden, G.I. (editor), *Geology and the Environment in Western Europe*. Oxford University Press, Oxford, 202–280.

Badgley, P.C., 1959. *Structural Methods for the Exploration Geologist*. Harper & Brothers, New York, 280 p.

Bailey, E.B., 1952. *Geological Survey of Great Britain*. Her Majesty's Stationery Office, London, 278 p.

Bailey, G., King, G. and Sturdy, D., 1993. Active tectonics and land-use strategies; a palaeolithic example from northwest Greece. *Antiquity*, vol. 67, 292–312.

Barnes, J.W., 1995. *Basic Geological Mapping*, 3rd edition. John Wiley, Chichester, 133 p.

Bartlett, R.A., 1980. *Great Surveys of the American West*. University of Oklahoma Press, Norman, Oklahoma, 410 p.

Basset, D.A., 1969a. Wales and the geological map. *Bulletin, National Museum of Wales*, vol. 3, 10–25.

Bassett, D.A., 1969b. William Smith, the Father of English geology and of stratigraphy: an anthology. *Geology, Journal of the Association of Teachers of Geology*, vol. 1, 38–51.

Bell, M. and Boardman, J. (editors), 1992. *Past and Present Soil Erosion. Archaeological and Geographical Perspectives*. Oxbow Monograph 22, Oxbow Books, Oxford, 250 p.

Benedict, J.B., 1976. Frost creep and gelifluction features: a review. *Quaternary Research*, vol. 6, 55–76.

Berg, R.C., Kempton, J.P. and Stecyk, A.N., 1984. Geology for planning in Boone and Winnebago Counties. *Illinois State Geological Survey Circular* 531, 64 p.

Bergstrom, R.E., 1980. Illinois State Geological Survey: its history and activities. *Illinois State Geological Survey Division, Urbana, Educational Series* 12, 37 p.

Bernknopf, R.L., Brookshire, D.S., Soller, D.R., McKee, M.J., Sutter, J.F., Matti, J.C. and Campbell, R.H., 1993. Societal value of geologic maps. *United States Geological Survey Circular* 1111, 53 p.

Bhagwat, S.B. and Berg, R.C., 1991. Benefit and costs of geologic mapping programs in Illinois: case study of Boone and Winnebago Counties and its statewide applicability. *Illinois State Geological Survey Circular* 549, 40 p.

Bishop, M.S., 1960. *Subsurface Mapping*. John Wiley, New York, 198 p.

Bonham-Carter, G.F., 1994. *Geographic Information Systems for Geoscientists*. Pergamon, Oxford, 415 p.

Boud, R.C., 1985. The Highland and Agricultural Society of Scotland and John MacCulloch's geological map of Scotland. *Cartographica*, vol. 22, 92–115.

Brooks, R.R., 1987. *Serpentine and its Vegetation*. Dioscorides Press, Portland, Oregon, 454 p.

Brunskill, R.W., 1982. *Traditional Farm Buildings of Britain: An Introduction to Vernacular Architecture*. Gollancz, London, 160 p.

Budd, P., Gale, D., Ixer, R.A.F. and Thomas, R.G., 1994. Tin sources for prehistoric bronze production in Ireland. *Antiquity*, vol. 68, 518–524.

Bugayevskiy, L.M. and Snyder, J.P., 1995. *Map Projections: A Reference Manual*. Taylor & Francis, Basingstoke, 352 p.

Bureau of Mineral Resources, Geology and Geophysics, 1989. *Symbols Used on Geological Maps*. Canberra, 74 p.

Butcher, N.E., 1983. The advent of colour-printed geological maps in Britain. *Proceedings, Royal Institution of Great Britain*, vol. 55, 149–161.

Catt, J.A., 1986. *Soils and Quaternary Geology: A Handbook for Field Scientists*. Clarenden Press, Oxford, 267 p.

Catt, J.A., 1988. *Quaternary Geology for Scientists and Engineers*. Ellis Horwood, Chichester, 340 p.

Challinor, J., 1970. The progress of British geology during the early part of the nineteenth century. *Annals of Science*, vol. 26, 177–234.

Chapman, C.A., 1968. A comparison of the Maine coastal plutons and the magmatic central complexes of New Hampshire. In: Zen, E-an, White, W.S. and Hadley, J.B. (editors), *Studies of Appalachian Geology: Northern and Maritime*. Wiley-Interscience, New York, 385–396.

Chikesa, A.G. (editor), 1965. *Plant Indicators of Soils, Rocks and Subsurface Waters*. Consultants Bureau, New York, 210 p.

Clarke, A.R., Morey, A.J., Gilbert, S.T. and Bacciarelli, R.E., 1987. Planning constraints and their influence on quarrying of Magnesian Limestone. In: Culshaw, M.G., Bell, F.G., Cripps, J.C. and O'Hara, M. (editors), Planning and engineering geology. *Geological Society Engineering Geology Special Publication* 4, 379–387.

Clayton, L. and Attig, J.W., 1993. Pleistocene mapping at the Wisconsin Geological and Natural History Survey. *Canadian Plains Proceedings*, vol. 25, 138–154.

Coe, C.J., 1981. The use of geologic, hydrologic and geochemical mapping techniques in environmental assessment. *Groundwater*, vol. 19, 626–634.

Compton, R.R., 1985. *Geology in the Field*. John Wiley, New York, 398 p.

Condry, W., 1973. *The Snowdonia National Park*. Fontana, London, 320p.

Conzen, M.P. (editor), 1994. *The Making of the American Landscape*. Routledge, New York & London, 433 p.

Cook, K.S., 1989. Artistic and scientific design sources in geological cartography: George Bellas Greenough's 'A geological map of England and Wales, 1820-1839'. *13th Conference on the History of Cartography*, Amsterdam 1989, 3 p.

Correll, R.L. and Dillon, P.J., 1993. Risk assessment as a framework for management of aquifers – a literature review. *Journal of Australian Geology and Geophysics*, vol. 14, 155–159.

Cox, L.R., 1942. New light on William Smith and his work. *Proceedings, Yorkshire Geological Society*, vol. 25, 1–99.

Crimes, T.P., Chester, D.K., Hunt, N.C., Lucas, G.R., Mussett, A.E., Thomas, G.S.P. and Thompson, A., 1994. Techniques used in aggregate resource analyses of four areas in the UK. *Quarterly Journal of Engineering Geology*, vol. 27, 165–192.

Culshaw, M.G., Bell, F.G., Cripps, J.C. and O'Hara, M. (editors), 1987. Planning and engineering geology. *Geological Society Engineering Geology Special Publication* 4, 641 p.

Darragh, T.A., 1977. The first geological maps of the continent of Australia. *Journal, Geological Society of Australia*, vol. 24, 270–305.

Davies, G.L.H., 1983. *Sheets of Many Colours: The Mapping of Ireland's Rocks, 1750–1890*. Royal Dublin Society, Dublin, 242 p.

Davies, G.L.H., 1995. *North from the Hook. 150 Years of the Geological Survey of Ireland*. Geological Survey of Ireland, Dublin, 342 p.

Davis, S.N. and DeWiest, R.J.M., 1966. *Hydrogeology*. John Wiley, New York, 462 p.

de Mulder, E.F.J., 1988. Thematic applied Quaternary mapping – a profitable investment or expensive wallpaper? In: de Mulder, E.F.J. and Hageman, B.P. (editors), *Applied Quaternary Research*. Balkema, Rotterdam, 105–117.

de Mulder, E.F.J., 1990. Engineering geology maps: a cost benefit analysis. *Environmental Geology and Water Science*, vol. 16, 23–28.

de Mulder, E.F.J., 1994. Preface to Special Issue: Engineering Geology of Quaternary Deposits. *Engineering Geology*, vol. 37, 3–4.

de Mulder, E.F.J. and Hillen, R., 1989. Preparation and application of thematic engineering and environmental geology maps in the Netherlands. *Engineering Geology*, vol. 29, 279–290.

de Smedt, P., de Breuck, W., Loy, W., van Autenboer, T. and van Dijk, E., 1987. Groundwater vulnerability maps. *Aqua*, vol. 5, 264–267.

de Villiers, J., 1983. Geology of the Helderberg. *Transactions, Geological Society of South Africa*, vol. 88, 175–177.

Dearman, W.R., 1987. Land evaluation and site assessment: mapping for planning purposes. In: Culshaw, M.G. et al. (editors), Planning and Engineering Geology. *Geological Society Engineering Geology Special Publication* 4, 195–201.

Deer, W.A., 1935. The Cairnsmore of Carsphairn igneous complex. *Quarterly Journal of the Geological Society of London*, vol. 191, 47–76.

Dennison, J.M., 1968. *Analysis of Geologic Structures*. Norton, New York, 209 p.

Dixon, H.R. and Lawrence, W.L., Jr., 1968. Structure of Eastern Connecticut. In: Zen, E-an, White, W.S. and Hadley, J.B. (editors), *Studies of Appalachian Geology: Northern and Maritime*. Wiley-Interscience, New York, 219–229.

Dobson, M.R. (compiler), 1995. *The Aberystwyth District*. Geologists' Association Guide 54, The Geologists' Association, London, 109 p.

Dunham, Sir K., Beer, K.E., Ellis, R.H., Gallagher, M.J., Nutt, M.J.C. and Webb, B.C., 1978. United Kingdom. In: Bowie, S.H.U., Kvalheim, A. and Haslam, N.W. (editors), *Mineral Deposits of Europe. Volume 1: Northwest Europe*. Institute of Mining and Metallurgy and the Mineralogical Society, London, 263–317.

Earp, J.R., 1938. The Higher Silurian rocks of the Kerry district Montgomeryshire. *Quarterly Journal of the Geological Society, London*, vol. 94, 125–160.

Edmonds, C.N., Green, C.P. and Higginbottom, I.E., 1987. Subsidence hazard prediction for limestone terrains, as applied to the English Cretaceous Chalk. In: Culshaw, M.G., Bell, F.G., Cripps, J.C. and O'Hara, M. (editors), Planning and Engineering Geology. *Geological Society Engineering Geology Special Publication* 4, 283–293.

Erdelyi, M. and Galfi, F.J., 1988. *Surface and Subsurface Mapping in Hydrogeology*. John Wiley, Chichester, 384 p.

Evans, J.W., 1921. The representation of stratigraphical divisions by shading. *Geological Magazine*, vol. 58, 40–41.

Eyles, J.M., 1967. William Smith: the sale of his geological collection to the British Museum. *Annals of Science*, vol. 23, 177–212.

Eyles, J.M., 1969. William Smith (1769–1839): A bibliography of his published writings, maps and geological sections, printed and lithographed. *Journal, Society of Bibliography of Natural History*, vol. 5, 87–109.

Eyles, V.A., 1937. John MacCulloch, F.R.S., and his geological map. *Annals of Science*, vol. 2, 114–129.

Eyles, V.A., 1950. The first national geological survey. *Geological Magazine*, vol. 7, 373–382.

Eyles, V.A. and Eyles, J.M., 1938. On the different issues of the first geological map of England and Wales. *Annals of Science*, vol. 3, 190.

Faul, H. and Faul, C., 1983. *It Began with a Stone. A History of Geology from the Stone Age to the Age of Plate Tectonics*. John Wiley, New York, 270 p.

Fergusson, C.L., Cas, R.A.F., Collins, W.J., Craig, G.Y., Crook, K.A.W., Powell, C.McA., Scott, S.A. and Young, G.C., 1979. The Upper Devonian Boyd volcanic complex, Eden, New South Wales. *Journal, Geological Society of Australia*, vol. 26, 87–105.

Flett, J.S., 1937. *The First Hundred Years of the Geological Survey of Great Britain*. Her Majesty's Stationery Office, London, 280 p.

Fleuty, M.J., 1964. The description of folds. *Proceedings, Geologists' Association*, vol. 75, 461–489.

Folger, P., 1996. The National Geologic Mapping Reauthorization Act of 1996. *GSA Today*, vol. 6, no. 8, 22–23.

Fortey, R., 1993. *The Hidden Landscape. A Journey into the Geological Past*. Pimlico, London, 310 p.

Fowler-Billings, K., 1993. *Stepping Stones. The Reminiscences of a Woman Geologist in the Twentieth Century*.

Connecticut Academy of Arts and Sciences, New Haven, Connecticut, 222 p.

Fracolli, D.L., 1985. An investigation of preparation and costs for a state bibliography of geology. In: Kidd, C.M. (editor), *Maps in the Geoscience Community*. Proceedings, Geoscience Information Society, Reno, Nevada, vol. 15, 117–124.

Frazer Darling, F. and Morton Boyd, J., 1970. *The Highlands and Islands*. Fontana, London, 405 p.

Freeman, T.J., Littlejohn, G.S. and Driscoll, R.M.C., 1994. *Has Your House got Cracks? A Guide to Subsidence and Heave of Buildings on Clay*. Institute of Civil Engineers, Building Research Establishment, London, 114 p.

Freund, R., 1971. The Hope Fault. A strike-slip fault in New Zealand. *New Zealand Geological Survey Bulletin*, vol. 86, 1–49.

Fulton, R.J., 1993. Surficial geology mapping at the Geological Survey of Canada: its evolution to meet Canada's changing needs. *Canadian Journal of Earth Science*, vol. 30, 232–242.

Geiger, C. and Barnes, K.B., 1994. Indoor radon hazard: a geographical assessment and case study. *Applied Geography*, vol. 14, 350–371.

Geikie, A., 1905. *The Founders of Geology*. Reprinted 1962, Dover Publications, New York, 486 p.

Giles, J.R.A., 1994. Geographical Information Systems – the key to the geological maps of the next century. *Episodes*, vol. 17, 73–74.

Goudie, A., 1993. *The Human Impact on the Natural Environment*, 4th edition. Blackwell, Oxford, 454 p.

Gozzard, J.R., 1985. Medium-scale engineering- and environmental-geology mapping of the Perth Metropolitan Region, Western Australia. *Engineering Geology*, vol. 22, 97–107.

Green, G.W. and Welch, F.B.A., 1965. Geology of the country around Wells and Cheddar. *Memoirs, Geological Survey of Great Britain*, 280, Her Majesty's Stationery Office, London, 225 p.

Greenly, E., 1938. *A Hand Through Time: Memories – Romantic and Geological; Studies in the Arts and Religion; and the Grounds of Confidence in Immortality*. Murby & Co., London, 2 volumes.

Greenly, E. and Williams, H., 1930. *Methods in Geological Surveying*. Thomas Murby, London, 420 p.

Groves, J.R., 1980. Step-by-step procedure for mounting maps on cloth. *Journal of Geological Education*, vol. 28, 141–143.

Hagan, W.W., 1994. The role of the state geological surveys in environmental geology. *Environmental Geology*, vol. 23, 166–167.

Hageman, B.P., 1989. The application of Quaternary geology in coastal areas. In: de Mulder, E.F.J. and Hageman, B.P. (editors), *Applied Quaternary Research*. Balkema, Rotterdam, 43–63.

Hardaway, J.E., 1968. Possibilities for subsurface waste disposal in a structural syncline in Pennsylvania. In: Galley, J. (editor), *Subsurface Disposal in Geologic Basins – A Study of Reservoir Strata*. American Association of Petroleum Geologists, Tulsa, Oklahoma, 93–127.

Harrison, J.M., 1963. Nature and significance of geological maps. In: Albritton, C.C. (editor), *The Fabric of Geology*. Freeman Cooper & Co., Stanford, California, 225–231.

Hayward, H.A., 1932. The geology of the Lower Greensand in the Dorking–Leith Hill district, Surrey. *Proceedings of the Geologists' Association*, vol. 43, 1–31.

Hiscock, K.M., Lovett, A.A., Brainard, J.S. and Parfitt, J.P., 1995. Groundwater vulnerability assessment: two case studies using GIS methodology. *Quarterly Journal of Engineering Geology*, vol. 28, 179–194.

Holmes, A., 1965, *Principles of Physical Geology*, 2nd edition. Ronald Press, New York, 1028 p.

Hudak, P.F., 1996. Distribution of indoor radon concentrations and uranium-bearing rocks in Texas. *Environmental Geology*, vol. 28, 29–33.

Hudson, R.G.S. and Dunnington, H.V., 1944. The Carboniferous rocks of the Swinden Anticline, Yorkshire. *Proceedings of the Geologists' Association*, vol. 55, 195–215.

Ichoku, C., Chorowiz, J. and Parrot, J.F., 1994. Computerized construction of geological cross-sections from digital maps. *Computers and Geosciences*, vol. 20, 1321–1327.

Ireland, H.A., 1943. History of the development of geologic maps. *Bulletin, Geological Society of America*, vol. 54, 1227–1280.

Jamieson, H.C., Brockett, L.D and McIntosh, R.A., 1980. Prudhoe Bay – a 10-year perspective. In: Halbouty, M.T. (editor), *Giant Oil and Gas Fields of the Decade 1968–1978*. American Association of Petroleum Geologists, Tulsa, Oklahoma, 289–314.

Jenkins, D.E., 1981. Geology of the Auburn Quadrangle, Carbon County, Idaho, and Lincoln County, Wyoming. *Brigham Young University Geology Studies*, vol. 28, 101–116.

Johnson, G.A.L. and Dunham, K.C., 1963. The geology of Moor House. *Monograph, Nature Conservancy*, 2, Her Majesty's Stationery Office, London, 182 p.

Jones, O.T., 1922. Lead and zinc. The mining district of North Cardiganshire and West Montgomeryshire. *Special Report 20, Mineral Resources of Great Britain*, 207 p.

Jones, O.T., 1949. The geology of the Llandovery district. Part II: the northern area. *Quarterly Journal, Geological Society of London*, vol. 105, 43–64.

Korn, M. and Martin, M., 1959. Gravity tectonics in the Naukluft Mountains of south-west Africa. *Bulletin, Geological Society of America*, vol. 70, 1047–1078.

Kupfer, D.H., 1966. Accuracy in geologic maps. *Geotimes*, vol. 10, 11–14.

Lemoine, M. (editor), 1978. *Geological Atlas of Alpine Europe and Adjoining Alpine Areas*. Elsevier, Amsterdam, 584 p.

Lemon, J.T., 1972. *The Best Poor Man's Country, a Geographical Study of Early Southeastern Pennsylvania*. Johns Hopkins Press, Baltimore, 295 p.

Levorsen, A.I., 1960. *Paleogeological Maps*. Freeman, San Francisco, 174 p.

Linton, D.L., 1947. The ideal geological map. *Advancement of Science*, vol. 5, 141–149.

Lloyd, J.W., 1985. Hydrogeology and beer. *Proceedings of the Geologists' Association*, vol. 97, 213–219.

Lobeck, A.K., 1939. *Geomorphology. An Introduction to the Study of Landscapes*. McGraw-Hill, New York & London, 731 p.

Lobeck, A.K., 1958. *Block Diagrams and Other Graphic Methods Used in Geology and Geography*, 2nd edition. Emerson-Trussell Book Co., Amherst, Maryland, 206 p.

Love, J.D., Christiansen, A.C. and Ploeg, A.J., 1993. Stratigraphic nomenclature chart of Wyoming. *Wyoming Geological Survey Map Series* 41.

Lumsden, G.I. (editor), 1994. *Geology and the Environment in Western Europe*. Oxford University Press, Oxford, 325 p.

Martin, E.L., 1973. Geological maps. In: Wood, D.N. (editor), *Use of Earth Sciences Literature*. Butterworth, London, 122–150.

McCall, J. and Marker, B. (editors), 1989. *Earth Science Mapping for Planning, Development and Conservation*. Graham & Trotman, London, 268 p.

McClay, K., 1987. *The Mapping of Geological Structures*. Open University Press, Milton Keynes, 161 p.

McLean, A.C. and Gribble, C.D., 1992. *Geology for Civil Engineers*, 2nd edition. Allen & Unwin, London, 314 p.

McMillan, A.A. and Browne, M.A.E., 1987. The use or abuse of thematic mining information maps. In: Culshaw, M.G., Bell, F.G., Cripps, J.C. and O'Hara, M. (editors), Planning and engineering geology. *Geological Society Engineering Geology Special Publication* 4, 237–245.

McPhee, J., 1986. *Rising from the Plains*. Farrar, Strauss, & Giroux, New York, 214 p.

Merrill, G.P., 1924. *The First One Hundred Years of American Geology*. Reprinted 1969, Hafner Publishing Co., New York, 773 p.

Moseley, F., 1979. *Advanced Geological Map Interpretation*. Edward Arnold, London, 80 p.

Moseley, F., 1981. *Methods in Field Geology*. Freeman, Oxford, 211 p.

Nickless, E.F.P., 1982. Environmental geology of the Glenrothes district, Fife Region: Description of 1:25 000 sheet 20. *Report, Institute of Geological Sciences*, 82/15, Her Majesty's Stationery Office, London, 53 p.

Nickless, E.F.P. and Jackson, I., 1994. Digital geological map production in the UK – more than just a cartographic exercise. *Episodes*, vol. 17, 51–56.

North, F.J., 1928. *Geological Maps. Their History and Development with Special Reference to Wales*. National Museum of Wales, Cardiff, 133 p.

Oldroyd, D.R., 1996. *Thinking about the Earth*. (Chapter 5: The Earth surveyed and geologically mapped.) Athlone Press, London, 410 p.

Page, B.M., 1966. Geology of the Coast Ranges of California. *Bulletin, California Division of Mines and Geology*, vol. 190, 255–276.

Parkinson, D.P., 1936. The carboniferous succession in the Slaidburn district, Yorkshire. *Quarterly Journal of the Geological Society of London*, vol. 92, 294–331.

Pavlinov, V.N., 1984. Evolution of geological mapping in connection with the development of natural sciences. In: Dudich, E. (editor), *Contributions to the History of Geological Mapping*. Akademiaikido, Budapest, 35–40.

Pearsall, W.H., 1968. *Mountains and Moorland*. The Fontana New Naturalist, Collins, London and Glasgow, 415 p.

Penoyre, J. and Penoyre, J., 1978. *Houses in the Landscape: A Regional Study of Vernacular Building Styles in England and Wales*. Faber & Faber, London, 175 p.

Perring, F.H. and Walters, S.M. (editors), 1976. *Atlas of the British Flora*, 2nd edition. EP Publishing Ltd., Wakefield, England, for the Botanical Society of the British Isles, 432 p.

Pitcher, W.S. and Berger, A.R., 1972. *The Geology of Donegal: A Study of Granite Emplacement and Unroofing*. Wiley-Interscience, New York, 435 p.

Pomerol, C. (editor), 1989. *The Wines and Winelands of France*. Robertson McCarta, London, 370 p.

Powell, C.McA., Edgecombe, D.R., Henry, N.M. and Jones, J.G., 1976. Timing of regional deformation of the Hill End Trough: a reassessment. *Journal, Geological Society of Australia*, vol. 23, 407–421.

Powell, D., 1992. *Interpretation of Geological Structures through Maps*. Longman Group, London, 176 p.

Powell, J.W., 1881. Sur la nomenclature générale, sur le coloriage et les signes conventionnels des Cartes géologiques. *International Geological Congress, Compte Rendu*, 2nd session, Boulogne, 627–641.

Powell, J.W., 1895. *The Exploration of the Colorado River and its Canyons*. Reprinted 1961, Dover Publications, New York, 400 p.

Rabbit, M.C., 1980a. A brief history of the US Geological Survey. *US Geological Survey Popular Publication*, 50 p.

Rabbit, M.C., 1980b. John Wesley Powell, soldier, explorer, scientist. *US Geological Survey Popular Publication*, 23 p.

Radbruch-Hall, D.H., Edwards, K. and Batson, R.M., 1987. Experimental engineering and environmental geologic maps of the conterminous United States. *Bulletin, US Geological Survey* 1610, 7 p. plus maps.

Ragan, D.M., 1985. *Structural Geology. An Introduction to Geometrical Techniques*, 3rd edition. John Wiley, New York, 393 p.

Ramsay, J.G. and Huber, M.I., 1987. *The Techniques of Modern Structural Geology. Volume 2: Folds and Fractures*. Academic Press, London, 700 p.

Rapp, G., Henricksen, E. and Allert, J., 1990. Native copper sources of artefact copper in pre-Columban North America. In: Lasca, N.P. and Donahue, J. (editors), *Archaeological Geology of North America. Geological Society of America Centennial Special Volume* 4, Boulder, Colorado, 479–498.

Rappaport, R., 1969. The geological atlas of Guettard, Lavoisier, and Monnet: conflicting views of the nature of geology. In: Schneer, C.J. (editor), *Toward a History of Geology*. MIT Press, Cambridge, Massachusetts, 272–287.

Richey, J.E., 1961. *Scotland: The Tertiary Volcanic Districts*, 3rd edition. British Regional Geology, Her Majesty's Stationery Office, Edinburgh, 120 p.

Robinson, G.D. and Spieker, A.M., 1978. Nature to be commanded . . . Earth-science maps applied to land and water management. *US Geological Survey Professional Paper* 950, 95 p.

Rowland, S.M. and Duebendorfer, E.M., 1994. *Structural Analysis and Synthesis*, 2nd edition. Blackwell, Oxford, 279 p.

Rudwick, M.J.S., 1976. A visual language for geology. *History of Science*, vol. 14, 149.

Schwartz, G.M., 1964. History of the Minnesota Geological Survey. *Minnesota Geological Survey Special Publication Series*, SP–1, 39 p.

Statham, I., Golightly, C. and Treharne, G., 1987. Thematic mapping of the abandoned mine hazard: a pilot study for the South Wales Coalfield. In: Culshaw, M.G., Bell, F.G., Cripps, J.C. and O'Hara, M. (editors), Planning and engineering geology. *Geological Society Engineering Geology Special Publication* 4, 255–268.

Stenestad, E. and Sustrac, G., 1992. The role of geoscience in planning and development. In: Lumsden, G.I. (editor), *Geology and the Environment in Western Europe*. Oxford University Press, Oxford, 281–301.

Strauss, G.K., Madel, J. and Fdez Alonso, F., 1977. Exploration practice for strata-bound volcanogenic sulphide deposits in the Spanish–Portuguese pyrite belt: geology, geophysics, and geochemistry. In: Klemm, D.D. and Schneider, H-J. (editors), *Time and Strata-Bound Ore Deposits*. Springer-Verlag, Berlin, 55–93.

Tearpock, D.J. and Bischke, R.E., 1991. *Applied Subsurface Geological Mapping*. Prentice-Hall, Englewood Cliffs, New Jersey, 648 p.

Topley, W., 1872. *The Geology of the Weald.* Geological Survey of Great Britain, London, 503 p.

Travis, R.B. and Lamar, D.L., 1987. Apparent-dip methods. *Journal of Geological Education,* vol. 35, 152–154.

Tsuya, H., Machida, H. and Shimozuru, D., 1988. *Explanatory Note for Geologic Map of Mt. Fuji* (2nd printing). Geological Survey of Japan, Tsukuba, 22 p.

Tyler, M.B., 1995. Look before you build: geologic studies for safer development in the San Francisco Bay area. *US Geological Survey Circular* 1130, 54 p.

van der Bark, E. and Thomas, O.D., 1980. Ekofisk: first of the giant oilfields in Western Europe. In: Halbouty, M.T. (editor), *Giant Oil and Gas Fields of the Decade 1968–1978.* American Association of Petroleum Geologists, Tulsa, Oklahoma, 195–224.

Varley, N.R. and Flowers, A.G., 1993. Radon in soil gas and its relationship with some major faults of SW England. *Environmental Geochemistry and Health,* vol. 15, 145–151.

Vigneaux, M. and Leneuf, N., 1980. Géologie et Vins de France. *Bulletin, Institute Géologique Bassin d'Aquitaine, Bordeaux,* vol. 27, 165–234.

Wallace, R.M., 1965. Geology and mineral resources of the Pico de Itabirito District Minas Gerais, Brazil. *US Geological Survey Professional Paper* 341-F, 68 p.

Willats, E.C., 1970. Maps and Maidens. *Cartographica,* vol. 7, 50.

Wilson, A.A., 1986. Geological factors in land use planning at Aldridge-Brownhills, West Midlands. In: Culshaw, M.G., Bell, F.G., Cripps, J.C. and O'Hara, M. (editors), Planning and engineering geology. *Geological Society Engineering Geology Special Publication* 4, 51–64.

Wilson, H.E., 1985. *Down to Earth. One Hundred and Fifty Years of the British Geological Survey.* Scottish Academic Press, Edinburgh, 189 p.

Winnock, E. and Pontalier, Y., 1970. Lacq Gas Field, France. In: Halbouty, M.T. (editor), *Geology of Giant Petroleum Fields.* American Association of Petroleum Geologists, Memoir 14, 370–387.

Winter, Th., Binquet, J., Szendroi, A., Colombet, G., Armijo, R. and Tapponier, P., 1994. From plate tectonics to the design of the Dul Hasti hydroelectric project, Kashmir (India). *Engineering Geology,* vol. 36, 211–241.

Withers, C.W.J., 1982. The Scottish Highlands outlined: an examination of the cartographic evidence for the position of the Highland–Lowland boundary. *Scottish Geographical Magazine,* vol. 98, 143–157.

Wolff, F.C., 1987. *Geology for Environmental Planning.* Norges Geologiske Undersokelse, Trondheim, 121 p.

Woodcock, N., 1994. *Geology and Environment in Britain and Ireland.* UCL Press, London, 164 p.

Woodward, L.A., 1984. Potential for significant oil and gas fracture reservoirs in Cretaceous of Raton Basin, New Mexico. *American Association of Petroleum Geologists Bulletin,* vol. 68, 628–636.

Wooldridge, S.W. and Goldring, F., 1953. *The Weald.* The New Naturalist 26, Collins, London and Glasgow, 276 p.

Index

Page numbers in **bold** type refer to main text entries; page numbers in *italics* refer to figures and tables